THE COMPOSITION OF KEPLER'S *ASTRONOMIA NOVA*

THE COMPOSITION OF KEPLER'S *ASTRONOMIA NOVA*

James R. Voelkel

⌘

PRINCETON UNIVERSITY PRESS

PRINCETON AND OXFORD

PUBLISHED BY PRINCETON UNIVERSITY PRESS, 41 WILLIAM STREET,
PRINCETON, NEW JERSEY 08540
IN THE UNITED KINGDOM: PRINCETON UNIVERSITY PRESS,
3 MARKET PLACE, WOODSTOCK, OXFORDSHIRE OX20 1SY

LIBRARY OF CONGRESS CATALOGING-IN-PUBLICATION DATA

VOELKEL, JAMES R. (JAMES ROBERT), 1962–
THE COMPOSITION OF KEPLER'S *ASTRONOMIA NOVA* / JAMES R. VOELKEL.
P. CM.
INCLUDES BIBLIOGRAPHICAL REFERENCES AND INDEX.
ISBN 0-691-00738-1 (ACID-FREE PAPER)
1. KEPLER, JOHANNES, 1571–1630. ASTRONOMIA NOVA. 2. KEPLER'S LAWS. I. TITLE.
QB355.3 .V64 2001
521′.3—dc21 2001036296

THIS BOOK HAS BEEN COMPOSED IN SABON

PRINTED ON ACID-FREE PAPER. ∞

WWW.PUP.PRINCETON.EDU

PRINTED IN THE UNITED STATES OF AMERICA

1 3 5 7 9 10 8 6 4 2

IN MEMORIAM AMANTISSIMAM

Robert T. Voelkel (1933–1987)

Victor E. Thoren (1935–1991)

Richard S. Westfall (1924–1996)

Herr, lehre doch mich,
daß ein Ende mit mir haben muß,
und mein Leben ein Ziel hat,
und ich davon muß,
und ich davon muß.
—Brahms, *Ein deutsches Requiem*

⌘ CONTENTS ⌘

⌘ ILLUSTRATIONS ⌘

⌘ ACKNOWLEDGMENTS ⌘

I have had the sad misfortune while working on this book, and the Ph.D. thesis on which it is based, of losing three important men. The first was my father, Robert T. Voelkel, who died during my second year of graduate school. I owe more of what I am to him than I realized or had a chance to acknowledge while he was alive. From him I learned a love and respect for knowledge, and I hope that I have inherited a sufficient measure of the intellectual rigor he employed to ward off the pretensions and prejudices of men. The second was Victor E. Thoren, who died shortly after the approval of my dissertation proposal. I count it a sad but honored distinction to have been his final thesis student. Finally, after the dissertation was finished, Richard S. Westfall, who had sat on my committee and commented thoroughly, thoughtfully, and enthusiastically on the thesis, passed away as well. To these three wise men, this work is dedicated.

To Owen Gingerich, who gathered me in, found support for me, and advised me on the overwhelming bulk of my thesis, as well as on subsequent drafts of this book, and on countless other things, I owe the greatest thanks of all. This work—and my career—would not exist without him. That he has done so much for me outside the bounds of institutional affiliation is but one manifestation of his extraordinary devotion to the history of astronomy. My debt to him as a teacher, advisor, and friend is incalculable.

I acknowledge the kind and skillful comments of Curtis Wilson and Bruce Stephenson, who read a draft of this book. William H. Donahue and Peter Barker have also provided valuable comments and consultations. All remaining errors are my own. Graeme Bird, Roger Ceragioli, and Lawrence Principe offered help at various times with Greek and Latin, though I would not dream of impugning their skill by associating them with my translations.

As much as we would wish it to be otherwise, scholarship depends on money. During the writing of my thesis, I was supported by a Victor E. Thoren scholarship from Indiana University, a Smithsonian Institution predoctoral fellowship, and a grant from the Josephine De Karman Fellowship Trust. Research on Kepler and Fabricius was supported by a grant from the Dudley Observatory. A postdoctoral fellowship at the Dibner Institute for the History of Science and Technology in Cambridge, Massachusetts, provided not only the funds to continue this work, but also an extraordinarily luxurious and congenial atmosphere. I extend my special thanks to then-director Jed Buchwald, executive director Evelyn

Simha, and to the ladies, Carla Chrisfield, Rita Dempsey, and Trudy Kontoff. My fellow fellows, especially David McGee and Noah Efron, also deserve my great thanks. The last stages of work on this book were supported by a postdoctoral fellowship at the Department of History of Science, Medicine, and Technology at the Johns Hopkins University.

Finally, to Kathryn Fogle, who continues to be a source of inspiration, affection, consolation, and exasperation, I give my love.

⌘ PREFACE ⌘

In 1964, still early in the computer age, Owen Gingerich set out to demonstrate the power of the mainframe computer by programming one to perform Kepler's laborious iterative derivation of the elements of the vicarious hypothesis.[1] Kepler complained he had repeated this tedious procedure seventy times. In addition to the great speed with which the computer sailed through the calculation, Gingerich discovered that it required the computer the minimum number of iterations—nine—to converge on the solution. From this, he could conclude only that Kepler's huge number of trials were due to his being horribly plagued by numerical errors.

Gingerich returned to this problem several years later, after he had succeeded in securing a microfilm of Kepler's Mars manuscripts, designated Pulkova XIV, in what was then Leningrad. He was surprised to discover that, contrary to the logical systematic approach of the *Astronomia nova*, the manuscripts contained a variety of approaches to the orbits of the earth and Mars almost indiscriminately mixed together. In addition to reassessing his conclusions regarding Kepler's large number of trials, he announced:

> Most commentators have assumed, because of Kepler's sequential and at times autobiographical style, that Kepler has spared no detail in the chronicle of his researches. Examination of the manuscript material . . . shows, on the contrary, that the book evolved through several stages and represents a much more coherent plan of organization than a mere serial recital of his investigations would allow.[2]

In retrospect, it is clear that some earlier scholars were fully aware that the *Astronomia nova* did not represent a straightforward chronicle of Kepler's researches. Max Caspar, the last century's leading Kepler scholar, declared in the 1937 Johannes Kepler *Gesammelte Werke* edition of the *Astronomia nova*, "Regardless of all its wrong paths and detours, the internal structure of the work which unveils itself upon deeper consideration is dictated by strict logic, and is accomplished clearly in a dramatic step-by-step process. A prelude and an epilogue in Parts I and V frame the main plot in Parts II–IV."[3] Reading between the lines, we see that Caspar understood the purpose of Kepler's dramatic narrative. And it is obvious from Caspar's account of the progress of Kepler's research from both the notes to the 1937 edition of the *Astronomia nova* and his biography, *Kepler* (1948; English ed., 1959), that he drew upon his knowledge of Kepler's correspondence and manuscripts to establish a sequence of events different from that described in the *Astronomia nova*. However, the biography was mysteriously published without notes, and it takes a very good familiarity with the supporting material to recognize how faithfully Caspar captured Kepler's life. Similarly, when editing the 1937 edition of the *Astronomia nova*, Caspar had his eye on this remarkable book. He knew and described the supporting manuscripts, including Pulkova XIV, but the variance of the manuscripts from the book, rather than being addressed as an issue of intrinsic interest, itself became the reason for ignoring them:

> This brief summary of the contents [of the Kepler Mars manuscripts] must suffice at this stage. Most of it, as the summary of contents shows, does not come into question for the publication [of the *Astronomia nova*]. The material that volume XIV contains is all used up, so to speak. Because in his work, as was already noted, Kepler himself does not present finished results but rather the story of the discovery of his results, and in fact on the broadest basis, it would not do to draw the work out even further by taking up drafts. For this reprint only that can be considered which serves for the correction of the printed text, which perhaps illuminates certain trains of thought from another side, or which further clears up the story of discovery of the present work.[4]

Caspar's failure to press the distinction between the *Astronomia nova* as history and as argument can be seen to belong to a historiographical tradition that treated the great book as a singular achievement without regard to its audience.

The first major biographer of Kepler in English was Arthur Koestler, who drew upon Caspar's *Kepler* for his own *The Sleepwalkers* (1959). For Koestler, who was wont to treat the genius as psychopath,[5] Kepler's narrative became a valuable example of the irrationality of scientific discovery:

> Fortunately, [Kepler] did not cover up his tracks, as Copernicus, Galileo and Newton did, who confront us with the result of their labours, and keep us guessing how they arrived at it. Kepler was incapable of exposing his ideas methodically, text-book fashion; he had to describe them in the order they came to him, including all the errors, detours, and the traps into which he had fallen. The *New Astronomy* is written in an unacademic, bubbling baroque style, personal, intimate, and often exasperating. But it is a unique revelation of the ways in which the creative mind works.[6]

As such a revelation, the truth of the account offered in the *Astronomia nova* had to be assumed. The fact that the book was written for an audience and that there was a rhetorical purpose to Kepler's account would have fitted only awkwardly with Koestler's assertion that Kepler was "incapable" of describing his discoveries methodically.

Elegantly edited excerpts of much of Kepler's correspondence had been published in Christian Frisch's nineteenth-century edition of Kepler's complete works, but the complete correspondence was not published in the Johannes Kepler *Gesammelte Werke* until 1959 (although all the correspondence through the composition of the *Astronomia nova* was published by 1954). Alexandre Koyré, in *The Astronomical Revolution* (1961; English ed. 1973), integrated correspondence more fully into his account of Kepler's discoveries, but more as a matter of expanding on Kepler's account in the *Astronomia nova* than of probing the dissimilarities. He also better understood the relation of the book's structure to Kepler's context and the intentionality of its narrative, but he too ultimately used the narrative of the *Astronomia nova* as a reflection of the nature of Kepler's mind rather than his audience:

> Whilst he assigned Kepler the task of studying the motion of this planet, Tycho Brahe, nevertheless, did not give him a free hand. He asked—and he renewed the request on his deathbed—that the motion should be treated according to his (Tycho Brahe's) principles, and not according to those of Kepler, or of Copernicus. Kepler fulfilled this request, without conforming to it exactly.
>
> This partly explains the unusual character and extreme difficulty of the *Astronomia nova*, which are responsible for the exceptional interest of the work. Indeed, in this book, which is unique among the great classics of science, and in which all astronomical problems are treated three, and even four, times after the manner of Ptolemy, Tycho Brahe, Copernicus and finally Kepler himself, Kepler does not restrict himself to setting forth the results, as did Copernicus and Newton: he relates at the time, intentionally as he did in the *Mysterium cosmographicum*—the development of his thought, his efforts, and his setbacks, Kepler's mind was so constituted that he was unable to find the way to truth without first having explored all the paths leading into error—but perhaps the mind of man in general is naturally framed in this manner [sentence structure *sic*].[7]

Thus again, the structure of the *Astronomia nova* came to be treated largely as an innate production of Kepler's mind rather than as a purposeful device.

Seminal articles on Kepler's methodology, such as Curtis Wilson's "Kepler's Derivation of the Elliptical Path" (1968) and Eric Aiton's "Kepler's Second Law of Planetary Motion" (1969), although continuing to cite letters occasionally, came to rely even more heavily on the *Astronomia nova* as a true account of Kepler's work.[8] In his article

"Keplerian Planetary Eggs, Laid and Unlaid" (1974), D. T. Whiteside acknowledged Gingerich's recent announcement of the confused state of the Kepler Mars manuscripts but continued to treat the correspondence similarly.[9] As in the other articles, when correspondence is incorporated at all, it is to elucidate the methodology of the *Astronomia nova*. In addition, much of this work can be faulted for focusing rather too narrowly on those aspects of Kepler's work that today are considered to be significant—the ellipse and the area law—and not, as in Koyré, the idea of a physical astronomy, or Kepler's purpose in writing the *Astronomia nova*.

The breakthrough for modern historiography of Kepler's work came with the publication of Bruce Stephenson's *Kepler's Physical Astronomy* (1987). Stephenson rightly understood physical astronomy to be the central feature of Kepler's work. He described its development in the *Mysterium cosmographicum* (1596) through the *Epitome astronomiae Copernicanae* (1618–1621). And, although he masterfully described the argument of the *Astronomia nova*, he showed how the ellipse and the area law were of significance only to the extent that they supported Kepler's physical ideas.

Stephenson was also the first to make an explicit statement of the rhetorical character of the *Astronomia nova*:

> This profoundly original work has been portrayed as a straightforward account of converging approximations, and it has been portrayed as an account of gropings in the dark. Because of the book's almost confessional style, recounting failures and false trails along with successes, it has in most cases been accepted as a straightforward record of Kepler's work. It is none of these things. The book was written and (I shall argue) rewritten carefully, to persuade a very select audience of trained astronomers that all the planetary theory they knew was wrong, and that Kepler's new theory was right. The whole of the *Astronomia nova* is one sustained argument, and I shall make what I believe is the first attempt to trace that argument in detail.[10]

However, having raised the issue of the *Astronomia nova*'s being rhetorical, Stephenson then followed its argument almost exclusively. While his work is an extremely valuable reading of the book, Stephenson failed to convince many readers of the validity of his claims.[11]

The evidence that the *Astronomia nova* was "written and rewritten," which Stephenson failed to provide, was swiftly produced in a spectacular way in William H. Donahue's "Kepler's Fabricated Figures: Covering up the Mess in the *Astronomia nova*" (1988). Donahue's intricate analysis revealed that Kepler had worked over at least one chapter of the *Astronomia nova* so many times prior to publication that it scarcely hung together, and that ultimately Kepler resorted to calculating positions with his final finished theory and passing them off as the results of an earlier, observational procedure. This was either "patent fraud," as Donahue somewhat recklessly declared,[12] or very good evidence of Kepler's narrative's didactic intention.

By the time Donahue published the first English translation of the *New Astronomy* in 1992, he weighed in heavily on Stephenson's side with regard to the rhetorical nature of the book:

> That is, although Kepler often seems to have been chronicling his researches, the *New Astronomy* is actually a carefully constructed argument that skillfully interweaves elements of history and (it should be added) of fiction. Taken as history, it is often demonstrably false, but Kepler never intended it as history. His introduction to the "Summaries of the Individual Chapters" makes his intentions abundantly clear. *Caveat lector!*[13]

With the growing acceptance that the *Astronomia nova* was in fact a painstakingly prepared, sustained argument, one major question remained: why? What had driven Kepler to these lengths? Here, Kepler's vast, unexploited correspondence promised an answer, and in particular, the part of Kepler's correspondence that has been almost entirely ignored by historians: letters written *to* Kepler. In this work, I have sought to pursue two questions. First, if the *Astronomia nova* is didactic or rhetorical, what can be reconstructed about the development of Kepler's physical astronomy without reference to the account offered in it? Second, and more important, to what extent did Kepler's interaction with the astronomical community affect the content and presentation of his work? In treating Kepler at last as a figure in time and context and not independent of it, this work provides the long-needed rhetorical context of the *Astronomia nova*.

THE COMPOSITION OF KEPLER'S *ASTRONOMIA NOVA*

⌘

⌘ INTRODUCTION ⌘

Johannes Kepler's *Astronomia nova* (1609) has long been recognized as one of the canonical works of the Scientific Revolution. Between Copernicus's *De revolutionibus orbium coelestium* (1543) and Newton's *Philosophiae naturalis principia mathematica* (1687), it occupies a position of central importance in the development of astronomy during the sixteenth and seventeenth centuries. Its significance is twofold. In terms of astronomical theory, it signifies the beginning of the end for a millennia-old tradition of mathematical astronomy, in which the motions of the planets were represented using only compounds of uniform circular motion. Kepler's elliptical orbits and his area law (the first two of what later came to be called his three laws of planetary motion) subsequently became essential elements in the Newtonian theoretical synthesis that was the culmination of the Scientific Revolution. Kepler's achievement also presaged Newton's in a second, more fundamental way. In place of the ancient tradition of mathematical astronomy, Kepler substituted a physical approach to astronomy—"celestial physics," as he named it—in which theories of planetary motion were derived from the physical consideration of the cause of their motion. He was able to derive his first two laws of planetary motion from a flawed but self-consistent set of physical principles. The unification of physics and astronomy in which Kepler played a leading role represents the most important conceptual change in science during the period.

The conceptual importance of Kepler's methodology of physical astronomy has been described by historians of astronomy, beginning with Alexandre Koyré and more recently by Bruce Stephenson. Their task of analyzing the manifest role it played in his discoveries was made possible by another highly unusual feature of the *Astronomia nova*. Unlike the traditional literary models for astronomical treatises, such as Ptolemy's *Almagest* or Copernicus's *De revolutionibus*, in which the exposition of planetary theory proceeded deductively with few clues regarding the ways those theories came into being, the *Astronomia nova* was a narrative odyssey through Kepler's development of his astronomical theory. Kepler did not hesitate to discuss the series of false starts, blind alleys, and failures he encountered on his road to eventual success.

Recent research, especially that of William H. Donahue, has shown that the account Kepler offers his readers is not a true history of the course of his research—something Kepler never claimed—but is rather a didactic or rhetorical pseudohistory. But until now, the question of why Kepler chose this form of exposition has not been addressed. My work finds the answer to this question in the context of the composition of the *Astronomia nova* and in Kepler's relation to the contemporary astronomical community. I argue that the unique conceptual and stylistic features of the *Astronomia nova* are intimately related: Kepler purposely chose this form of exposition precisely because of the response he knew to expect from the astronomical community to the revolutionary changes in astronomical methodology he was proposing.

This interpretation also resolves a broader tension in our view of Kepler's intellectual achievement. Throughout his life, Kepler's astronomical work was devoted to showing that the Copernican heliocentric system of the world was true. Yet some of his works are very different in character. His youthful *Mysterium cosmographicum* (1596) argued for heliocentrism on the basis of metaphysical, astronomical, astrological, numerological, and architectonic principles. By contrast, the *Astronomia nova* was far more tightly argued on the basis of only a few dynamical principles. The contrast in the works seems to embody a transition from Renaissance to early modern science; in Arthur Koestler's characterization, Kepler seems to have passed over a "watershed." However, Kepler did not subsequently abandon the broader approach of the *Mysterium cosmographicum*. Similar metaphysical arguments reappeared in his *Harmonice mundi* (1619), and he reissued the *Mysterium cosmographicum* in a second edition in 1621, in which he qualified only some of his youthful arguments. Given the persistence of these ideas in Kepler's work, it is clear that he himself did not experience some sort of conversion experience and become a modern scientist. We must ask instead how it was that the *Astronomia nova* in particular was written in the style it was. One of the conclusions of my work is that the *Astronomia nova* is only accidentally modern—that is, that the particular context in which the book was composed forced Kepler to rein in his broader arguments for heliocentrism, leaving only a subset of his physical reasoning that appears distinctly modern in retrospect.

Two interrelated questions arise from the fact that Kepler's *Astronomia nova* does not provide an entirely faithful account of his research on the theory of Mars nor of his broader approach to the physical truth of heliocentrism. First, if it is not a true account of his Mars research, how did Kepler actually proceed? Second, what was Kepler's motivation for presenting his findings in the form of a narrative, and for obscuring his broader conception of physical reasoning? I argue in my work that answers to both these questions can be found in the development of Kepler's research and his interaction with the astronomical community.

This book covers the evolution of Kepler's thought through the publication of the *Astronomia nova*. My argument is twofold. First, I establish the breadth of Kepler's notion of physical reasoning and the continuity of research from the *Mysterium cosmographicum* to the *Astronomia nova*. I describe how the conditions of his work under Tycho Brahe strictly limited Kepler's research, but that it nevertheless proceeded along lines that came forth from the *Mysterium cosmographicum*. Second, I address the composition of the *Astronomia nova*. I argue that Kepler intentionally obscured the continuity between the *Mysterium cosmographicum* and the *Astronomia nova* in the face of the negative response his physical reformation of astronomical theory faced from within the astronomical community. And I show how his rhetorical narrative was meant to convince his readers of the necessity of his approach and to lead them through difficult and contentious material.

Part 1 covers the period from Kepler's days as Michael Maestlin's student at the University of Tübingen up until he began his research with Tycho Brahe. In chapter 1, I introduce the prevailing attitude among astronomers to Copernicus's work and show how Kepler deviated from it. Although we regard Copernicus's work as significant for putting forward the idea that the earth travels around the sun, sixteenth-century astronomers largely ignored that claim, which violated Aristotelian physics and apparently contradicted the testimony of Holy Scripture. Instead, they were attracted by Copernicus's novel form of mathematical planetary theory, which eliminated Ptolemy's equant, a mechanism that caused the center of a planet's epicycle to travel nonuniformly around its eccentric and thus violated the precept that planetary theories should be composed of compounds of uniform circular motion.[1]

In Kepler's earliest writing on Copernicus, a fragment of a student disputation from 1593, he ignored conventional astronomers' interpretation of heliocentrism and disregarded Copernicus's detailed mathematical arguments. Instead, and apparently in the face of resistance from his audience, Kepler argued for the physical truth of heliocentrism on the basis of what he called "cosmographical" reasons. These were largely conventional metaphysical arguments for heliocentrism, taken from either Copernicus or Rheticus's *Narratio prima* (1540), but Kepler also introduced one highly significant innovation. He expanded a conventional claim that the sun was the source of all heat, light, and motion in the solar system to suggest that one might derive the planets' periods from their distances from the sun, the source of their motive power. Copernicus had noted the correlation but had never quantified it.

In chapter 2, I recount how Kepler, after having been forced to leave seminary to assume the position of mathematics teacher at the Protestant school in Graz, returned to the ideas of his student disputation and to his defense of Copernicus based on physical reasoning. He sought to redirect his religious aspirations into astronomy by arguing that the heliocentric system of the world made plain the glory of God in His creation of the world. Thus he made the establishment of the physical truth of heliocentrism a religious vocation.

To the problem of accounting mathematically for the relationship between the planets' distances and their periods, Kepler now added the questions of accounting for the number of planets and their particular distances from the sun. He promptly hit upon an explanation for the latter problems. In his "polyhedral hypothesis," he reasoned that God had used the five perfect Platonic solids as archetypes when constructing the solar system. By interpolating the five solids between inscribed and circumscribed spheres, Kepler was able to derive values for their distances and to provide an explanation for the number of planets.

In addition, Kepler began to develop the notion he had expressed in his student disputation into a quantitative "motive force" hypothesis relating the planets' periods and distances. Reasoning that planets' periods increase with distance both because the planet-moving force is weaker and because the circumference of their orbits are longer, he combined the effects to come up with an expression for the relationship between the planets' distances and periods that was somewhat less accurate than the polyhedral hypothesis.

The polyhedral hypothesis became the centerpiece of Kepler's first book, the *Mysterium cosmographicum* (1596), which I discuss in chapter 3. The polyhedral hypothesis proved to be a very fertile source of ideas, and Kepler buttressed the argument with numerous auxiliary arguments based on the astrological, numerological, and metaphysical appropriateness of the arrangement he was proposing. Kepler considered all of these to be elements of his "physical" argument for Copernicus. In the preface, he refers to the arguments of his student disputation as "physical, or if you prefer, metaphysical." His conception of what constituted physical arguments corresponded roughly to Aristotelian causes, and especially to the formal cause of the world.

Although arguments of formal cause based on the polyhedral hypothesis had swelled to constitute the bulk of the *Mysterium cosmographicum*, Kepler did not lose sight of the significance of the sun as the source of motion in the solar system, and he included an additional argument based on this motive force hypothesis toward the end of the book. In a highly significant application of the idea of motive force, he considered what effect the change in a planet's distance from the sun would have on its motion around its own orbit. He came to the conclusion that the physical motion of a planet around an eccentric orbit would be the same as that described in classical mathematical astronomy by either Ptolemy's bisected equant or Copernicus's eccentric epicyclet arrangement. He thus concluded that both these theories were merely mathematical models for the physical motion whose cause he had described.

Kepler conceived his physical "proof" of the reality of heliocentrism in the *Mysterium cosmographicum* as an affirmation of faith. However, this aim of the book was subverted by prepublication censorship. In chapter 4, I describe how the theologians at the University of Tübingen arranged to suppress a chapter of the book intended to address the reconciliation of the Copernican system with Holy Scripture. In doing so, they urged Kepler to "play the part of the pure mathematician" and desist from arguing for the physical truth of heliocentrism. Their view that mathematics had no claim to physical truth reflected a common fictionalist stance toward the status of astronomical hypotheses, which Kepler endured at that moment but ultimately could not accept.

The response of the astronomical community toward the *Mysterium cosmographicum*, which I also describe in chapter 4, was mixed. On the one hand, there were those who embraced Kepler's finding that the dimensions of the solar system could be found from the inscribed polyhedra. Georg Limnaeus, for instance, lavishly praised Kepler for reviving the *prisca philosophia* of the ancients. And Michael Maestlin even suggested that the polyhedral hypothesis could be used to derive better values for the planetary distances than could be found from observation. Even Tycho Brahe said that Kepler's scheme was ingenious, in spite of the fact that some expected Tycho to take the leading role in refuting Kepler's pro-Copernican argument.

However, there was one point to which astronomers reacted uniformly negatively: they all agreed that Kepler's attempt to account for the function of the equant on the basis of his planet-moving force was ill-conceived. They considered it inappropriate—even dangerous—to apply physical reasoning to mathematical planetary theory. I argue that the distinction between the fairly positive reaction to the book as a whole versus the critical reaction to Kepler's explanation of the equant was based on a rigid division within astronomy between cosmography and planetary theory. The former addressed broader questions about the form of the world and was closely allied to physics; thus Kepler's physical arguments were acceptable. The latter, however, was considered part of mathematics and did not admit physical reasoning. Thus to the mathematical astronomer Johannes Praetorius, Kepler's work was more aligned to physics, and "cannot be of use to the astronomer in almost any way."

The *Mysterium cosmographicum* had the fateful consequence of bringing Kepler into contact with Tycho Brahe. In part 2, I cover the period from Kepler's collaboration with Tycho Brahe until the publication of the *Astronomia nova*. During this time, Kepler's qualitative explanation of planetary motion based on his planet-moving force acquired a quantitative exactness. With the help of Tycho's unprecedentedly accurate observations, his earlier physical insight led him to his first two laws of planetary motion. During the same period, he also became definitively aware of the resistance this new kind of physical astronomy would face.

The portentous encounter between Tycho Brahe, the aged observer, and Johannes Kepler, the young theorist, is so convenient that it can seem inevitable. In chapter 6, I try to take an unbiased view of their collaboration in the light of recent scholarship that has suggested that Kepler was more desirable to Tycho as a pawn in his legal struggle with Nicholas Reimers Ursus than as an assistant. From the terms of their agreement, Kepler does not seem to have occupied a particularly favored position in his first few months with Tycho, but was probably rather low in the hierarchy of assistants. Nor does Tycho appear to have overseen his work too closely.

Despite Kepler's express hope of receiving from Tycho improved values for the planetary distances with which to test and improve the polyhedral hypothesis, Tycho would not provide this information. Instead, he assigned Kepler to work on the theory of Mars and gave him observations for just that planet. Despite being barred from developing the primary argument from the *Mysterium cosmographicum*, Kepler could still pursue his motive force hypothesis. And during his first few months with Tycho, Kepler experienced some remarkable successes in his research with Mars. First, he discovered that the theory of Mars seemed to require being referred to the true sun—the source of its motion, to Kepler—rather than the center of the earth's orbit (the "mean sun"), as Copernicus and Tycho had done. Second, he discovered that the eccentricity in the theory of the earth needed to be bisected, just as Ptolemy had bisected the eccentricities in the theories of the planets. Ever since the time of Hipparchus, up to and including Tycho's successful solar theory, the earth had always been assigned a simple, unbisected eccentricity. But to Kepler, the earth's simple eccentricity had been an unsatisfying qualification in the motive force hypothesis in the *Mysterium cosmographicum*, for it had not been amenable to Kepler's explanation in terms of the planet-moving force. In addition to bringing the theory of the earth into line with the theories of all the other planets, the bisection of the earth's eccentricity also eliminated an annual variation in Mars's eccentricity that Tycho had raised as an objection to the planet-moving force hypothesis.

The bisection of the earth's eccentricity later became an important element of the argument of the *Astronomia nova*, where it was presented in part 3 and provided the justification for Kepler to change from a purely mathematical to a physical approach to finding Mars's orbit. But at the time Kepler found it, he had not yet completed the research presented in part 2. Moreover, the continuity between the *Mysterium cosmographi-*

cum and Kepler's Mars research makes it clear that he had pursued a physical approach to planetary theory from the beginning of his collaboration with Tycho.

A clue to Kepler's reorganization of the account of his research comes from Tycho's reaction to Kepler's resort to natural (physical) principles. The available evidence shows that Tycho objected "vehemently" to this kind of research. In chapter 7, I show how the direction of Kepler's research after Tycho's death, though always motivated by the physics of the planetary orbit, took an abruptly more physical turn, as he began for the first time to employ a version of his area law and to experiment with oval orbits.

Under the circumstances prevailing just after Tycho's death, the *Astronomia nova* would probably never have been published at all. In chapter 8, I explain how a struggle between Kepler and Tycho's heirs over the right to profit on Tycho's astronomical inheritance led to Kepler's losing responsibility for the *Rudolphine Tables*. At the same time, he was ordered to name what works he would produce to justify his recent appointment as imperial mathematician. Placed in this bind, Kepler named as one of the works he would produce his *Commentaries on Mars*—that is, the *Astronomia nova*. He was thus forced to conceive the book as a preliminary announcement of the fruits of his physical astronomy as applied to the orbit of Mars. It would contain his important finding regarding the bisection of the earth's eccentricity, which vindicated his physical account of the cause of the equant as well as clearing up certain problems in the orbit of Mars (and the orbits of Mercury and Venus as well). At that time, however, he had no clear idea of what the eventual solution to Mars's orbit would be. Although he was employing a form of the area law, the discovery of Mars's elliptical orbit was still two and a half years away.

Kepler's struggle with Tycho's heirs also led to Tycho's son-in-law, Franz Tengnagel, gaining the right to censor any of Kepler's work based on Tycho's observations, and this outcome had serious consequences for the composition of the *Astronomia nova*. Tengnagel did indeed censor some of Kepler's work because it strayed too far from Tycho's intention. And when Kepler received letters of criticism from Christian Severin Longomontanus, Tycho's longtime chief assistant, imploring him to give up his ill-conceived program of physical astronomy, Kepler had reason to fear that a conspiracy among Tycho's legal and scientific heirs—whom he called "the Tychonics"—might threaten his philosophical freedom. In response, he justified the course of his research on the theory of Mars in a long letter to Longomontanus, whose rhetorical narrative is similar to the argument of the *Astronomia nova*. I argue that pressure from the Tychonics, including ridicule of the *Mysterium cosmographicum* from

Longomontanus, influenced Kepler to restrict the range of his physical arguments to only those dynamical arguments that were essential for the *Astronomia nova*.

At the same time as his problems with the Tychonics were developing, Kepler learned that even a sympathetic friend and correspondent could raise serious objections to his work. In chapter 9, I describe how Kepler's correspondence with David Fabricius, an East Frisian pastor and amateur astronomer, acted as a kind of peer review for the *Astronomia nova*. In a friendly and forthright manner, Fabricius demanded certain demonstrations in order for Kepler win his assent to the radical innovations he was proposing in the theories of the earth and Mars. I demonstrate how Fabricius's queries formed the framework for numerous specific chapters in the *Astronomia nova*.

As the complexity of Kepler's Mars work increased, Fabricius became more and more skeptical about the course of Kepler's research and began to offer him alternative theories of Mars based on compounds of uniform circular motion in the classical style. When the third of these reproduced an ellipse that seemed to differ insensibly from Kepler's own theory, Kepler viciously attacked it and broke off their correspondence. The threat to the argument of the *Astronomia nova* was clear. Kepler was arguing on the basis of the fact that only by the use of physical reasoning had he arrived at the correct solution of Mars's orbit. His solution of the problem of Mars's orbit would justify physical astronomy and, consequently, also the physical truth of the heliocentric system. He knew the argument could not succeed if an alternative in the classical instrumentalist form were available.

In the concluding chapter 10, I offer a reading of the *Astronomia nova* as rhetoric. I show how the argument of the book was a response to the various criticisms he had encountered during the course of his research. To the charge that his physical astronomy was an unjustified aberration, he responded by constructing his argument to make it appear as though he resorted to a physical approach to planetary theory only after a comprehensive failure of the most general kind of model in the classical form (which he presented in part 2, even though he actually completed the research only after parts of the research presented in part 3). He countered the charge that his radical innovations were themselves the source of the difficulties he had encountered by repeating many of the demonstrations in the book (as with the repeated demonstrations involving the true and the mean sun). And in order to justify his unprecedented innovation of bisecting the earth's eccentricity, he offered numerous redundant demonstrations.

Moreover, I argue that many of Kepler's failed attempts served a valuable didactic function. For instance, the faulty orbit of Mars called the *via buccosa* was the result of Kepler's mistaken construction of Mars's position on the ellipse. His experience with Fabricius had shown that when he omitted the explanation as to why the construction did not work, Fabricius was suspicious of the complexity of the true construction. Thus, many features of the *Astronomia nova* become comprehensible only when they are viewed in the context of Kepler's experience in writing the book as elements of an elaborate and purposefully-constructed rhetorical argument.

This work analyzes Kepler's composition of the *Astronomia nova* in a detail never attempted before. By viewing the account of his research that Kepler offers in his published work with skepticism and by attempting to reconstruct the actual course of his research from contemporary sources, it adds a new depth to our appreciation of this canonical text. In particular, it establishes the meaning of this text within the context of late sixteenth- and early seventeenth-century astronomy and against the backdrop of Kepler's contemporaries' view of his work.

When viewed in context, the meaning of the *Astronomia nova* becomes clear. By observing the persistence with which Kepler pursued long-held convictions deep into the investigation of Mars's orbit and emerged triumphant, we understand what in the book was important to him. By examining how the situation in which he found himself after Tycho's death determined this announcement of his results, we understand why it was written. Most important, by knowing the suspicion with which Kepler's introduction of physics into astronomy was viewed and the incomprehension his work elicited, we can at last explain its curious structure.

PART 1

THE *MYSTERIUM COSMOGRAPHICUM*

⌘

⌘ CHAPTER 1 ⌘

THE COPERNICAN CONTEXT

When Johannes Kepler entered the University of Tübingen in 1589, the Copernican revolution was far from complete. Nearly fifty years earlier, an aged Nicolaus Copernicus (1473–1543) had published his *De revolutionibus orbium coelestium* [*On the Revolutions of the Celestial Orbs*] (1543). In this monumental work, Copernicus had put forth the first developed proposal that the earth travels around the sun. However fundamental the heliocentric system may seem from our modern vantage point, even by Kepler's time there were relatively few astronomers who believed that it was actually true. And yet Copernicus was widely recognized as one of the most brilliant astronomers of the time. To reconcile Copernicus's astronomical renown with the extreme skepticism with which the heliocentric system was received, it is necessary to examine his work within the contemporary context. Heliocentrism faced serious ambiguity within Copernicus's work itself and from views of the relationship between astronomy and physics.

In the prefatory letter to *De revolutionibus*, Copernicus himself delineated two separate concerns with the state of astronomical theory that had led him to the composition of his heliocentric system:

> I was impelled to consider a different system of deducing the motions of the universe's spheres for no other reason than the realization that astronomers do not agree among themselves in their investigations of this subject. For in the first place, they are so uncertain about the motion of the sun and the moon that they cannot establish and observe a constant length even for the tropical year. Secondly, in determining the motions not only of these bodies but also of the other five planets, they do not use the same principles, assumptions, and explanations of the apparent revolutions and motions. For while some employ only homocentrics, others utilize eccentrics and epicycles, and yet they do not quite reach their goal. For although those who put their faith in homocentrics showed that some non-uniform motions could be compounded in this way, nevertheless by this means they were unable to obtain any incontrovertible result in absolute agreement with the phenomena. On the other hand, those who devised the eccentrics seem thereby in large measure to have solved the problem of the apparent motions with appropriate calculations. But meanwhile they introduced a good many ideas which apparently contradict the first principles of uniform motion.

> Nor could they elicit or deduce from the eccentrics the principal consideration, that is, the structure of the universe and the true commensurability of its parts. On the contrary, their experience was just like some one taking from various places hands, feet, a head, and other pieces, very well depicted, it may be, but not for the representation of a single person; since these fragments would not belong to one another at all, a monster rather than a man would be put together from them.[1]

Copernicus related a complex state of affairs in the astronomy of his time. But broadly, there were two separate strands of concern, one having to do with planetary theory and the precise prediction of planetary positions, and the other with a larger, cosmological concern for the system of the world. Both were in essence physical problems, but physical problems of a fundamentally different kind.

Planetary Theory

In terms of prediction and planetary theory, Copernicus outlined two different camps. On the one hand, there were those who used homocentric spheres but who could not accurately reproduce the phenomena. On the other hand, while those who used eccentrics and epicycles had adequately solved the problem of modeling the apparent motions, at the same time Copernicus expressed his discomfort that in doing so they violated the "first principles of uniform motion." This physical problem in astronomical theory he alluded to was well known and stretched back in time to Aristotle.

According to the cosmology set forth by Aristotle, the heavens were a fundamentally different substance than the terrestrial elements in the sublunary realm. This heavenly substance—aether—alone had the property of unending uniform circular motion. This physical account of the nature of aether had been appropriate for the astronomical theory of Aristotle's time, Eudoxus's system of homocentric spheres. In Eudoxus's system—the details of which are debatable—the irregular motions of the planets were the result of the compound motion of various spheres centered on the earth. Although the system of homocentric spheres became a fundamental tenet of Aristotle's cosmology, it was sufficient only for a qualitative account of planetary motion. Only much later in antiquity did planetary theories acquire observational exactness, and at this time it became necessary to violate, or at least stretch the tenets of Aristotelian cosmology.

The acme of classical astronomy was Ptolemy's *Almagest*, composed around A.D. 150. In providing quite accurate geometrical theories of the

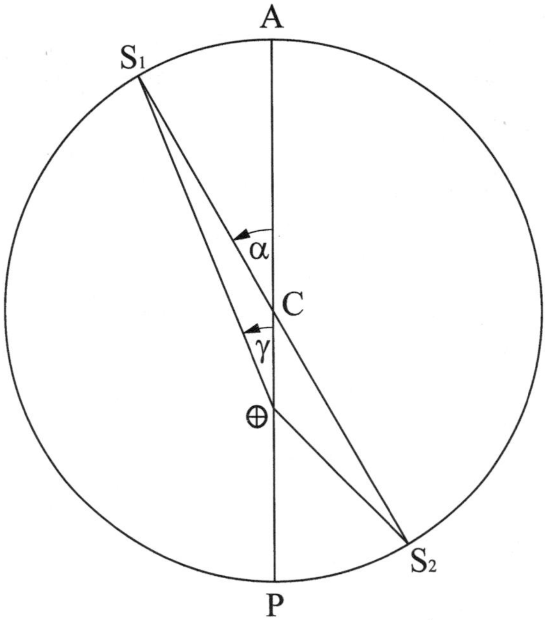

Figure 1.1 A simple eccentric

In a simple eccentric planetary theory, such as that for the sun, the earth (⊕) is displaced from the center (C) of a circular orbit around which a planet (S) moves uniformly. The planet's closest approach to the earth is at perigee (P) and its farthest distance is at apogee (A). The line of apsides connects A and P and passes through the center and the earth. Although the planet moves uniformly around the circle, its apparent motion viewed from the earth is nonuniform, as can be seen from the two locations of the planet S_1 and S_2. The angles ACS_1 and PCS_2 are equal, meaning that the planet has moved the same amount of time since crossing the line of apsides, but the angular motion as viewed from the earth ($A\oplus S_1$ and $P\oplus S_2$) is different, with the planet appearing to move slower at apogee and faster at perigee.

planets' motions, Ptolemy made use of a number of mathematical developments in astronomical theory from the preceding few centuries that were difficult to reconcile with the older Aristotelian cosmology. One of the most elementary was the simple eccentric orbit, in which the center of the orbit is displaced from the earth (figure 1.1). The simple eccentric was an elegant geometrical solution to modeling a planet's observed nonuniform motion around the zodiac, known as the first anomaly of planetary motion or the zodiacal anomaly. If a planet moves uniformly around an eccentric, it will appear from our off-center vantage point to spend more time in one part of the sky than in another. The simple eccentric was employed in Hipparchus's solar theory from the second century B.C.,

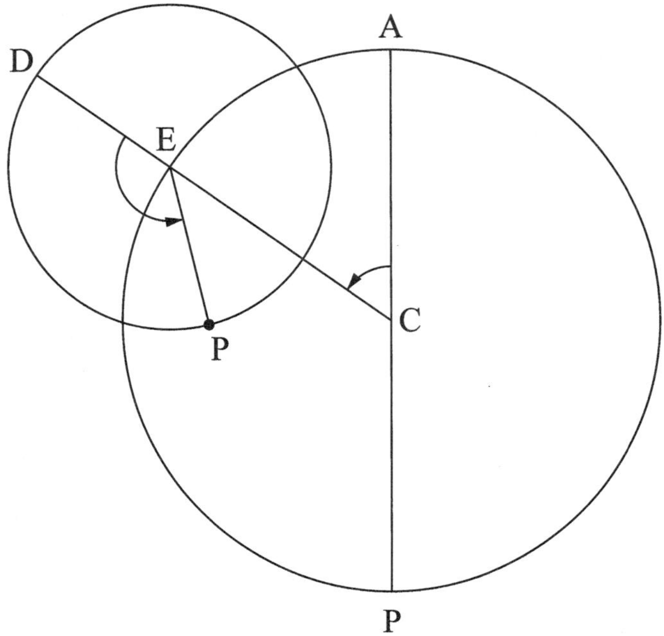

Figure 1.2 The epicycle and deferent
In an epicycle-deferent theory, a planet (P) moves on the circumference of a small circle (DP) called an epicycle, whose center (E) moves in turn on the circumference of a "carrying circle" or deferent (AEP). By adjusting the periods of the epicycle and deferent and the ratio of their sizes, the system can be used to cause a planet to move temporarily backward, when the backward motion of the planet on the epicycle briefly overcomes the continuing forward motion of the deferent.

and adopted by Ptolemy in the *Almagest*. However, although the simple eccentric was an adequate tool for more precisely reproducing a planet's observed motion, it did represent a violation of Aristotle's cosmological scheme, for, of course, with eccentrics, the orbits of the planets will no longer be homocentric with their centers on the earth.

Ptolemy's use of epicycles and deferents represented an even more severe violation of Aristotelian principles. In an epicycle-deferent system, first used in astronomical theory by Apollonius of Perga in the third century B.C., the planet rides on the circumference of a small circle, the epicycle, whose center is carried in turn on the circumference of a larger "carrying circle," the deferent (figure 1.2). Suitable adjustments of the sizes and speeds of the epicycle and deferent allow for a straightforward explanation of the second anomaly of planetary motion, that is, the retrograde motion of planets. During retrograde motion, planets cease their ordinary eastward motion with respect to the stars and move briefly westward. In an epicycle-deferent model, retrograde motion occurs when the planet is in the bottom half of the epicycle and its backward motion temporarily overcomes the continuing forward motion of the deferent. Again, the epicycle-deferent system represented a challenge to Aristotelian physics; although the motions might be uniform and circular, one element of the theory—the epicycle—no longer even surrounded the earth.

It was Ptolemy's greatest innovation in planetary theory that posed the largest challenge to Aristotelian physics and that Copernicus singled out as "contradict[ing] the first principles of uniform motion." A simple eccentric epicycle-deferent is only approximately capable of modeling the motion of a planet. If the eccentricity is made sufficiently large to model the first anomaly of planetary motion, the epicycle at apogee becomes apparently too small to replicate accurately the second anomaly—that is, a planet's retrograde loops become smaller than observed. Likewise, at perigee, the epicycle will be apparently too large, and the retrograde motion will be larger than observed. It seems that it was this problem that led Ptolemy to introduce the equant, a point situated equidistant across the center of a planet's deferent from the earth around which the center of the epicycle moved uniformly (figure 1.3). By dissociating the center of uniform motion of the deferent from its geometrical center, Ptolemy was able to accentuate the nonuniformity of motion of an eccentric without exaggerating the apogeal and perigeal distances of the epicycle, which would have resulted in too-large or too-small retrograde episodes.[2] Because the orbits of the planets are so nearly circles, the equant nearly reproduces Kepler's area law, and the epicycle is a good approximation of the earth's orbit,[3] Ptolemy's theories were quite accurate and remained the foundation of astronomical theory until the sixteenth century.

But there was and there remained a lingering inconsistency between Ptolemaic astronomy and Aristotelian natural philosophy. The fundamental element in the conflict between the two had to do with Aristotle's requirement that aether have only uniform circular motion. Ptolemy adhered to the spirit of this principle, but while the angular motion of the epicycle center around the equant point was uniform, its physical motion around the eccentric deferent could not be. It was this nonuniform motion of the epicycle center that Copernicus referred to when he mentioned that those who used eccentrics and epicycles violated the first principles of uniform motion.

Ptolemy's violation of the principle of uniform motion did not go unnoticed by Islamic commentators as early as the eleventh century, and the revision of planetary theory undertaken by the so-called Marāgha school of astronomers in the thirteenth and fourteenth centuries was directed at recasting Ptolemaic theory into some other compound of uniform epi-

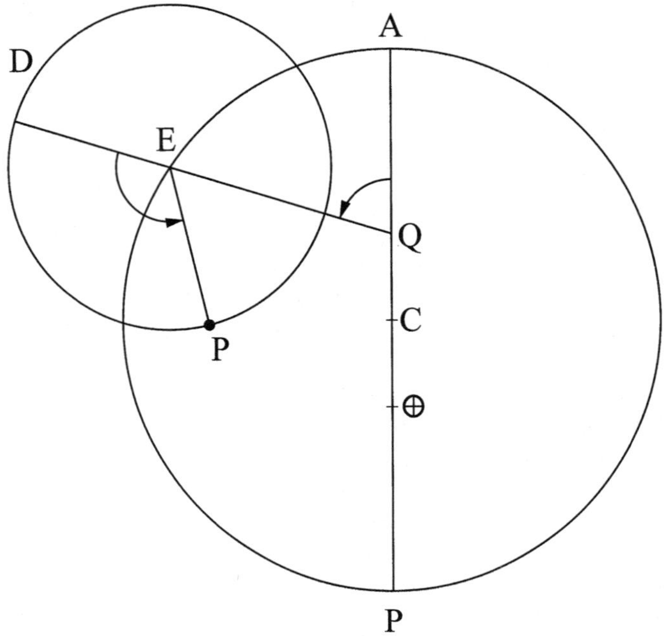

Figure 1.3 A Ptolemaic planetary theory
In the general form of his planetary theory, Ptolemy modified an epicycle deferent system by incorporating a "point of equal motion" or equant (Q). The center of the epicycle (E) still moves along the circle centered at C, but it moves uniformly around an eccentric point (Q). At the same time, the earth (⊕) is moved to an eccentric point equal and opposite the center from the equant. The equant has the effect of accentuating the nonuniformity of motion of a simple eccentric without exaggerating the apogeal and perigeal distances, which would cause retrograde motion near perigee to be too large and too small near apogee.

cyclic motion that did not entail this violation of uniform circular motion. The evidence strongly suggests that Copernicus drew on this work, though the channels of transmission have not been definitively established.[4] For instance, Copernicus's technique for reproducing the effect of Ptolemy's equant without its violation of uniform circular motion in the models of the superior planets was identical to that employed by Mū'ayyad ad-Dīn al-ʿUrḍī and Ibn ash-Shāṭir. The model involved replacing Ptolemy's bisected eccentric deferent with a deferent of different eccentricity and a small epicycle, which I shall henceforth refer to as an "epicyclet" to distinguish it from the large epicycles in Ptolemy's theory that accounted for retrograde motion (figure 1.4). As Copernicus's deferent rotated counterclockwise, the epicyclet rotated counterclockwise at exactly

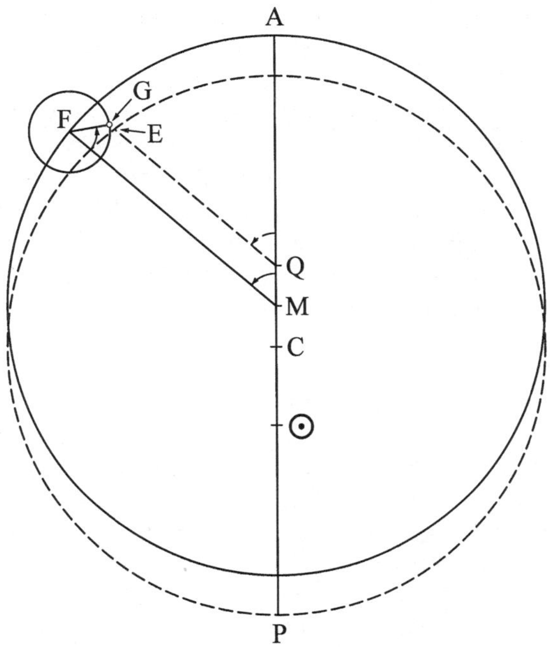

Figure 1.4 A Copernican eccentric epicyclet
Copernicus replaced Ptolemy's eccentric deferent with an equant in the theories of the superior planets (shown here in dotted lines) with a deferent with an eccentric epicyclet (shown solid). Copernicus's deferent is centered at a point M midway between the center (C) and the equant (Q), such that QM = MC = ½ e. The epicyclet—so called to distinguish it from the large epicycles in Ptolemy's theory that account for retrograde motion—rotates in the same sense as the deferent at the same speed. The radius FG = QM = ½ e. The angles AMF and MFG are always equal; thus the line GQ is always parallel to MF. Therefore, the point G rotates uniformly around the equant, perfectly replicating that aspect of Ptolemy's theory. The difference between the theories can be seen in the slight difference between the point G and the point E from Ptolemy's theory, which is negligible.

Note that the large epicycle DP from Figure 1.3 has been omitted from this figure. In Copernican theory, retrograde motion was explained by the motion of the earth. To recast a Copernican theory into geocentric form, simply attach the same epicycle DP to the point G.

the same the speed, which ensured that a point on the epicyclet always moved around uniformly with respect to the equant point in Ptolemy's theory. The only deviation from Ptolemy's theory was that the point on the epicyclet did not trace Ptolemy's deferent circle exactly, though the observational difference was negligible.

The aspect of Copernicus's critique of contemporary astronomy in the prefatory letter to *De revolutionibus* addressing Ptolemy's violation of "the first principles of uniform motion," then, was not especially revolutionary. And the reformulations of Ptolemaic theories he offered in the book found a ready and appreciative audience among sixteenth-century astronomers, to whom Ptolemy's violation of the principle of uniform motion was likewise troubling. Nor can it be said that Copernicus's motivation was to be particularly more accurate than Ptolemy had been. He set his goal to reformulate astronomical theory to replicate Ptolemy's empirical success, rather than to exceed it.[5]

Cosmology

As discussed thus far, nothing of Copernicus's reform of planetary theory relies on the motion of the earth around the sun. This innovation was the product of his second critique of astronomical theory, that it failed in "the principal consideration, that is, the structure of the universe and the true commensurability [*symmetria*] of its parts." This was a physical concern of a different kind, one of a more metaphysical, cosmological nature.

Ptolemy put his astronomical theories forward in the *Almagest*, where they were presented more or less individually. His explanation of how all the theories could be thought of together as a cosmological system was reserved for his *Planetary Hypotheses*. Although the *Planetary Hypotheses* was lost in the West, the sense of Ptolemy's account of his cosmological system was known to Islamic commentators and passed to the West, where it was published notably in Georg Peurbach's *Theorica novae planetarum* (Nuremberg, 1473).[6]

In his physics, Ptolemy was an Aristotelian. To accommodate his epicycle-deferent planetary theories to Aristotelian cosmology, it was necessary to explain how they might be thought of in terms of aether spheres. His solution was to describe each planetary theory in its entirety, from a planet's farthest apogeal to nearest perigeal distance, as one homocentric sphere centered on the earth. Within this thickest sphere was another eccentric sphere, whose motion corresponded to the motion of the deferent. Within this second sphere, touching its outer and inner surfaces, was a spherical cavity, in which another aether sphere could rotate independently. This third sphere represented the epicycle.

In accordance with Ptolemy's cosmological axiom that the universe is a plenum, it was now necessary to arrange the planetary theories with no wasted space in between. To do so, Ptolemy scaled the planetary models so that the farthest distance of one corresponded to the closest approach of the next. First, he used a measurement of the moon's diurnal parallax to determine its absolute distance in terms of the size of the earth. His theory of the moon's motion then indicated that its furthest distance would be 64 earth radii. He scaled Mercury's theory so that its closest approach would be 64 earth radii, which required in turn that its farthest distance would be 115 earth radii, and so on through Venus, the sun, Mars, Jupiter, and Saturn, out to the sphere of the fixed stars, which finally ended up being about 20,000 earth radii away.[7] Significant correlating evidence came when Venus's farthest distance fell within 100 earth radii of the closest distance of the sun as determined through an independent observational demonstration.

Copernicus's critique of Ptolemy's way of assembling the planetary models into a cosmological system was that it had no commensurability—that is, that this arrangement of the universe had no common measure. In the Copernican system, retrograde motion was explained as the result of the earth's motion around the sun. This simplifying assumption allowed Copernicus to collapse Ptolemy's epicycles in the theories of Mars, Jupiter, and Saturn, and the deferents in the theories of Mercury and Venus, into a single orbit, that of the earth around the sun. In the process, however, it was necessary to retain the previous proportions between the planets' epicycles and deferents, and so the proportions of all of the planetary orbits were locked into a single specific relation with the orbit of the earth. Thus the sizes of all the orbits were known in proportion to the size of the earth's orbit and to one another. It was, as Copernicus touted, commensurable.

It is not clear, however, that Copernicus's metaphysical principle of commensurability is any less idiosyncratic than Ptolemy's principle of plenitude. One might ask, as Kepler would, why were the planets located at precisely those distances and not any other? And why was there so much useless intervening space? But these were not issues Copernicus addressed. He did offer a cogent correlation between the distances of the planets and their periods. When appropriately scaled to the size of the earth's orbit, the planets were necessarily arranged so that Mercury, the fastest planet, was closest to the sun and so on all the way out to Saturn, with the largest, slowest orbit. "Therefore," Copernicus wrote, "in this ordering we find that the world has a wonderful commensurability and that there is a sure bond of harmony for the movement and magnitude of the orbital circles such as cannot be found in any other way."[8]

We noted at the beginning of this chapter that the status of heliocentrism occupied an ambiguous place within Copernicus's own work, for these cosmological arguments in favor of heliocentrism are only brought up in the first part of Book I of *De revolutionibus*, and take up only twelve of the entire treatise's 196 folios. The rest of the treatise is a formidably technical astronomical work in the style of Ptolemy's *Almagest*. It is easy, therefore, to see *De revolutionibus* primarily as a reformation of astronomical theory and secondarily as a proposal that the universe is heliocentric. And this was in fact how the book was received.

The Reception of Copernicus's Work

Copernicus's work was taught and used in Germany starting shortly after the publication of *De revolutionibus*. It was particularly well received at the University of Wittenberg, where a fostering of the mathematical sciences was being overseen by the Protestant educational reformer Philipp Melanchthon (1497–1560). But the adoption of Copernicus was accompanied by an adaptation illustrative of the disjunction between astronomical theory and cosmology. During the 1550s, Melanchthon and an influential group of astronomers, including Erasmus Reinhold (1511–1553) and Caspar Peucer (1525–1602), crafted an institutionalized interpretation of Copernicus, wherein his improved astronomical theories were seriously investigated, while his cosmological claims, which contradicted Aristotelian physics and Holy Scripture, were rejected or treated with indifference. This "Wittenberg interpretation," as Robert Westman named it, subsequently spread to many German universities.[9] Among the most influential vehicles for spreading this interpretation was Reinhold's *Prutenicae tabulae coelestium motuum* (1551, 1562), a reorganized and recalculated system of tables based on Copernicus's theories but one that made no reference to the cosmological claims.[10]

Nor was this adaptation of Copernicus peculiar to Germany, or to Lutherans. The most prominent Italian astronomer of the age, Giovanni Antonio Magini (1555–1617), in his *Novae coelestium orbium Theoricae congruentes cum observationibus N. Copernici* [*New Theory of the Celestial Spheres in Agreement with the observations of N. Copernicus*] (1589) wrote:

> The observations of Nicholas Copernicus, a very well-informed man and most outstanding specialist, agree with the heavens marvelously, and I shall follow his observations. On the other hand, I shall ignore his hypotheses about the multiple motions of the terrestrial globe, and about the sun's immobility in the center of the universe and the immobility of the sphere of the stars. He devised these hypotheses most ingeniously in order to avoid a multitude of spheres. But his hypotheses are rejected by practically everybody as being absurd and too far away from the truth. For this one reason alone so great a man, to whom astronomy will forever owe a gigantic debt, is merci-

> lessly assailed by many people, even though nobody has been able thus far to produce a theory of the motions that is more suitable and more reliable than this theory of Copernicus. . . . I shall cling so closely to Copernicus . . . as to seem to have formulated in accordance with his thinking . . . all the phenomena. There is, however, this one exception: in opposition to the truth and all philosophy, Copernicus regards the earth as mobile, with the sun and the eighth sphere stationary, whereas I make the opposite assumption.[11]

In addition to those who used Copernicus's theories while explicitly rejecting his cosmological claims, there were those who attempted to reformulate heliocentrism to maintain the immobility of the earth.[12] The most notable of these was the great Danish astronomer Tycho Brahe (1546–1601), whose Tychonic system, in which the sun with the planets orbiting around it orbited the earth, was published in 1588. This idea was evidently so attractive that it was plagiarized by Nicholas Reimers Ursus (1551–1600), who published it, albeit slightly flawed, in the same year.[13] Another of Europe's most prominent astronomers, Johannes Praetorius (1537–1616), a professor at the University of Altdorf, also toyed with the idea in the early 1590s.[14]

Sixteenth-century astronomers were capable of an easy indifference to the cosmological truth of heliocentrism because of a long-standing ambiguity in the status of astronomical hypotheses and in the relationship between astronomy and physics. Aristotle himself had addressed the awkward status of mathematics in the investigation of nature in the *Physics*, asking, "Is astronomy different from physics or a department of it?"[15] On the one hand, the tools of the mathematical astronomer were abstract mathematics; on the other hand, the object of his inquiry was obviously part of the physical world. The well-known equivalence of a simple eccentric and a properly contrived epicycle and deferent was emblematic of the skepticism toward astronomers' claim to demonstrative certainty. Though they might be able to reproduce the phenomena of a planet—that is, model and predict its motion—that was no guarantee that their particular theory, in its combination of epicycles and deferents, actually represented the true configuration of the heavenly realm.[16]

This ambiguity corresponded also to a more or less standard view of the relationship between astronomy and physics.[17] Astronomy in the sixteenth century consisted of two related yet distinct parts. The first was mathematical astronomy, the technical discipline of calculating the positions of the planets. Mathematical astronomy was in general ambivalent about the physical truth of astronomical theories. Magini was a representative case. Though eagerly embracing elements of Copernicus's *De revolutionibus*, especially Copernicus's much improved lunar theory, he immediately rejected heliocentrism and recast Copernicus's theories of the superior planets into the geocentric form. To be sure, mathematical astronomy was not independent of physical concerns. Magini's rejection of heliocentrism was based on its physical absurdity. And Copernicus's own reformation of Ptolemy's theories to rid them of nonuniform motion along the deferent was based on the physical principle that planetary motions could be only uniform and circular, and astronomers such as Reinhold and Peucer did embrace them for this reason. Nevertheless, most questions of a physical nature were left to the other, natural philosophical part of astronomy. The physical part of astronomy included questions about the nature of the heavens, their substance, and so on. As such, the larger cosmological question of whether the earth was at the center of the universe fell to natural philosophy rather than mathematical astronomy. Conversely, the mathematical details of the theories of the planets were understood to belong exclusively to mathematics.

Barker and Goldstein have cogently analyzed the status of astronomy in terms of the status of other disciplines and the different types of demonstrations astronomers and natural philosophers were able to bring to bear. The standard of reasoning in natural philosophy was the demonstration *propter quid*, that is, the determination of the one and only cause of a phenomenon, which was achieved by the elimination of possible but not actual causes, usually through an appeal to a higher discipline. Astronomy was stymied by the well-known equivalence of different possible arrangements of spheres. The one true arrangement of spheres could not be determined, and thus the winnowing of the different possible arrangements could not proceed and astronomy was prevented from achieving *propter quid* demonstrations. As Barker and Goldstein put it, "sixteenth century astronomers might be depicted not as fictionalists but as perpetually frustrated realists."[18] This difficulty maps onto the distinction between reasoning a priori and a posteriori, as the terms were understood in the sixteenth century.[19] A posteriori reasoning was understood as reasoning from effects to causes, and a priori from causes to effects. Astronomers were understood to be limited to a posteriori reasoning, coming up with theories that would account for the observed phenomena of the planets. Deriving the phenomena a priori was not considered possible since the unique cause could not be determined. The only recourse of true, convinced Copernicans was to claim that heliocentrism offered a priori reasons for effects that were not explained in the Ptolemaic system, such as that a planet is always at apogee in its epicycle when it is at opposition to the sun. However, Reinhold and his followers could just as well reject this claim.[20]

The fact that, as a discipline, astronomy was subordinate to physics or natural philosophy clarifies the attitude of sixteenth-century astronomers noted above. In terms of Copernicus's recasting of Ptolemy's equant into uniform circular motion, the doctrine that aether is limited to uniform circular motions came from physics. Thus, although there were alternate arrangements of Copernicus's constructions (such as an eccentric epicyclet versus a double epicyclet), at least the Ptolemaic form could be eliminated because of its violation of physics.

In the generation of astronomers at the end of the sixteenth century, a true Copernican, one who believed not only that Copernicus's planetary theories were superior but also that the earth was truly in motion around the sun, was rare. It is highly significant therefore that one such astronomer was Michael Maestlin (1550–1631), Kepler's teacher of astronomy at the University of Tübingen.[21] Maestlin was an heir to the Wittenberg interpretation; he had himself published a corrected edition of Reinhold's tables in 1571. But he was also among the first astronomers who were compelled, particularly by the evidence provided by the comet of 1577, to take Copernicus's cosmological claims seriously.[22] And Maestlin had the distinction of being virtually the only convinced Copernican teaching at a university in Europe when Kepler was a student.[23] Maestlin's belief in the truth of Copernicus's claims was essential, for Kepler would become the most fervently realist promoter Copernican heliocentrism had ever known.

⌘ CHAPTER 2 ⌘

THE DEVELOPMENT OF THE *MYSTERIUM COSMOGRAPHICUM*

Kepler's Early Espousal of Copernicanism: His 1593 Student Disputation

Within the context described in the previous chapter, Kepler's receiving his astronomical education from Michael Maestlin, a convinced Copernican, was unusual and significant. Furthermore, the nature of Kepler's interest in Copernicanism was even further removed from that of the astronomical community. For whereas most astronomers were attracted to the technical superiority of Copernicus's theories—i.e., the fact that his reformulation of the equant did not entail physically nonuniform motion along the deferent[1]—and Maestlin himself had become convinced of the truth of heliocentrism on astronomical bases (from the evidence provided by the comet of 1577), Kepler immediately seized on its physical significance. He acknowledged the importance of this in a significant statement in the preface to the *Mysterium cosmographicum* (1596):

> At the time, six years ago, when I was studying under the distinguished Master Michael Maestlin at Tübingen, I was disturbed by the many difficulties of the usual conception of the universe, and I was so delighted by Copernicus, whom Mr. Maestlin often mentioned in his lectures, that I not only frequently defended his opinions at the disputations of candidates in physics but even wrote out a thorough disputation on the first motion, arguing that it comes about by the Earth's revolution. I had then reached the point of ascribing to this same Earth the motion of the sun, but where Copernicus did so through mathematical arguments, mine were physical, or if you prefer, metaphysical. And for this purpose I collected together little by little, partly from the words of Maestlin, partly by my own efforts, the advantages which Copernicus has mathematically over Ptolemy.[2]

At least a fragment of the latter disputation ascribing the annual motion of the sun to the earth has apparently survived and has recently been republished.[3] It is Kepler's earliest extant astronomical writing. It shows both that many of his fundamental physical principles were nascent while he was still a student at Tübingen and also that he recognized the resistance Copernicanism faced.

The disputation fragment apparently consists of the continuation of an argument begun during an earlier session, and it suggests that Kepler had been having a rough time. He began by reminding the members of his audience that he had previously asked that they should receive him more humanely, and he reproached them for having attacked him unfairly. It appears that theological objections had been the most troublesome, for Kepler bemoaned the "hostility toward a new and impious idea,"[4] yet his attitude was conciliatory. Invoking the metaphor of a battle, Kepler wrote that Copernicus asked that they not send Holy Scripture against him, "[f]or he is convinced that, although they [i.e., the Holy Scriptures] may not be so bitter that he could not hope to come to terms with them easily in private, still it would come to pass that having been roused by the shouts of the spectators they would indeed want to fight."[5] But Kepler did not suggest that Copernicus could not vanquish them. Rather, he stated that Copernicus would not fight "as one who would slay a defenseless woman with a double-edged sword or slay an opponent from horseback for sport."[6]

Kepler continued to stake out the area he was prepared to defend by disarming Copernicus in turn. Despite his testimony in the preface to the *Mysterium* that he had been collecting the mathematical advantages that Copernicus had over Ptolemy, for the sake of this disputation Kepler eschewed Copernicus's two great weapons, "that extraordinary and hardest mathematical club of his along with the most powerful sling, which threw entire earths."[7] In doing this, Kepler again did not concede anything. There is no doubt, he wrote, that with one blow and one swing he would have crushed the epicycles, but Copernicus was concerned for his followers that they should have to witness this carnage.

Kepler was thus left with the following arguments:

> You [will] see light, trifling reasons, and some thinly spun cosmography, wrapped in which our Copernicus wished to appear today, so everyone would notice that he wanted to play, not to fight.[8]

Kepler did not characterize his arguments beyond calling them "cosmographical." Later, in the *Mysterium*, he called the arguments of his disputation, of which this is apparently a fragment, "physical, or if you prefer metaphysical." At the time, *physica* referred broadly to natural philosophy, and Kepler's arguments in the disputation can certainly be classified as natural philosophical/metaphysical.

Having focused the debate on the physical or natural philosophical issues, Kepler enumerated the arguments in favor of heliocentrism. The initial arguments were perhaps more poetic or rhetorical than scientific. Even in them, however, we see that Kepler identified the sun as the source of light, heat, and motion in the universe, and also that he conceived light as the medium of motion and change:

> In the first place, not even a blind man would deny that of all the bodies in the universe the most superior is the sun, whose entire essence is nothing other than the purest light, than which no star is greater, which is the one and only, the effector, preserver, and warmer of all things; the source of light, the wellspring of abundant heat; most beautiful, clearest, and purest to the sight; the origin of vision, the expressor of all colors, itself lacking any color; called the king of the planets as regards motion, the heart of the world in strength, and the eye with regard to beauty; and which we would all judge to be the one worthy thing, which the best and greatest God, if He would delight in a corporeal home and could be held in a place, would inhabit with the blessed angels. For who would push Him, who is said to inhabit unapproachable light (Jo:15),[9] out into those outer and supercelestial shadows? In the respect that the Germans elect as emperor him who is the most powerful in the whole empire, who would doubt that the celestial motions would cast their ballots for him who already governs all the other motions and alterations by the benefit of light, all of which he possesses.[10]

Kepler's reasons clearly derived from Copernicus's in *De revolutionibus*.[11] But Kepler greatly expanded the role of the sun, adding that the sun is also the source of the planets' motion. This modification was a harbinger of essential elements of Kepler's physical astronomy.

Continuing, Kepler examined more closely the idea that the sun is the source of motion for the planets. He acknowledged the traditional Aristotelian distinction between terrestrial and celestial matter, but at the same time he asserted that one should not abandon the expectation of similarity between celestial and terrestrial matters. Kepler's belief that celestial physics ought in principle to be treated like terrestrial physics was a highly unusual and significant aspect of his thought. It is seen here in his heliocentric reinterpretation of the prime mover:

> Certainly we perceive at other times a thing and, in particular, strength in any degree, to be stronger where there is more in the same place, [and] to vanish more quickly [when] diffused into separate parts. Therefore, although I do not ignore [that] the matter is otherwise with that purest essence than with the accidents of bodies, still it is impossible in this matter that he who does not have better reasons should abandon every expectation of similarity. Accordingly, since it is not appropriate for the prime mover to be spread out spherically, but rather to go forth from one certain source, as if from a point, no part of the orb and also not even any star is worthy of so much honor. It

> should most justly go to the sun, which alone in both worthiness and strength seems suitable and deserving of this manufacturing of motion, and which could hold even God Himself, not to mention the prime mover. Since it is glorified with so many honors, burdened by so many gifts, Copernicus thought the Sun should be situated so that it sits foremost in the center of the world as the mover itself. Just as it is as such necessarily immobile, so too it should be fixed in an immobile dwelling, whence as if from the center and the certain heart of the world it would extend itself through its effect most equally to all surrounding orbs.[12]

Here, Kepler introduced the idea that the strength that moves the planets is concentrated in the sun and extends from there into the orbs. Next, he expanded the idea, reasoning that since the sun distributes motion equally, one should be able to derive the periods of the planets from their distances. His treatment at this time was not rigorous, but it was mathematical. He clearly posited a function by which the circulating motion is nearly infinite at the center and falls off with distance to nothing at the sphere of the fixed stars. He continued:

> From which consequently the ratio of the speed in the planets can be easily devised. Without doubt, the one, only, first, and ultimate mover in the very center of the world, in which it is the smallest—that is, it has no orbit—executes a kind of thing that is like the fastest motion, which you cannot perceive except in its lack of motion and from the disposition of the moving principle, because the motion is nearly infinite rather in the parts here, where its principal action is, than where it is secondary.[13] From the point there, undoubtedly from the common center of the sun and the world, motion extending itself most equally into all neighboring orbs makes an impression into them with equal effort and hastens their particular revolutions accordingly in proportion to the orbit. First from the sun, Mercury has the smallest orbit. Therefore there is not another, faster sphere, whose motion is less than three months. Venus follows, whose period increases with the distance of the orb and is nine months. Following this, in the middle among the planets, is the Earth, a planet and itself wrapped in the monthly lunar orb, as is clear from the moon, makes twelve months. Mars above has 23 months; Jupiter, five times more remote from the sun than the Earth, has 12 years; and Saturn, nine times as remote, has 30 years. Certainly it was fitting for the fixed stars, since they are distant from the sun by an almost infinite interval, also to be that much slower. And since they cover all things, and produce the relationship of motions and spaces in the world, neither of which certainly can be intelligible without a cover, it is agreed that they are not moved at all.[14]

Again, there are clear similarities to Copernicus, who used both the progression in the periods of the planets and the role of the immobile sphere of the fixed stars as a definitive container to argue for his cosmology.[15] But Kepler introduced motive power into his period-distance function.

Interestingly, this bold departure may itself have derived from Copernicus, who maintained that only uniform circular motion was appropriate for describing the path of heavenly bodies,

> because a simple heavenly body cannot be moved by a single sphere nonuniformly. For this nonuniformity would have to be caused either by an inconstancy, whether imposed from without or generated from within, in the moving force or by an alteration in the revolving body. From either alternative, however, the intellect shrinks.[16]

This is the only mention of such a "moving force" in *De revolutionibus*. In rejecting the inconstancy of the motor virtue, Copernicus suggested the existence of a constant motor virtue. Perhaps Kepler seized on Copernicus's vague statement to develop his own idea that there is a planet-moving force coming forth from the sun that moves the planets around their orbits.

A more concrete expression of the idea that the sun is the source of motion in the solar system would have been Georg Joachim Rheticus's *Narratio prima* (1540), but Kepler's debt to Rheticus at this stage in his thought is difficult to establish. Kepler mentioned in the preface to the *Mysterium* that at the stage at which he was collecting the "mathematical advantages" Copernicus had over Ptolemy, he would have been greatly helped had he read Rheticus, which apparently he had not yet done.[17] But by the time he wrote the *Mysterium cosmographicum*, he was explicitly drawing on Rheticus. The place of the composition of his fragmentary oration on the motion of the earth in this timeline is unclear. Nevertheless, the parallel with the *Narratio prima* is worth pointing out, for Rheticus goes much further than Copernicus in addressing the cause of the planets' motion in the heliocentric system. After addressing the problematic Aristotelian doctrine that motion is passed down from the prime mover through the planetary spheres, Rheticus writes:

> However, in the hypotheses of my teacher, which accept, as has been explained, the starry sphere as boundary, the sphere of each planet advances uniformly with the motion assigned to it by nature and completes its period without being forced into any inequality by the power of a higher sphere. In addition, the larger spheres revolve more slowly, and, as is proper, those that are nearer to the sun, which may be said to be the source of motion and light, revolve more swiftly. Hence Saturn, moving freely in the ecliptic, revolves in thirty years, Jupiter in twelve, and Mars in two . . .[18]

and so on. Rheticus's suggestion is clearly akin to Kepler's idea that the sun is the source of motion. Even so, Kepler had already developed the idea much further in his student disputation by beginning to see what can be derived from it as a physical principle.

From Kepler's description of the sun's motion at the center of the system as being nearly infinite yet imperceptible, we can infer that it is rotatory, as it is in his later writings. For Kepler, motion extended from the sun uniformly into space, and diminished predictably with distance. This idea of a planet-moving force became the most operative principle in Kepler's astronomical work. It extended from this, his earliest writing, through the *Mysterium cosmographicum* and the *Astronomia nova* to the *Harmonice mundi*, and was responsible for all three of his laws of planetary motion.

At the same time, however, Kepler's account in the disputation was inconsistent and betrayed certain Aristotelian presuppositions. First, he made the sun a kind of inverted prime mover, justifying its rest by appealing to the Aristotelian distinction between the unmoved mover and the moved. At the same time, even though the sun was immobile in a translatory sense, it moved with nearly infinite motion (in what we infer to be a rotatory sense). Moreover, it did so because of its proximity to the center of the world, which therefore retained some of its Aristotelian significance.

The last section of the disputation was devoted to a further elaboration of the theme of the sun as governor and visible god, a theme that, as we have seen, was expressed in *De revolutionibus*.[19] While Kepler did not use the elegant analogy of the Trinity to the sphere that he later used in the *Mysterium*, there are hints of the geometrical bases of that too in his suggestion that the spherical surface of the sphere of the fixed stars serves to reflect all rays back to the origin at the center, thus reflecting the glory of creation back on its Creator.

Finally, Kepler's fragment closed with a brief statement on the problems posed by a moving earth:

> And we should not fear for the Earth, lest it should be too weak and weighty, so that it and its soundness of body could not bear this motion. The center is of too little importance; it is not heavy. It is easily agreeable to the orb to be moved to whatever place. All heavy things naturally follow the center of the earth.[20]

This statement, too, could easily have been derived from some of Copernicus's remarks on this problem, for example:

> For my part I believe that gravity is nothing but a certain natural desire, which the divine providence of the Creator of all things has implanted in parts, to gather as a unity and a whole by combining in the form of a globe.[21]

As measured against Copernicus's remarks, Kepler's dismissal of the problem seems laconic. However, this brevity may be artificial; the disputation fragment ends here, perhaps in the middle of his remarks. But if this were all that he had to say on the subject, it would not be surprising, for Kepler never seems to have viewed Aristotelian arguments against the motion of the earth as serious obstacles to heliocentrism.

The interpretation of the "Fragmentum orationis de motu terrae" is problematic. It represents only a part of one of Kepler's two Copernican disputations, and therefore cannot necessarily be assumed to be representative. But it clearly expressed some of the most influential themes of Kepler's thought, many of which he repeated in the *Mysterium cosmographicum* and the introduction to the *Astronomia nova*.[22] That Kepler held these ideas even as a student is significant, for it casts doubt on the assertion that, like Maestlin, his primary motivation for the adoption of Copernicus was the astronomical evidence from the comet of 1577,[23] and provides additional evidence that it stemmed from a unified consideration of philosophical, theological, and astronomical factors.[24] Most important to this study is that it shows that, from his student days, Kepler was arguing for Copernicus by means of physical reasons (however ill-defined) and that even then he was aware that he faced strong resistance.

Kepler's Work in Graz

Early in 1594, Kepler was assigned to teach mathematics at the Protestant school in Graz, Styria, a district of Inner Austria.[25] The assignment was a pivotal event in his career, for it forced him to shift his attention from theology to mathematics, a substantial aspect of which was astronomy. Kepler was initially distressed by his assignment. He was not bothered by the remoteness of Graz; he had criticized his comrades for shunning foreign assignments for such cowardly reasons. He was repelled rather "by the low opinion and contempt in which this kind of function [i.e., astronomy teacher] is held, and the sparsity of erudition in this part of philosophy."[26] This view is a fundamental element of Kepler's attitude toward astronomy. He had wanted to be a theologian. Astronomy as a purely mathematical discipline was not compelling for him. Only as a greater natural philosophical pursuit, in which he could ascertain the traces of the Creator, did it come to captivate him. Later, after making the discoveries he would publish in the *Mysterium*, he understood how to make astronomy his calling, as he explained to his teacher Maestlin:

> I am in haste to publish, dearest teacher, but not for my benefit . . . I am devoting my effort so that these things can be published as quickly as possible for the glory of God, who wants to be recognized from the Book of Nature. . . . Just as I pledged myself to God, so my intention remains. I wanted

> to be a theologian, and for a while I was anguished. But now, see how God is also glorified in astronomy, through my efforts.[27]

As a result of Kepler's new assignment, he refocused his attention on Copernican astronomy, developing the arguments he had outlined in his disputation at Tübingen. As he wrote in the preface to the *Mysterium cosmographicum*:

> In the meantime, while I was rolling that rock [the mathematical advantages of Copernicus over Ptolemy], but as a sideline apart from theology, by lucky chance I came to Graz . . . and there the duties of my post obliged me to attend more closely to those studies. For expounding the principles of astronomy there, everything I had previously worked out for myself was of great value. And as in Virgil "the report grows by traveling and gains strength as it goes," so for me the careful contemplation of these topics was the cause of further contemplation. . . .
>
> There were three things about which I persistently sought the reasons why they were such and not otherwise: the number, the size, and motion of the circles.[28]

As we have seen, of the three concerns he mentioned, only the last—the precise mathematical relationship determining the periods of the planets as a function of their distances—can definitely be dated to his student days.

Kepler's posting to Graz had another beneficial result—this one for the historian—for it was only due to his isolation in what he called "this obscure and gloomy wilderness"[29] that we have his letters to Maestlin, in which we have the opportunity to study the development of his thought. We can determine, among other things, that the idea of finding the cause of the distances was most probably added to his interest in the Copernican system after leaving Tübingen. In a letter to Maestlin of 2 August 1595, Kepler asked excitedly whether,

> you have ever heard or read there to have been anything, which devised an explanation for the arrangement of the planets? The Creator undertook nothing without reason. Therefore, there will be reason why Saturn should be nearly twice as high as Jupiter, Mars a little more than the Earth, [the Earth a little more] than Venus, and Jupiter, moreover, more than three times as high as Mars.[30]

Surely, had he already been interested in the question of the reason for the planets' particular distances at Tübingen, he would have asked his teacher then. But his newfound interest in the distances had not superseded his interest in the motions, for he added immediately, "I want the same statement of the proportion of motions."[31]

The cause of Kepler's excitement was the discovery he had made on 20 July 1595 of the distance scheme he published later in the *Mysterium cosmographicum*. In the preface of the *Mysterium*, he related the series of models he tried over that summer. The first were numerical ratios, in which he went so far as to postulate invisible planets to fill empty spaces in the sequence, but these were unsuccessful, and provided no answer for the number of planets. He then described a sine model (described below), which did not, in fact, derive the distances but took them as given for deriving the periods. Finally, on 19 July, while lecturing on the progression of conjunctions of Jupiter and Saturn, he noticed that the ratio of the circle circumscribing an equilateral triangle to the circle inscribed in it was approximately the same as the ratio of Saturn's orbit to Jupiter's. He must have run through various arrangements of planar figures quickly, for the following day he hit upon the idea of using the Platonic solids instead.[32] As he wrote in the *Mysterium*:

> I give you the proposition, conceived in words just as it came to me and at that very moment: "The Earth is the circle which is the measure of all. Construct a dodecahedron round it. The circle surrounding that will be Mars. Round Mars construct a tetrahedron. The circle surrounding that will be Jupiter. Round Jupiter construct a cube. The circle surrounding that will be Saturn. Now construct an icosahedron inside the Earth. The circle inscribed within that will be Venus. Inside Venus inscribe an octahedron. The circle inscribed within that will be Mercury."[33]

This model—which I will refer to as the "polyhedral hypothesis" (see plate 2.1)—provided a reasonably accurate derivation of the distances of the planets with respect to one another. The model had another compelling feature, for it also provided a reason for the number of planets. Since there are only five regular solids, there can only be six circumscribing and circumscribed spheres, and hence there are six planets.

Kepler's polyhedral hypothesis is well known to historians of science from Kepler's *Mysterium cosmographicum*, where the structure was buttressed with arguments from the mathematical, numerological, and astrological significance of the solids. But in some sense, the very fecundity of the polyhedral hypothesis in suggesting ancillary arguments obscured Kepler's account of the cause of the motions.

The letter to Maestlin of 2 August 1595 was the first in a series of three letters in which Kepler described the material that was to appear in the *Mysterium*. In the first letter, Kepler did not describe his new system, but asked a number of questions directed toward the uncertainty in the distances of the planets as derived from observations. For instance, he asked whether the values of the *prosthaphaereses*—the corrections added to the mean motions—were accurate. The size of the *prosthaphaeresis* was di-

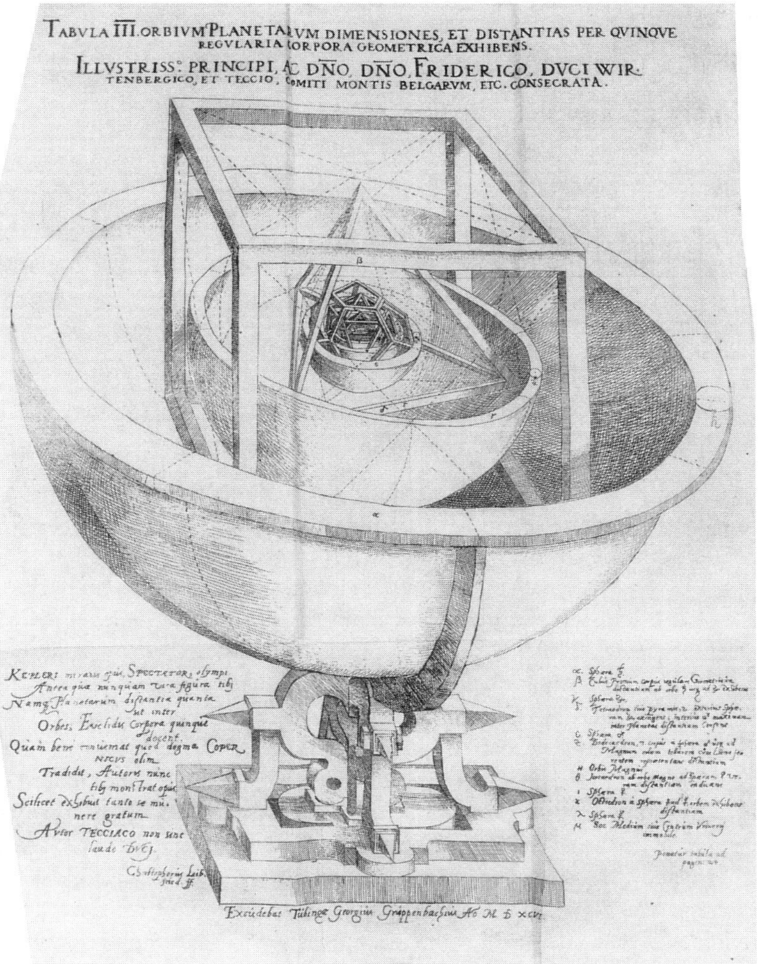

Plate 2.1 The polyhedral hypothesis from Kepler's *Mysterium cosmographicum* (1596)

rectly related to the size of a planet's orbit relative to the earth's orbit. Kepler was interested in whether planetary theories based on the distances derived from his model would differ sensibly from those derived from observation. It is clear that he was asking with reference to the polyhedral hypothesis, for he gave the distances he had derived from it and asked whether they were too much in disagreement with the observations.

Only in his next letter to Maestlin of 14 September 1595, after he had determined that the polyhedral hypothesis did not disagree significantly with observations, did Kepler give the details of the system. And after a

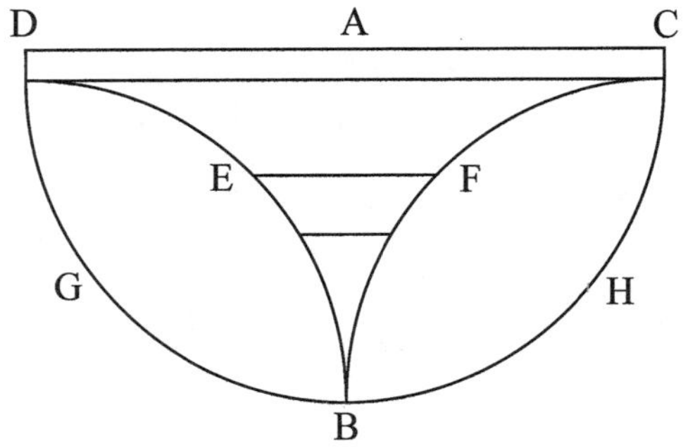

Figure 2.1 Kepler's Sine Model

single sentence describing the order of the solids, he proceeded to a different subject, the cause of the motions:

> Next, there is in the Sun a moving spirit [*anima movens*] and an infinite motion. In the mobiles [there is] a twofold diminution of motion. The first [comes from] the unequal revolution that is caused by the unequal size of the orbs, assuming that the strength of the motion [*vigor motus*] were the same in all orbs. But now, secondly, that strength of motion (as light in optics), which is further from the source is that much weaker. And certainly this weakness does not diminish proportionally but thus as you see in the attached diagram [see figure 2.1], where A is the Sun and through it an infinite line DC, that is just tangent to the quadrants DEB and CFB. DGBHC is the fixed stars. Therefore there is not any motion at B. The remaining lines indicating the proportion of the moving force [*vigoris moventis*] at different distances are determined from the complement of the sines by means of the distance of the planets along the line AB, or some parallel, as you see. But I have not fully investigated this latter [point] on account of the most tedious calculation. However would that these things so please you, as they please me, who regards them as stupendous miracles of God.[34]

This is the sine model that Kepler discussed in the preface to the *Mysterium* as a precursor to the polyhedral hypothesis, though it does not provide distances and was apparently developed after that model had been discovered.[35] There, he also related that, after having carried through that "most tedious calculation," he abandoned the model. It had two faults. First, although the ratio of distances of planets was known, the total distance to the fixed stars AB was not. It would therefore have to be determined by fitting the known ratio of the distances and the periods to the curve. Second, "the strengths of the motions were expressed only in terms of a ratio of one to another," by which Kepler perhaps meant that, although it might have provided the relationship of distance to motion, it provided no account of the distances themselves.[36] This interpretation would agree with Kepler's erroneous account of the sine model as an unsuitable precursor distance model.

What is most interesting about this model is that it represents a working out of Kepler's line of thought in his student disputation. There, he had implied that the motions of the planets could be derived mathematically from the distances of the planets from the sun, their source of motion. In his further development of the idea, Kepler had noted that change of the period of the planets with distance from the sun has two elements. First, there was the simple fact that the length of the orbit increases proportionally to the length of the radius. However, there was a second element by which the moving strength itself diminished with distance, "as light in optics." Kepler had already written in the disputation that movement was conferred by light. Here, he introduced optics, implying that the moving force can be investigated using the same principles as terrestrial science, although the use in this letter was only as an analogy.

Whereas Kepler's first letter to Maestlin had been a cautious query that gave away little and the second contained only the briefest mention of the polyhedra, in the third letter he began to elaborate his vision toward the complexity that finally appeared in the *Mysterium*. The final and most complete of the series of three letters in which Kepler introduced Maestlin to the material that was to appear in the *Mysterium* was written on 3 October 1595. Kepler had promised in the second letter to send his entire "little work," and the third letter represented an outline of the book as he then planned it:

> In the second letter I promised to send you the entire little work at the first opportunity. However, I have not yet finished it. . . . Therefore, you will get a brief précis of those things that are not geometrical. I will begin with several theses I discuss from Holy Scripture, and show how at the same time both its authority would be consistent with them and yet that Copernicus, if he says something inconsistent, cannot be refuted by them. Next, I offer three reasons for which I always believed in Copernicus. The first is astronomical, wherein I respond to those who say that truth sometimes follows from falsehood [and that] therefore it is possible that Copernicus proposed a falsehood even though he demonstrates truth most beautifully from it. I deny that this is possible. Another is physical [*Physica*], wherein I maintain that the whole mass of the Copernican hypothesis is supported from the fount of nature far better than the contrary. The third is your [reason from] the comet of Venus.[37]

There are several points of interest here. First, Kepler wrote that he would begin by defending Copernicus against Holy Scripture. This theological defense was suppressed and did not appear.[38] Second, Kepler briefly mentioned the philosophical point—that in astronomical hypotheses, truth cannot be derived from falsehood—which appeared in chapter 1 of the *Mysterium* and was further developed later in his *Apologia pro Tychone contra Ursum*.[39] And third, this passage contains Kepler's first known use of the term *physica*.

In the rest of the letter, Kepler did not elaborate the theological argument, the astronomical (philosophical) argument, or the argument from the comet of 1577.[40] We can infer that all the reasons that Kepler brings forward fall under the idea of *physica* that he held at that time. Most of his arguments were more metaphysical, in accordance with his statement in the preface to the *Mysterium*, that the arguments he brought forward in his disputation were "physical, or if you prefer metaphysical."[41] It is clear that at this time Kepler used the term *physica* loosely. In later years, he was more precise about the distinction.

Many of the arguments for his new construction of the world that Kepler used in this letter also appeared in the *Mysterium*. God, he wrote, created a certain number of heavenly bodies. But number is an accidental quantity; before the creation, there was no number other than the Trinity God embodies. Linear and plane geometry do not determine number, so God must have begun with solid geometry, in which there are six perfect bodies: the sphere and the five Platonic solids. The sphere represents God himself, since in its center, surface, and volume it embodies the Father, the Son, and the Holy Spirit. The sphere also represents the world; the sun is the center, the outermost sphere of the fixed stars the surface, and the air or aether the volume. The five Platonic solids can distinguish no more than the six spheres between which they are interpolated, and so they determine the number and the spacing of the planets.

Kepler then continued with the argument "that I held already at Tübingen,"[42] that the sun, immobile in the center of the world yet the source of all motion, is the image of God, the Creator of all things. And he elaborated the idea in the disputation that one can derive the periods of the planets from their distances from the sun. Since the sun disperses motion through the medium in which the planets are situated,

> from the necessity of the presuppositions, it follows that the motion is proportional to the distances. So therefore I concluded that that opinion which would preserve the same order with the motion and distances would be correct; that which does not [would be] false.[43]

Kepler did not immediately continue in this vein of thought, preferring evidently to remain with those things of which he was more certain. After a brief description of the properties of the solids, their associated planets, and the significance of their arrangement, he paused to plead for Maestlin's judgment and to distinguish the surety of the conclusions already reached from those that follow:

So far, there is no doubt [that I am right]. But because many doubts occur to me in what follows, I wanted to give you a full account and to ask your advice. You will find many things that are partly based on a false foundation, and partly on none; in other places you will see me get bogged down because of my inadequate knowledge of Copernican astronomy. You may polish, change, edit, advise me or give me orders. I'll be most grateful for any letter you can possibly write.[44]

Only after this plea did Kepler turn to those things that were more problematic. One of these was the problem of the motions:

Now I would add the whole of those things that follow from the stated theses. There is, as I said, a moving spirit [*motrix anima*] in the Sun. If equal motion and the same strength came from the Sun into all orbs, one would still circulate more slowly than another on account of the inequality of the orbs. The periodic times would be as the circles. For quantity measures motion. However, circles [go] as the radius, namely as the distance. Thus from the certainly-known mean motions we could easily construct also the mean distances. But another cause enters which makes the more remote slower. Let us take the experience [*experimentum*] of light. For as both light and motion are connected in their origin so also [are they connected] in their actions, and perhaps light itself is the vehicle of motion. Therefore, in a small orb and also in a small circle near the Sun, there is as much light as there is in a large and more remote sphere. Therefore the light is thinner in the large, and denser and stronger in the narrow. And this strength is in inverse proportion to the circles, or the distances. Now if there is the same relation of motion (for which matter certainly nothing neater can be conceived), it would follow that the distances act doubly to induce slowness of motion. Whereby half the difference of the two motions added to the smaller motion will be the distance of the more remote: the smaller motion itself will be the distance of the nearer.[45] Look what follows in this table:

			From the motion		*From your table*	*Difference*[46]
	Saturn	is 1000,	Jupiter	574	563	+11
If the	Jupiter	then the	Mars	274	294	–20
mean	Mars	mean	Earth	694	658	+36
distance	Earth	distance	Venus	761	719	+42
	Venus	will be	Mercury	563	549	+14

Behold how near the truth I might have come! And so far you question that I shed tears abundantly whenever some such thing occurs. For by God this is the a priori way to correcting the motions; and every hope, if others, to whom observations are available, exert themselves.[47]

This development represents a clear continuation of Kepler's idea dating back to the disputation that there was a moving force (or spirit) issuing forth from the sun. As in the earlier sine model, he used the analogy of the force with light to explain how it is weaker farther from the source. But this time, he went so far as to suggest explicitly that light is the vehicle through which motion is conveyed. However, whereas light propagates spherically, Kepler confined his attention to the plane of the orbit, reasoning that because the total amount of motion is the same in all orbs (thought of as circles), it is spread more thinly in larger orbits, and thus decreases in strength in direct proportion to the radius. Only much later did he reconsider the spherical propagation of the motive virtue and address the problem of whether the strength ought, as light, to decrease as the square of the distance.[48] His notion of the relationship between force and motion, incidentally, was thoroughly Aristotelian: motion was directly proportional to force.

The distance also contributed to the length of a planet's period simply by virtue of the fact that the planet had to traverse a longer orbit. Again, the effect was proportional to the radius. Kepler thus concluded that the distance contributes twice to the slowing of the motion, and by *adding* (rather than multiplying) these two factors, he concluded that difference in the motion of a more distant planet would be the change in distance divided by two, or as he put it, "half the difference of the two motions added to the smaller motion will be the distance of the more remote." Thus in general terms, Kepler made the ratio of the distances of a superior planet to an inferior planet:

$$\frac{P_S - P_I}{2} + P_I : P_I ,$$

where P_S and P_I are respectively the periods of the superior and inferior planets.

Kepler went on to describe the errors produced by this scheme. Still absorbed in the elaborate justification of his arrangement of the solids, he noted that the errors were similar for the solids that were similar.[49] But he was dissatisfied:

> Here I am up to now perplexed. For I reckon to make up for [the error] by adding or subtracting so much by way of correction in the mean distances. If you reckon that this is permitted, it follows that we can have the most exact motions. For the mean places are from this method.[50]

The errors were larger than those for the polyhedral hypothesis, and they were troubling. In principle, the theory of the motions should have produced precise values for the planets' mean distances, for the observed data—the periods—were very well known. Observations stretching back for millennia had produced highly accurate values for the periods.[51] Kepler clearly believed that if the precise relationship of the periods to the distances were known, a more certain value for the distances could be derived than that found from uncertain theories and imprecise instruments. From his earliest letter regarding the polyhedral hypothesis, Kepler had been concerned with the uncertainty in observationally derived distances. Moreover, the theory of the motions promised to proceed directly from an accurate observed quantity to the relative distances. Kepler had in the meantime been juggling with the distances in the polyhedral hypothesis. Because the polyhedra determined the distance between the maximum distance of one planet and the minimum distance of the next, the eccentricity of each planet—another series of derived distances—had to be introduced. The theory of the motions was in principle much simpler, if more unforgiving.

By this time, Kepler had discarded the sine model, ruefully noting, "No calculation ever occupied more time or exhausted more spirit than that so empty and so purposeless one."[52] But still he thought there was something to it, for it predicted that there would be no motion at the sphere of the fixed stars, whereas in the new model, there would always be some moving strength at any distance. But he dismissed the problem. The sphere of the fixed stars could not be moved because it did not occupy a place but rather defined place in the world.[53] It was, if you wish, the framework of absolute space. Kepler concluded that the moving force that encounters the sphere of the fixed stars is simply reflected.[54]

Maestlin's Involvement with the *Mysterium*

Michael Maestlin was involved to an unusual degree in the publication of his student's first book. Not finding a printer in Graz competent to publish a serious astronomical book, Kepler chose to have the book printed in the familiar academic environment of Tübingen. There were also possibilities for patronage to be pursued in his native Württemberg; even though he now resided in Styria, Kepler continued to regard himself primarily as a subject of the Duke of Württemberg. As Württemberg's principal astronomer, Maestlin was unfailingly called on to give his opinion of Kepler's new discovery, whether by university authorities in Tübingen or the ducal court in Stuttgart. Maestlin's participation extended even further as he became intimately involved in seeing the book through press. But despite his steadfast support and loyalty to his student, we must bear in mind that Maestlin was a sixteenth-century astronomer—no matter how unusual a one—and his reaction to the work was shaped by his perception of how the work fit within contemporary definitions of physical and mathematical astronomy. He would be supportive of the cosmographical features of the work, those that dealt with the overall structure and dimensions of the planetary system, but critical of the extension of Kepler's physical analysis to the determination of the details of planetary theory.

To begin with, and to the extent with which we will deal with Maestlin in this chapter, he overlooked Kepler's analysis of the cause of the planets' motions. In the context of Kepler's entire letter of 3 October 1595, the section on deriving the distances from the motions was but a part. In his first letter in response to Kepler of 27 February 1596 (o.s.), Maestlin did not mention the moving force or the motion model. But Maestlin was very supportive of the polyhedral hypothesis and well disposed toward Kepler's project of deriving the dimensions of the world a priori rather than from observations, especially since he thought that this information could be used to overcome those who opposed Copernicus:

> Whenever I think about your demonstrations, in which you most skillfully prove Copernicus's hypotheses a priori, I am exceedingly glad that finally some learned man is found, who dares to assail the utterances of that throng of ignorant mathematicians speaking against these hypotheses of Copernicus. And so much more, that he even lays open the plan of the Creator in the creation of the world, and with the utmost ingenuity the measure of the spheres, which even for Copernicus himself had to be tracked down for computing prosthaphaereses, only a posteriori by means of observations. It seems now to have an open access a priori.[55]

Maestlin's only reservation was that Kepler's values for the distances of the planets were ones that Maestlin evidently had given him at Tübingen, perhaps in a lecture, that were merely pedagogical and did not take into account the whole size of the apparatus of Copernicus's eccentric epicyclet (see figure 2.2). He consequently appended to his letter a document titled "The Dimensions of the Spheres of the World, according to the calculations of the *Prutenic Tables*, following the hypotheses of Copernicus," which contains a detailed account of how the values are calculated and a table giving the maximum and minimum distances from the mean sun to the planets.[56]

Ironically, Maestlin's support of Kepler's venture in early 1596 was due in some part to Kepler's failure to derive a more adequate physical theory of the motion of the planets and the resulting emphasis given to the elaboration of the polyhedral model. Maestlin ignored his discussion of a moving spirit entirely, and consequently he did not recognize at this time the extent of Kepler's program of physical astronomy. Instead, Maestlin saw Kepler's work to derive the distances of the planets a priori from the Platonic solids as eminently reasonable from from a cosmological point

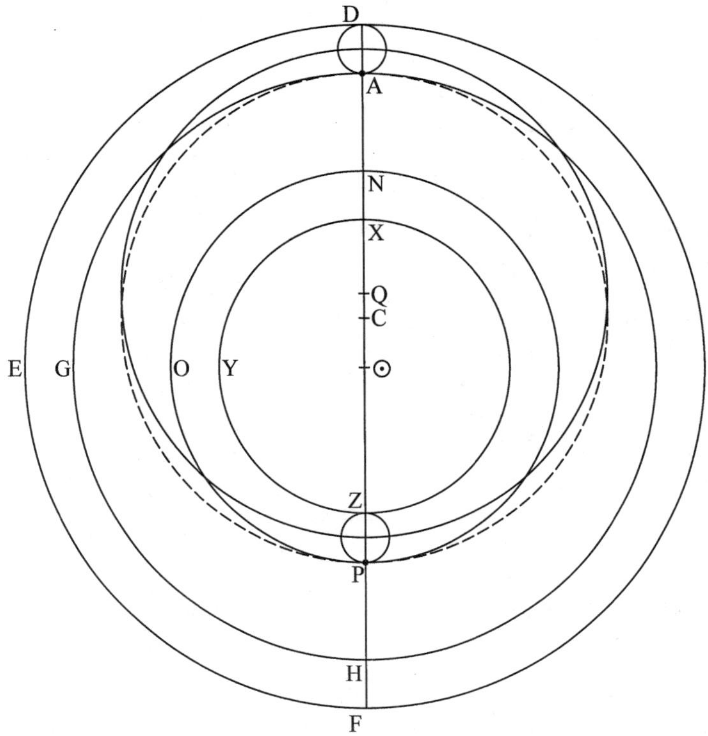

Figure 2.2 The greatest and least distances of a planet's orb in Copernican theory

In determining the planets' greatest and least distances for the *Mysterium cosmographicum*, Kepler was faced with a technical wrinkle. In Copernican theory, a planet's greatest distance from the sun (⊙) at aphelion (A) and least distance at perihelion (P) do not correspond to the farthest and least extent of the theory's apparatus. If the extent of the planet's orb were calculated using these positions of the planet (similar to the way it would be done with Ptolemaic theory), its inner surface would correspond to the circle NOP and its outer surface to circle AGH. To take into account the Copernican epicyclet, the extent of the planet's orb must be measured from the circle DEF to the circle XYZ. The larger size of the planet's orb would have vitiated the planetary distances in Kepler's polyhedral hypothesis.

of view; for a Copernican such as himself, the attraction of heliocentrism lay in the possibility of such a priori demonstrations.[57] It is along these lines that Maestlin let his support of Kepler be known to Duke Friedrich von Württemberg.[58] Kepler had applied to the duke for printing costs for the *Mysterium cosmographicum*, and the commissioner had solicited Maestlin's opinion.[59] Maestlin answered that until Kepler, astronomers, believing the plan of the Creator to be inscrutable, had derived the proportion of the celestial spheres from observation, not knowing that their theories could be helped by such a priori reasoning:

> Up to now everything that has been disputed and written about the height and size of all celestial spheres has its basis only from astronomical observations, that is a posteriori. But no one had ever understood that astronomy might have some help a priori, and that or how the observations themselves and then the hypotheses, quantities, and magnitudes of the celestial orbs were to be regulated and examined by means of a natural, geometrical and right proportion, as is to be hoped from this invention. For formerly it was sure that no certain proportion was to be found in the common hypotheses, rather it was thought that the Creator had created the order and height of the spheres according to His omniscient, but to all men unfathomable, counsel. So also Copernicus himself, who used other hypotheses, and on whom this invention is based, did not know or notice any other means than only to justify observations from observations (though in this he doubts not a little whether they are altogether correct). But now that M. Kepler has finally noticed that the planets' intervals have very nearly the proportion that the spheres have to the inscribed and circumscribed five regular bodies, I do not doubt that by this means astronomy will be helped in the same fundamental way.[60]

Maestlin felt that Kepler's was a worthy contribution to astronomy. He stated later that all who were skilled in astronomy believed Copernicus, and he clearly believed that the polyhedral hypothesis was another significant argument for Copernicanism. But Maestlin also elaborated the idea that Kepler's work could make a fundamental contribution to improving astronomical theory a priori. He had in mind the determination of distances from observation, which he recognized was problematic. He wrote that at first he had thought Kepler's theory was in error, but upon reflection decided that the error probably lay in observations and previous theory:

> I indeed found that the proportions of the regular bodies did not hold true very accurately with the spheres of the world (inasmuch as they are already known and demonstrated from Copernicus). However after painstaking consideration of all the facts, I saw that such things were to be ascribed much more to the imperfections of astronomy, since the tables often deviate from observations and the motions are not yet sufficiently known. For that reason also I have undertaken observations for some years up to now with the intention, God willing, of reforming the motions and from that to calculate certain ephemerides . . . and suchlike. For which this new invention of M. Kepler's will be very helpful. . . .

> For this reason, gracious Prince and Lord, I deem this invention of M. Kepler's to be a lofty and, in astronomical studies, a very useful work, for which this time of ours when it came to light is well to be congratulated. As I too well know to make use of it, if God wills, in reforming the motions from my observations.[61]

Of course, Maestlin's statement that he would use Kepler's work in the reformation of astronomical theory was overgenerous. His letter was in essence a letter of recommendation to a patron for a former student. I have no knowledge that Maestlin seriously undertook any reformation of planetary theory based on distances derived from Kepler's polyhedral hypothesis.

This letter does show that Maestlin's early, favorable attitude toward Kepler's work was due to its support for Copernicanism by means of a priori mathematical proof. And thus his support for Kepler was in some sense based on a misunderstanding: the mathematical and metaphysical aspects of the polyhedral hypothesis had eclipsed Kepler's efforts to derive the motion of the planets from physical considerations. When the full extent of Kepler's physical ideas later became clear to Maestlin, he became critical.[62]

⌘ CHAPTER 3 ⌘

THE *MYSTERIUM COSMOGRAPHICUM*

The first thing about the *Mysterium cosmographicum* that indicated its author's intention was the title. In full, it read, *The Forerunner of Cosmographical Essays, Containing the Cosmographical Secret: On the Marvelous Proportion of the Celestial Spheres, and on the True and Particular Causes of the Number, Size, and Periodic Motions of the Heavens, Demonstrated by Means of the Five Regular Geometric Bodies.*[1] Kepler explicitly chose to emphasize in the title that this was not a mathematical but a cosmographical work. As he explained in a letter to Maestlin written while he was preparing for publication:

> Still, lest it should wholly deceive the community of mathematicians, I want to be careful with the title. It should be openly declared not a Mathematical but a Cosmographical work (where the mathematical things are borrowed and not of the greatest accuracy, just as in Aristotle's book *De caelo*), and not the *Cosmography* itself but merely a forerunner, from which men should understand that a more certain and better work will be published someday.[2]

Kepler planned the *Mysterium* as the first of a series of cosmographical books which would defend heliocentrism by means of physical arguments. After the publication of the *Mysterium*, he gave a full outline of this plan and his reasons for it in a letter to Herwart von Hohenburg:

> It has happened to me several times already that when I run into those who certainly understand the matter but cannot believe the motion of the earth, I hear that *the unusual is imagined*. That is, if this one hypothesis regarding the motion of the earth were indeed true, everything else would be most beautiful to them. In the title [of the *Mysterium*] itself I promised the material which you deem worthy of some attention. For that little book of mine was supposed to be a forerunner [*Prodromus*] of upcoming Cosmographical dissertations. My plan is this: In the new cosmography physical arguments for the motion of the earth will be brought up, and many things will be collected, in which taken separately there would be diverse things accorded little credence. And so that this invention of mine of the 5 solids would not lie hidden among those things like chaff, and among uncertain things be rejected in one thought along with the rest, I wanted to publish it separately, [and see] in this way if it should deserve credence. For it is this which ought to secure some authority for the plausible upcoming arguments drawn from physics. I plan four little cosmographical books:

> 1. On the universe, and especially on the quiescent parts of the world, the place and rest of the sun, the arrangement and rest of the stars, the world's unity etc.
>
> 2. On the mobiles, in which there is a repetition of the business of the 5 solids, a disputation on the motion of the earth, the proportion of motions, Pythagorean music, etc.
>
> 3. On the globes themselves in particular, but principally on the globe of the earth, causes of mountains, rivers, etc. In general terms, however.
>
> 4. On the relation of heaven and earth due to action and on passions, wherein [I address] light, aspects, meteorology, and astrology, by means of physical principles. Thus the first three correspond to Aristotle's books *On the Heaven*, and the fourth to the books *On Generation*.[3]

Kepler spoke in his letter to Maestlin of "deceiving the community of mathematicians," and in both letters he referred to Aristotle's *De caelo*. Kepler recognized that the classification of his work was somewhat problematic. He did not want it to be confused with technical astronomy, which was then classified as a branch of mathematics.[4] By making clear his intention to introduce physical issues into his discussion of what might otherwise have been seen as a mathematical work, Kepler was defining his work as cosmography. And yet, as is clear from references scattered throughout the book, he hoped to convince astronomers, as well as natural philosophers, to accept his arguments.

The introduction of physical arguments was the foundation of Kepler's defense of Copernicus, although the precise relationship between mathematical and physical astronomy was addressed somewhat obliquely in the *Mysterium*. From remarks scattered throughout the book, Kepler's focus on physical issues would have been obvious to the reader. At least one critic dismissively characterized Kepler's approach as the use of "physical or metaphysical reasons (as the author says)," referring to Kepler's statement in the preface regarding the type of arguments employed in his student disputation.[5] It will not be my intent to give a detailed account of the entire *Mysterium*.[6] Instead, I will give a general outline of the book and examine in detail only those arguments in which Kepler explicitly discussed physics or employed the planet-moving force, since these presage the physical astronomy of the *Astronomia nova*.

Kepler came closest to defining the place of physics in astronomy in chapter 1, where he gave the philosophical argument for Copernicus that he mentioned in the letter to Maestlin of 3 October 1595. Kepler's target was those who would maintain that Copernican theory was false although it was observationally adequate, i.e., that this was a case in which truth followed from false premises.[7] This skeptical stance was an old conundrum for astronomers. The time-honored example was that the motion of a planet on simple eccentric and on a properly contrived epicycle and deferent was completely geometrically equivalent, and thus astronomers could not definitively determine which construction was true. The basis for Kepler's defense of the truth of the Copernican system was that "that which is false by nature betrays itself as soon as it is considered in relation to other related matters."[8] Hence the absence of any limit to the applicability of the Copernican hypothesis showed that its success was not fortuitous. He then counterattacked, applying this philosophical principle to the traditional theories:

> I reply, to start with, that the ancient hypotheses clearly fail to account for certain important matters. For example, they do not comprehend the causes of the numbers, extents and durations of the retrogradations, and of their agreeing so well with the position and mean motion of the sun. Since in Copernicus their regularity is made so beautifully apparent, there must be some inherent cause of all these things.[9]

In the second part of his reply, Kepler presented Ptolemy and Copernicus as different "species" of the same "genus" of theories, with the genus apparently consisting of the relative motion of the earth and sun.[10] Given that heliocentric and geocentric systems are in principle observationally equivalent,[11] by Kepler's philosophical principle the theories can be differentiated only by applying them to "another related matter," which is where support must be sought from physics:

> Yet for Copernicus's taking the species as the genus, and in addition setting the Sun at the center of the universe, and the Earth in motion round it, there were other reasons. For, to turn from astronomy to physics or cosmography, these hypotheses of Copernicus not only do not offend against the Nature of things, but do much more to assist her. She loves simplicity, she loves unity.[12]

It was precisely in this philosophical argument that Kepler articulated the need for physical astronomy and for the *Mysterium* itself, for Kepler saw the *Mysterium* as providing the arguments from physics that would show the truth of heliocentrism.

In the second chapter, Kepler continued with arguments he had outlined in the letter to Maestlin.[13] Before the creation of the world, he argued, there was no quantity, so God relied on geometrical archetypes. The most perfect, the sphere, whose center, surface, and volume correspond to the Father, Son, and Holy Spirit, represented the sphere of the fixed stars. The planets were also assigned spheres, with their number and relative sizes determined by the interpolation of the remaining perfect solids, the five Platonic polyhedra (cube, tetrahedron, dodecahedron, icosahedron, and octahedron). Chapters 3 through 8 went into the classification of the polyhedra, its significance in the order of the solids, the place of the earth in this arrangement, and the significance of the placement of individual solids.

Chapters 9 through 12 concerned astrological topics. In the notes to the 1621 edition, Kepler called chapter 9 "merely an astrological game, [which] should be considered not a part of the work but a digression," and wrote that chapter 11 "could be omitted, for it carries no weight."[14] But it is noteworthy that he began chapter 9 by writing, "I cannot avoid here abstracting a little from that part of physics which concerns the properties of the planets, to make it apparent that their natural powers also observe this order and keep this proportion to each other," and he remarked in chapter 11, "I shall have the physicists against me in these chapters."[15] Kepler's comments that his excursions into astrology departed from the part of physics that concerned the qualities of the planets confirm that he saw the rest of the work as physics, which corresponds with Maestlin's views about the physical part of astronomy.[16] Furthermore, they show that Kepler recognized that some would object that astrology did not rightfully belong to physics.

Chapters 13 through 19 were on the observational sufficiency of the polyhedral hypothesis, i.e., based on parameters taken from Copernican theory. Kepler made it clear from the beginning that his hypothesis would stand or fall depending on the evidence. At the beginning of chapter 13, he wrote:

> So far all that has been said is that certain signs agree with the theorem proposed and make it probable. Let us now pass to the distances between the astronomical spheres and the geometrical derivations: if they do not agree, the whole of the preceding work has undoubtedly been a delusion.[17]

Determining the ratios of the spheres to the polyhedra was straightforward, but transferring this to planets introduced the problem of the thickness of each planet's sphere. Kepler noted that traditionally, i.e., in Ptolemaic cosmology, the extent of the entire mechanism of spheres for a planet determined the thickness of a planet's sphere and that the outermost surface of a planet's sphere was in contact with the innermost surface of the superior planet's sphere. Here, he again brought up the difference between physics and astronomy. It is the physicist who was concerned with nesting these spheres; the astronomer, presumably, was concerned only with the theories themselves.[18] This distinction corresponds to a division in Ptolemy's work between the *Almagest* and the *Planetary Hypotheses* (represented in Kepler's time in *Theorica* literature). In the *Almagest*, the planetary theories were all presented independently in the same standard scale. The relative scaling of the planetary theories and their nesting as a cosmological system was taken up in the *Planetary Hypotheses*. Thus as Kepler saw it, astronomers were concerned with the spare mathematics of planetary theories, and physicists or natural philosophers with the physical details of the system as a whole.

It had become clear to Kepler that he would have to depart from the traditional method of calculating the thickness of a planet's sphere when Maestlin sent him the manuscript "The Dimensions of the Spheres of the World, according to the calculations of the *Prutenic Tables*, following the hypotheses of Copernicus."[19] He realized then that any attempt to incorporate the entire extent of Copernicus's theoretical mechanism into the space between the polyhedra would ruin his hypothesis. He also recognized, however, that the extent of the mechanism was larger than the area actually traversed by a planet; the planet is always in the bottom of its epicycle at aphelion and in the top of its epicycle at perihelion (see figure 2.1). He consequently concluded that he need only make allowance for the maximum and minimum distances of the planet itself, reasoning that because the epicycles were incorporeal they should not constrain his model. He added chapter 16, "A Particular Comment on the Moon, and on the Material of the Solids and Spheres," to this effect,[20] and referred the reader to chapter 22 for a justification of using only the limits of the planets' motion.[21]

Kepler accounted for the discrepancy between the polyhedral hypothesis and the distances of the planets from Copernicus in chapter 15. Here again, the distinction between astronomy and physics came into consideration. He began the chapter by stating:

> In case, friendly reader, I should offer you any occasion for rejecting the whole of this enterprise because of a trifling dispute, I must here mention to you something which I should like you to remember carefully: Copernicus's purpose was not to deal with cosmography, but with astronomy. That is, he is not much concerned whether there is a mistake relating to the true proportion of the spheres, but only with establishing from the observations the values which are best suited for deriving the motions of the planets and computing their positions, as far as possible. But if anyone should try to give better suited values and rectify Copernicus's in such a way as to upset in the process nothing, or very little, in the system of equations, that will readily be permitted as far as Copernicus is concerned.[22]

With this, Kepler introduced a proposition that galled some contemporary astronomers: if the polyhedral hypothesis did not agree with Copernican theory, then Copernicus was liable to alteration.[23] Although he did allow that the changes must be insensible or very small, cosmography evidently took precedence over astronomy. Kepler had to propose forcing the polyhedral hypothesis and Copernican theory into precise agreement, or his effort to provide supporting physical arguments for Copernicus adequate to astronomers would fail.

In his effort to assess the accuracy of the theory, Kepler, with Maestlin's help, constructed the distances of the planets anew based on the true sun at the center of the world. Copernicus had used the position of the mean sun as the center, the mean sun being the center of the earth's orbit. Since the theory of the sun (or, equivalently, the earth in the heliocentric system) had always been a simple eccentric since the time of Hipparchus, the sun was not at the center of the earth's orbit but slightly eccentric, which placement accounted for the nonuniform motion of the sun with respect to the stars. Kepler recognized that Copernicus had used the mean sun for ease of calculation, and in addition to making Copernicus's theory only pseudoheliocentric (i.e., that the center of the earth's orbit rather than the body of the sun was the center), it had the undesirable consequence of giving no thickness to the sphere of the earth (by definition). In the light of Kepler's later research, this work is significant, for one of the first major steps in the *Astronomia nova* is Kepler's shift from the mean to the true sun. After carrying out these calculations in chapter 15 of the *Mysterium*, Kepler presented a table of the distances with values from Copernicus, Copernicus revised to the true sun, the polyhedral hypothesis, and the polyhedral hypothesis with the moon included in the thickness of the sphere of the earth.

Kepler considered the meaning of the discrepancy between his hypothesis and observations in chapter 18, "On the Disagreement between the Equations Derived from the Solids and those of Copernicus in General, and on the Precision of Astronomy."[24] He pointed out again the distinction between the physicist and the astronomer; the physicist would already be satisfied by the agreement of his theory, the astronomer would not:

> Consequently those who wish to scrutinize everything too exactly will think that because the reckoning from the solids does not agree to the last detail with the theories of Copernicus, and with his values, the whole of my work has been made ridiculous. On the other hand, unless I remove this charge, in my opinion I shall have lost the case. To the physicists, indeed, or cosmographers, according to the role which I maintain in this little book, I owe no explanation of the difference. For although they borrow their arguments for their theories from the astronomers, yet they do not check them by calculation as precisely as the astronomers, and they are not so acute or so critical as to be influenced by this trivial difference. Consequently I have won my case among the cosmographers.
>
> Now I rightly fear the mass of astronomers; yet since it is right that the practitioners should take precedence in giving judgment, I do not despair of victory, even in the face of that. First, I tell them that they should be hopeful about their calculations.[25]

He suggested that the discrepancy that existed between his hypothesis and Copernican theory was due to the shortcomings of Copernicus and the *Prutenic Tables*. By going into detail about the various factors that contributed to error in Copernicus, he attempted to engage and win over the astronomers. This indicates Kepler's intent if not to merge, then at least to satisfy, the two disciplines.

In order to do this, he discussed four aspects of Copernican theory that could contribute to an error in the distances. First, although he conceded that Copernican theory was far better than previous theories, he pointed out that the tables were often in error. "I do not say this to show contempt for Astronomy," he wrote, "but to prevent anyone rashly putting a more serious construction on this disagreement, and while aiming at me, and the 5 solids, scoffing at the very foundations of astronomy."[26] Second, there were suspect values for the eccentricities. The size of the eccentricity determined the thickness of the sphere Kepler put between two solids, and his experience with the problematic thicknesses had shown how significant this could be. Not only did he lay the blame on these values, but he also suggested that his work could aid in determining them. Perhaps in reference to Maestlin's surprising position in his letter to the Duke of Württemberg,[27] Kepler wrote:

> For it is my opinion that after the discovery of this proportion in the heavens everything which still prevents us from attaining exact knowledge of the heavens is to be attributed to errors in the eccentricities; and if those were removed, I think that the five solids would be of great assistance to the practitioners for the correcting of the motions, which not a few of them in various places are contemplating.[28]

Third, he faulted both Reinhold, for having uncritically taken parameters for the *Prutenic Tables* directly from Copernicus, and Copernicus's haphazard derivation of those parameters. And fourth, he criticized Copernicus for having adopted eccentricities directly from Ptolemy. Finally, perhaps to justify and atone for his sustained criticism of Copernicus, Kepler appended an excerpt of a letter by Rheticus, in which Rheticus explained that Copernicus was content with a larger view of his work and limited his concern with the details to 10′ of error.

The Motive Force Hypothesis

After 19 chapters that focused exclusively on the polyhedral hypothesis, Kepler addressed his period-distance relation and the planet-moving force in chapters 20 through 22. With only three chapters out of a total of twenty-three on these topics, one might be tempted to conclude that Kepler placed more emphasis on the polyhedral hypothesis. Apart from the evidence offered above that Kepler was equally concerned from the beginning with a relationship determining the motions, there is evidence for his placing equal emphasis on the polyhedral hypothesis and the period-distance relation in a letter to the then imperial mathematician, Nicholas Reimers Ursus (1551–1600).

While in his letter to Maestlin of 3 October 1595 Kepler had not proceeded to the motive force hypothesis until after he had elaborated on the properties of the solids and their arrangement and had explicitly asked for Maestlin's judgment of what followed, in the letter to Ursus, the importance of the motive force hypothesis is nearly equal to that of the interposition of the solids. In a letter of 15 November 1595, Kepler informed Ursus of the discoveries he would publish in the *Mysterium*. After a succinct account of his polyhedral hypothesis, in which he related only that the number and sizes of the orbs are determined by the five Platonic solids and gave their order, he continued with an account of the period-distance relation of roughly equal length. He mentioned that the periods have a double relation to the distances, and that "a weakness is added in the exterior [parts], such as happens in the extenuation of light rays."[29] He gave only the slightest indication that the polyhedral hypothesis might be preferable, writing, "It disagrees very little with Copernicus in either case, more, however, if the distances are determined from the mean motions than from the solids."[30] This statement can be taken as evidence that Kepler found the polyhedra and the motive force the two most significant elements of the *Mysterium*, even if the latter was overshadowed by ancillary arguments regarding the solids.

Kepler introduced the relationship between the planets' periods and distances in chapter 20 of the *Mysterium*, where, like the polyhedral hypothesis, it was presented as an additional argument for Copernicanism:

> So far the argument by which I think a great deal of strength has been added to the novel hypotheses has run smoothly, and it has shown that the distances of the orbits in the hypotheses of Copernicus use the ratios of the five regular solids. Let us now see whether also from a second argument drawn from the motions both the new hypotheses and the Copernican dimensions of the orbits themselves can be verified, and for the ratio of the motions to the distances a more accurate account can be obtained from Copernicus than from the customary hypotheses.[31]

According to the philosophical tenet that he introduced earlier, that "that which is false by nature betrays itself when it is considered in relation to other related matters," the argument from the motions would be valuable confirming evidence that the Copernican system was true, for here, a different and unrelated physical derivation would again produce the Copernican dimension of the orbs.

Kepler began by showing how in the Copernican system, there is a steady progression in periods from Mercury to Saturn and how the increase in the planets' periods is greater than in direct proportion to their distances. He then proceeded to his physical explanation of this proportion, noting:

> . . . one of two conclusions must be reached: either the moving souls are weaker the further they are from the Sun; or, there is a single moving soul in the center of all the spheres, that is, in the Sun, and it impels each body more strongly in proportion to how near it is. In the more distant ones on account of their remoteness and the weakening of its power, it becomes faint, so to speak.[32]

The affinity of this idea to those in Kepler's student disputation is manifest in the lines that follow:

> Thus, just as the source of light is in the Sun, and the origin of the circle is at the position of the Sun, which is at the center, so in this case the life, motion and the soul of the universe are assigned to that same Sun; so that to the fixed stars belongs rest, to the planets the secondary impulses of motions, but to the Sun the primary impulse. In the same way the Sun far excels all others in the beauty of his appearance, and the effectiveness of his power, and the brilliance of his light. Consequently the Sun has a far better claim to such noble epithets as heart of the universe, king, emperor of the stars, visible God, and so on.[33]

Kepler's calculation of the ratio by which the periods increase with distance was the same as that which he sent to Maestlin in the letter of 3 October 1595. The increase in the periods is greater than that which would be caused merely by the increase of the length of the orbit with distance, he reasoned, so there must also be some weakening of the moving spirit. "Let us suppose, then, as is highly probable," he wrote, "that motion is dispensed by the Sun in the same proportion as light."[34] He repeated his conclusion that, because the amount of light in a circle becomes weaker in proportion to the radius and the lengths of the orbits increase in proportion to the radius, an increase in distance causes a twofold decrease in the period. And he gave the same statement of the ratio of the distances of a superior planet to an inferior planet as $1/2\ (P_S - P_I) + P_I : P_I$.[35] Kepler updated the table of distances with the new distances Maestlin had provided and added an alternative but algebraically equivalent phrasing of the equation, but otherwise the argument is the same as in the letter.[36]

When Kepler came to addressing the error of the period-distance relation in chapter 21, he was faced with a problem. He was pleased with his unprecedented period-distance relation, and allowed, "That indeed should have been the sole intention of this treatise on motion."[37] However, the errors for the period-distance relation were somewhat greater than those from the polyhedral hypothesis. Moreover, they were not identical and thus implied that the two theories were incompatible. Nevertheless, he still found the period-distance relation more significant than the solids:

> It seems to me unlikely that anyone will give any other answer than that this fitting of the motions to the spheres is very neat, a wonderful piece of handiwork by God the craftsman. Consequently, if one or other argument must be accepted, they will assent to the second argument rather than to the one from the solids, as being more obviously acceptable, even though the values still have a slight discrepancy from the Copernican ones.[38]

He then tried to turn this argument around, using a greater willingness to accept the less accurate period-distance hypothesis to argue for the acceptance of the more accurate polyhedral hypothesis.

Kepler's attempt to give an account of the errors in the period-distance relation was at the same time an argument for the polyhedral hypothesis and for the compatibility of the two. He did this by preparing a table of the absolute differences of the distances derived from the period-distance relation from those taken from Copernicus, and noting the similarities in the differences to the solids that determine the same spacing in the polyhedral hypothesis. Thus, for example, only in the case of the Jupiter-Mars distance was the difference negative, and that corresponded to the tetrahedron. After adjusting the mean distance of Mercury, the errors for the Saturn-Jupiter and Venus-Mercury distances were of comparable size (2 and 4 units) and the errors for the Mars-Earth and Earth-Venus distances were of comparable size (36 and 43 units).[39] This, Kepler noted, corresponded to the similarity of the solids involved, the cube and octahedron for the former pair, and the dodecahedron and icosahedron for the latter pair. He concluded from this pattern in the errors that "we must therefore consider that these values hint at the truth, but we have not yet completely achieved it."[40] And he suggested because of this that both hypotheses may be manifestations of some other as yet undiscovered regularity. Finally, by means of a complicated table, he argued that even if one disregarded the polyhedra in favor of the period-distance relation, it did not exclude the interpolation of the solids.

In chapter 22, Kepler took the very significant step of applying the reasoning behind the period-distance relation to another problem: why a planet moves uniformly around the equant. He used the idea that the strength of the planet-moving force diminishes with the distance from the sun to explain how the planet moves more slowly when farther from the sun and more quickly when closer to the sun. In light of the centrality of just this concern in his later determination of the first two laws of planetary motion, this chapter is perhaps the most interesting in the *Mysterium cosmographicum*.

As noted above, when Kepler received Maestlin's "The Dimensions of the Spheres of the World," he was forced to conclude that regardless of the extent of Copernicus's theoretical mechanism, he could only make allowance in the polyhedral hypothesis for the maximum and minimum distances of the planet itself. He justified this conclusion in chapters 16 and 22, reasoning that because the epicycles were incorporeal they should not constrain his model. But while he was scrutinizing Copernicus's eccentric-epicyclet, he apparently recognized (or believed he recognized) that though Copernicus had explicitly reconstituted his astronomical theories to avoid the mechanism of Ptolemy's equant, Copernicus's eccentric-epicyclet did produce uniform motion around the point formerly occupied by the equant.[41] This feature was, indeed, a necessary element of Copernicus's agreement with Ptolemy.[42] The incorporeity of Copernicus's spheres that was necessary in order to fit the polyhedra correctly between the planets may have given Kepler cause to wonder about the planets' nonuniform motion through space. At any rate, when he considered carefully that the planets move more slowly in their orbits when they are farther from the sun, he certainly thought back to his period-distance relation, and concluded that the same relation explained both the differences in the periods of the planets from one to another *and* the nonuniform motion of a single planet over its own orbit:

> [T]he reason and the means [of the irregularity in the planets' motion] are shown more clearly by these writings of our own, that is, if the cause of the retardation and acceleration is the same for the spheres of the individual planets as it was above in Chapter 20 for the whole universe, in the following manner. The path of the planet is eccentric, and it is slower when it is further out, and swift when it is further in. For it was to explain this that Copernicus postulated epicycles, Ptolemy equants. Then describe a concentric circle equal to the eccentric path of the planet. . . . Therefore at the middle part of the eccentric path where it projects above the concentric circle, the planet will be slower, because it moves further away from the sun and is moved by a weaker power; and in the remaining part it will be faster, because it is closer to the Sun and subject to a stronger power. . . . You know the cause of this slowness; let us now look at the measure of it as well.[43]

Kepler's quantitative argument, as corrected by Maestlin, was as follows: The sun is at A, with a reference circle NOPQ (see figure 3.1). The line of apsides passes through ABCD, with AB = BD. The center of the planet's orbit is B, with the orbit EFGH. And the equant is at D, with its reference circle IKLM. All circles are of equal radius. When the planet is

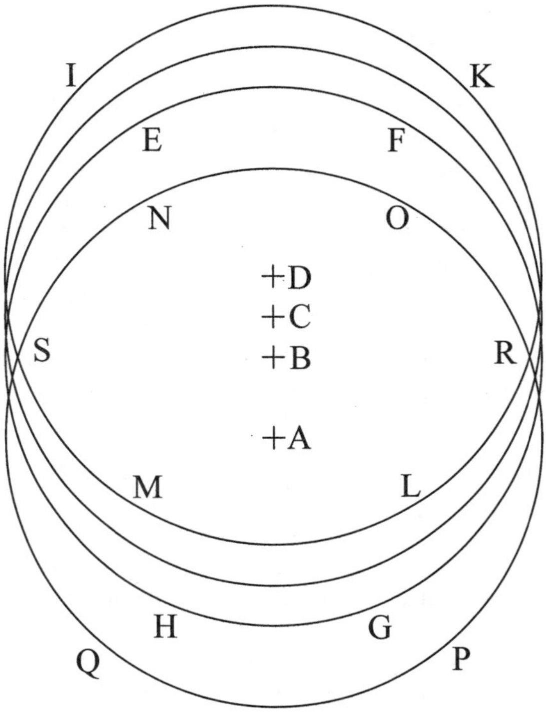

Figure 3.1 Kepler's motive force hypothesis from the *Mysterium Cosmographicum*, chapter 22

Kepler used this figure to illustrate his motive force hypothesis in the *Mysterium cosmographicum*. The planet's orbit is EFGH with center B. The sun is at A, and Ptolemy's equant at D.

around EF, it is further from the sun than NO by the length AB. Therefore, since, according to Kepler, the motive spirit decreases according to twice the radius, its motion with respect to how it would have moved at the constant distance NO is as slow as if it were traveling twice as far out, on IK. Similarly, when the planet is around HG, it is farther in than the constant circle QP by the amount AB, and it will move as though it were twice as far in, on ML. Thus, even though the planet travels on EFGH, it moves as though it were traveling with constant velocity on the circle IKLM around the equant D. This is a pleasant, if not rigorous, argument, valid only at apsides.

Kepler went on briefly to note that although Copernicus constructed an eccentric around C, his epicycle very nearly traces the path EFGH, and the planet's motion is also uniform around the equant D. Kepler did not offer a rigorous proof of this equivalence, however.

Kepler must have found it particularly compelling that his argument accounting for the equant was valid for both the Ptolemaic and Copernican models. He had already abandoned the notion that the epicycles in planetary models had any reality in order to make the polyhedral hypothesis fit with the minimum and maximum Copernican distances. Immediately before his account of the equant, he cited Tycho Brahe's abandonment of solid spheres on this point. But his model of the equant cast the different arrangement of spheres in Ptolemy and Copernicus in a new light. Not only were the spheres incorporeal in the sense that they would not interfere with the spacing of Kepler's polyhedra, but they were also actually nonexistent. With his argument, Kepler cast both the Ptolemaic and Copernican theories simply as models for the motion he explained from the physical hypotheses of a moving spirit extending forth from the sun:

> Then, naturally, let the whole universe be full of a spirit which whirls along any stars or comets it reaches, and that with the speed which is required by the distance from the Sun of their positions and the strength of its power there. Next let there be in each planet an individual spirit, by the impulsion of which the star goes up in its circuit; and even without the spheres the same results will follow.
>
> Anyone who reads this passage on the equant will, I know, rejoice. For if the astronomers are surprised that Ptolemy assumed this same measure of the center of the equant without proof, some people will now be all the more surprised that there was an explanation for it, and Ptolemy did not suspect it, since he assumed the fact to be as it is, and as if by divine guidance arrived blind at the proper destination.[44]

Here, then, was the foundation statement of Keplerian dynamics: There is a radial force surrounding the sun that pushes any planet around it such that closer objects are moved around more quickly than further ones. In addition, each planet has its own individual motive force by which it navigates itself now closer, now further from the sun. As a consequence, planets are moved around their orbits with a physically nonuniform motion, faster when they are closer and more slowly when they are further away.

Kepler's conclusion that the reigning mathematical theories of the planets' motion were both merely closely equivalent descriptions of an underlying physical process did not have mathematical rigor at this time. He apparently accepted that the shape of these orbits would be circular, or very nearly so (bear in mind that Copernicus's orbit was not a true circle). And he provided little more than a hand-waving mathematical demonstration that his formula for the strength of the planet-moving force would produce a similar motion to that produced by Ptolemy's equant.

In his enthusiasm, he does not seem to have had an inkling of just how intractable his physical analysis of planetary motion would turn out to be. But given the centrality of this idea in the *Astronomia nova*, as well as the general kinship of the *Mysterium* to the *Harmonice mundi*, one can easily see why Kepler wrote in the dedication to the 1621 edition of the *Mysterium*, "Almost every book on astronomy which I have published since that time could be referred to one or another of the important chapters set out in this little book."[45]

⌘ CHAPTER 4 ⌘

RESPONSES TO THE *MYSTERIUM COSMOGRAPHICUM*

Kepler had accepted the assignment to teach mathematics in Graz more or less unwillingly. Once there, he conscientiously turned his attention to astronomy and realized that he could redirect his religious calling to glorify God through his astronomical discoveries. With the publication of the *Mysterium cosmographicum*, he hoped to realize this ambition. For Kepler, the Copernican heliocentric system was a manifest symbol of God in His creation and a sign of His intelligent design. It was essential, therefore, to establish its physical truth. Anything less would make Kepler's new vocation meaningless.

He faced two distinct resistances, however, both of which would tend to return him to the conventional role of the mathematical astronomer. The first exerted its effort before publication. It was the successful effort of theologians at the University of Tübingen to suppress any discussion of how to reconcile the heliocentric system with apparently contradictory passages from the Bible. In so doing, they urged him to adopt the stance of an abstract mathematician, a position that would undermine his goal of proving the physical truth of heliocentrism. The second resistance came from astronomers themselves. Kepler's novel insistence that physical reasons must be employed to establish the truth of heliocentrism and—more unsettling—that physical reasoning was prior to mathematical astronomical theories trod on the convenient boundary between natural philosophy and mathematical astronomy. The response of astronomers to the publication of the *Mysterium* was varied inasmuch as some saw it as cosmography rather than astronomy proper, but their disapprobation of Kepler's importation of physics into mathematical astronomical theory seems to have been universal.

Theological Objections to the *Mysterium*

As we saw above, neither in his student disputation nor in his plans for the *Mysterium* had Kepler conceded that theological objections to Copernicus were insurmountable. However, the first chapter of the *Mysterium* as it was published began with this brief disclaimer:

> Although it is right to consider right from the start of this dissertation on Nature whether anything contrary to Holy Scripture is being said, nevertheless I judge that it is premature to enter into a dispute on that point now, before I am criticized. I promise generally that I shall say nothing which would be an affront to Holy Scripture, and that if Copernicus is convicted of anything along with me, I shall dismiss him as worthless. That has always been my intention, since I first made the acquaintance of Copernicus's *On the Revolutions*.[1]

This statement is disingenuous. In the letter of 3 October 1595 in which Kepler gave Maestlin the first outline of the *Mysterium*, he had expressed his intention to begin the work "with some theses I discuss concerning Holy Scripture, and I show at the same time both how Its authority would be consistent with them, and yet that Copernicus, if he says consistent things otherwise, cannot be refuted by them."[2] Yet instead of these "several theses," Kepler substituted a brief, weak disclaimer.

The reason for this substitution was that, during the publication of the *Mysterium*, Kepler had run afoul of the theological authorities in Tübingen.[3] When Kepler was in Tübingen in 1596, he approached the printer Gruppenbach about publishing his work. The printer expressed his willingness, on the reasonable condition that it be done with the university's cognizance and permission. Kepler sought permission in a letter of 1 May 1596 (o.s.).[4] The university then sought Maestlin's opinion, which he evidently submitted sometime that month.

Maestlin responded much along the same lines as he had the previous month in his letter of support to Duke Friedrich von Württemberg, again expressing the hope that "Then with this a priori foundation as it were having been given, there is no doubt, that those who collect the observations, having devoted themselves to the reforming of the motions of the celestial bodies, will have the greatest aid."[5] His only reservation was that Kepler's work was written for those who were already familiar with Copernicus and Euclid. However, too few whom this work would interest, he wrote, would have the leisure to learn these authors first, so Kepler should be asked to present his discovery more plainly and popularly, and he should include a preface with diagrams explaining the Copernican hypotheses. This mandate, along with the unanimous approval of the senate was passed to Kepler in a letter of 6 June 1596 (o.s.) from the prorector, Matthias Hafenreffer (1561–1619).[6]

Such was the official face of the university's oversight of the publication of the *Mysterium*. However, there had evidently been private pressure as well, as becomes clear in a series of letters written after publication. Immediately afterward, in a letter of 9 April 1597, Kepler expressed to Maestlin his relief that defenders of Holy Writ had raised no objections to the *Mysterium*.[7] He had had his originality questioned by a local, provincial critic, but "I know of no one in the brilliant light of your academy who would intentionally assail my doctrines."[8] Not enough time had elapsed for Kepler to hear the response of readers, but he was optimistic that he would be treated well in Tübingen, where he had met their criticism before publication.

But his youthful enthusiasm and the perspective from remote Graz gave Kepler an unrealistic view of the reception of his work in Tübingen. He was reminded of the true situation when later that year Maestlin reported back in a letter of 30 October 1597 about the theological atmosphere in Tübingen with respect to Copernicus. Maestlin, who had been a Copernican at Tübingen for many years, had a more profound understanding of the potential for conflict between Copernicans and theologians. He also understood that Kepler was in greater danger of a theological than a scientific refutation:

> I was easily and soberly able to persuade myself that there would not be lacking those who would defame your book, but to defame is not to refute. He who wants to quash this discovery of yours and the Copernican hypotheses must attack them with other considerations and stratagems. There were rumors being thrown about among us about two books published against your discovery, one of them in fact published by Dr. Hunnius (which I am simply not able to believe). But in truth nothing of them came to us in Tübingen; still I will take pains to obtain [a copy] if something is published. At the same time (which notwithstanding I would want to confide to you now) it also offends our theologians somewhat. They make no overt move, however, because they are deterred by the prestige of our duke, to whom the key diagram is dedicated. Time and again Dr. Hafenreffer has assailed me (jokingly, to be sure, although serious tones too seem to be intermingled with the jests). He wants to debate with me, while defending his Bible, etc. By the same token not so long ago in a public evening sermon he expounded Genesis, Chapter I: "God did not hang the sun up in the middle of the universe, like a lantern in the middle of a room" etc. However, I usually reply humorously to those jokes as long as they remain jokes. If the matter were to be treated seriously, I too would respond differently. Dr. Hafenreffer acknowledges your discovery to be wonderfully imaginative and learned, but he regards it as completely and unqualifiedly in conflict with Holy Writ and truth itself. Yet with these men (who are otherwise fine and very scholarly, but) who have no adequate grasp of the fundamentals of these subjects, in like manner it is better to act jokingly while they accept jokes.[9]

Kepler was inclined to dismiss Hafenreffer's comments. He felt that Hafenreffer was a Copernican at heart but hewed to the theological line to appease his colleagues. In a letter of 6 January 1598 responding to Maestlin's warning, he expressed this:

> With regard to what you write about a work written against me by Hunnius, I too certainly believe it is crazy that he would be associated with this. Unless perhaps it were a preface for something written against me. For not so long ago, when somebody would have published a joking dissertation—that women are not men—he brought forward the severe reprimand that a youth ought to restrain himself from those kinds of mockeries and profanations of Scripture. If something of these matters is brought to you, I beg you, do not keep it from me. By Dr. Hafenreffer's authority I changed the beginning of my book. He is the promoter of the printing, the discovery's eloquent praiser. That is why I wrote to him, concealing what I know from your letter. And I really cannot believe that he is opposed to this doctrine [Copernicanism]. He pretends [to be so] in order to appease his colleagues, whom perhaps he displeased by furthering my book. And this he must be granted since for him peace with his colleagues is more important than peace with me.[10]

Kepler's abrupt comment about having changed the beginning of the book on Hafenreffer's authority provides evidence that Kepler had experienced theological pressure during the process of getting approval to publish the *Mysterium*. It also reveals Kepler's feeling that having altered the book at Hafenreffer's suggestion he had thereby absolved himself from theological criticism.

Hafenreffer's response to Kepler in a letter of 12 April 1598 (o.s.) is more specific about the changes Kepler made to the *Mysterium*. Kepler's letter to Hafenreffer, which he mentioned to Maestlin in the passage above, has been lost, but Hafenreffer's response shows that Kepler was discreetly seeking advice about how to reconcile Copernicanism with Scripture. Kepler was clearly beginning to appreciate more fully the delicacy of the situation. In his letter to Maestlin of 3 October 1595, he had brashly proposed to show in the *Mysterium* that Copernicus could not be refuted by Scripture. Now, while receiving clandestine information from Maestlin about the atmosphere at Tübingen, he was ostensibly seeking "brotherly advice," and evidently going so far as to attach some arguments for Hafenreffer's perusal. Hafenreffer responded:

> In that you finally attach some of those attempts of yours to marvel in the new demonstrations, I discerned that you were wavering about what was to be done in this domain, whether to pass over the matter in silence or to try to make clear to everybody that those hypotheses of yours agree in all respects with Holy Writ. As I understood your letter, in that part of it you do not conceal your request for my brotherly advice. What I think about this subject,

most illustrious sir and dearest brother, I shall reveal to you frankly and openly. Between those hypotheses and Holy Writ there must of course be a clear distinction. This I have always felt without any question, even from the time when I first saw the hypotheses, as you could observe when you were here with us [at Tübingen in May 1596] and can still remember perfectly well. For this is the reason why I recommended, not only in my own name but also in the name of my colleagues, the omission from your treatise of the chapter (I think it was No. 5) which dealt with this harmonization [of Copernicanism with the Bible], lest those very [theological] disputes arise therefrom. That indeed was omitted, except that right from the start[11] some—but the briefest—mention of this matter is made.[12]

Hafenreffer's letter clarifies the fate of Kepler's attempt to reconcile Copernicanism with the literal truth of Scripture. He had in fact composed a chapter along the lines he had outlined to Maestlin, in which he confronted arguments from Scripture against Copernicus. (Material apparently from this chapter later surfaced in the introduction to the *Astronomia nova.*) Hafenreffer, prorector of the university, had "recommended" in the name of his colleagues that it be cut. From the last sentence of the passage, it is clear that not even Kepler's bland introductory paragraph escaped their scrutiny.

The *Mysterium* nevertheless turned out far more provocative than the authorities at Tübingen had wished. Hafenreffer's intention was consistent with the "Wittenberg interpretation" of Copernicus: he wanted Kepler to present Copernicanism and his own ideas as ingenious mathematical constructs with no claim to physical reality. However, he did not foresee that after Kepler's material had passed review Maestlin would supplement it. Maestlin, still not content with Kepler's account of Copernicus, added as an appendix Rheticus's *Narratio prima* (1540), the most avidly realist pro-Copernican work in existence. Kepler had, in fact, frequently mentioned the possibility of appending the *Narratio prima.*[13] But Hafenreffer had indicated that Kepler should insert his own exposition of Copernicus in place of the *Narratio* "which is more longwinded, has certain different views from your *Prodromus* [i.e., the *Mysterium*], and has certain shortcomings."[14] Maestlin's presumptive addition clearly incensed Hafenreffer:

> We cannot be blamed for the other material that was added later, especially the forward by M[aestlin], since none of these later additions were seen by us before they were sent to the printer. In the present situation, however, I advise and caution you in a brotherly way not to try even to propound the aforementioned harmonization in public and to fight for it. For if you do, many good men might be incensed, not without cause, and the whole business might be blocked or bespattered with the noxious stain of dissension. For I

have no doubt that if this opinion were openly advocated and defended, it would have its opponents, perhaps not lightly armed. Therefore, if there is any place for my brotherly advice (as I firmly hope), in demonstrating such hypotheses hereafter you will play the part of the pure mathematician, without worrying at all whether the hypotheses do or do not conform to the things in creation. For the mathematician accomplishes his purpose, I believe, if he produces hypotheses to which the phenomena correspond as closely as possible. And you yourself, I suppose, will yield to the man who could discover better hypotheses. Yet it does not follow that the truth of things instantly agrees with the hypotheses conceived by each and every specialist. I do not wish to adduce the irrefutable passages I could cite from Holy Writ. For in my judgment what is needed is not disputes but brotherly advice. If you heed it (as I confidently trust) and play the part of the abstract mathematician, I have no doubt that your thinking will be highly pleasing to very many persons (as it certainly is to me too). But if (may Almighty God forbid this) you should wish to harmonize those hypotheses with Holy Writ openly and fight for this harmonization, I am certainly afraid that this matter may result in dissension and strain. In that case I would wish that I had never seen your thoughts, although in themselves and considered mathematically they are splendid and lofty. For in God's church there has long been more strife than is good for the weak. However, I don't know where I am being carried along by my pen, or rather by my brotherly affection for you. Were it not as very strong and very sincere as it is, I would not have permitted my pen such unfettered freedom. But this very affection makes two demands upon you: For us, play the role of the strict mathematician, and constantly foster that peace in the church which I know you have favored up to now.[15]

Kepler discussed the gist of Hafenreffer's warning in a letter to Maestlin of 1/11 June 1598. Despite his clear and ominous warning, Hafenreffer's recommendation to continue pursuing mathematical astronomy reinforced Kepler's hope that Hafenreffer was truly a Copernican trying to maintain tranquility among his theological colleagues. Nevertheless, he understood the message: "I am to refrain from mentioning Holy Writ in public."[16] His response was both theologically and economically pragmatic. Having himself aspired to service in the church, he sincerely declared, "What are we to do? The whole of astronomy is not important enough to offend any of Christ's little ones."[17] On the other hand, as one who now made his living as an astronomer he also knew that:

The whole world is full of men who are ready to throw astronomy, if it sides completely with Copernicus, off the earth, and to forbid the specialists an income. But specialists cannot live off themselves or on air. Therefore, let us act in astronomical affairs in such a way that we hold on to the supporters of astronomy and do not starve.[18]

Finally, Kepler was concerned for his beloved teacher. He sensed that he might become the flash point of a dispute between Maestlin and his "joking" colleagues, and hastened to tell Maestlin, "I would not want you to make enemies or suffer setbacks for my sake or even for the sake of the truth itself (which affects nobody's salvation or income)."[19]

Later, in the *Astronomia nova*, Kepler would be able to express his opinion about the proper interpretation of Scripture and its relation to scientific knowledge. When he became imperial mathematician to Rudolf II, the authorities in Tübingen would no longer control his publication. And at that time, having experienced the vicissitudes he would suffer at the hands of Catholic authorities, he would disregard the qualms of his remote former professors. But in 1598 he knew where he stood: he was to "play the part of the abstract mathematician," to desist from mentioning Scripture, and thereby to refrain from claims to the physical truth of Copernicanism. Though the premises were different, the objections of astronomers to the *Mysterium* would reach the same conclusion. But Kepler could not abandon his uniquely passionate belief in the reality of Copernicanism; it was his primary scientific motivation. Instead, in the same letter to Maestlin, he formulated a plan whose consequences should not be ignored in the analysis of Kepler's thought: he resolved that they should emulate the secrecy of the Pythagoreans, "If anybody approaches us privately, let us tell him our opinion frankly. Publicly let us be silent."[20]

The Response of Astronomers to the *Mysterium*

After the publication of the *Mysterium* in early 1597, Kepler was anxious to hear the opinions of men of learning. He distributed copies of his new book and solicited responses. When he became aware later that year that the Bavarian Chancellor Herwart von Hohenburg was seeking some help with his astrological interpretation of a passage in Lucan's *Civil War*, he painstakingly worked out an astrological analysis of the passage and sent it to Herwart along with a covering letter, in which he asked in return that Herwart "solicit the judgments of mathematicians known to you about my newly published little book, no matter how favorable to me, and have them sent to me (if I do not ask presumptuously) on your authority."[21]

It is difficult to assess Kepler's reaction to the responses he received. In the 1621 edition of the *Mysterium*, he wrote that "just as if it had been dictated to my pen, an oracle fallen from heaven, every chapter of the little book was recognized at once, by those who understood it, as important and quite true (as the manifest works of God usually are)."[22] And he cited the letters he had received from Tycho Brahe and professors of mathematics, Galileo, Ursus, and Georg Limnaeus.[23] Kepler's perception of support is understandable. Maestlin criticized only a small aspect of the book. Galileo, who had not yet read the book, in expressing his support of Copernicus, supported Kepler by implication. Tycho's criticism was mitigated by his desire to cultivate his relationship with Kepler. And there is no evidence that Kepler was aware of the opinion of his most scathing critic, Johannes Praetorius.

The support that Kepler did receive was probably more a reflection of support for Copernicus than for his own a priori speculation. Furthermore, those elements of physical astronomy in the *Mysterium* that are directly related to Kepler's work in the *Astronomia nova* received little or no support.

Michael Maestlin

As we have seen, Maestlin was generally supportive of Kepler's work in the *Mysterium*. He wrote letters of support to Duke Friedrich von Württemberg and the University of Tübingen, and devoted no small effort to seeing the book through the press. There was, however, one element in the *Mysterium* about which Maestlin was critical. This was Kepler's idea of the planet-moving force (or "moving spirit," as Maestlin called it), which Kepler used in chapter 22 of the *Mysterium* to explain the motion of a planet around the equant.

It was only in one of Maestlin's final letters to Kepler preceding publication that he expressed this criticism, for Kepler had sent the material for that chapter only late in the process.[24] What is more, Kepler had made a geometrical mistake in the argument that required Maestlin to correct it for publication.[25] Maestlin explained the correction in a letter of 11 July 1597 (o.s.). After pointing out the error and his correction, he added:

> I do not reject this speculation of spirit and motive virtue. Nevertheless I fear that it is not very subtle if it is extended too much. . . . I am really afraid that if too specific it would become more than a method, [and] it would lead to the loss or at least the ruin of all of astronomy. I deem that this speculation is to be used altogether sparingly and with great moderation. And as I would truly say what I think, I do not reject, but my assent is really weak, indeed very many objections stand before me.[26]

This was Maestlin's only serious criticism of all of Kepler's work in the *Mysterium*.[27] Why was he supportive of the polyhedral hypothesis and the rest of the book, and critical of chapter 22 on the reason a planet moves uniformly around the equant?

The answer might be found in Maestlin's own astronomical textbook, the *Epitome astronomiae*, an edition of which was published in 1597.[28] In it, Maestlin made a conventional distinction between the astronomical and physical parts of astronomy. After a number of propositions on the structure of the world, for example, that the heavens are spherical, their motion is uniformly circular, the earth is at rest in the center, and so on[29] he wrote:

> These propositions, since the science of astronomy borrows them from physics, demand not only astronomical but sometimes even physical demonstrations (although demonstrations of this kind are not to be admitted into other parts of astronomy so as to avoid confusion of the sciences).[30]

And elsewhere, in pointing out the distinction between two elements in the answer to the question "What is the world?," he made this distinction clear:

> The former is astronomical, for it specifies in plain words parts of the world with which astronomy is occupied, that is to say, the heaven and the earth (from the earth the astronomer observes the motions of the heaven), notwithstanding that with these words it encompasses other simple and mixed bodies, and all remaining natures not pertaining in the same way to astronomical science, including those things that are confined to these natures. Furthermore, with the word *systematos*, which signifies the composition, constitution, or the coming together into the form of a globe, it expresses not so much the essential as the accidental round form of the world, properly subject to astronomical consideration. However, there should be no mention of efficient and final causes on the grounds that they are alien to astronomy.
>
> The other is physical, for it specifies not so much the material (of the whole, of which the world as it now is consists) and essential forms of the world (the arrangement and distribution, the most perfect and complete essence of the world, the elegant order, etc., whence the world is called *kosmos*), but also the efficient (by God) and final cause (for the sake of God; all things are certainly made for the sake of man—just as was piously said by the Stoics—but man himself was made for the sake of God, and therefore the cycle is finished in God). However, these things are within the realm of physics. But in truth, that definition, if the word "created" is added, is amended, and also entirely satisfies us in our sacred religion. Thus [modified it reads] . . . "The cosmos is the arrangement and distribution of everything created and preserved by God and for the sake of God."[31]

From Maestlin's comments, we can perhaps reconstruct his attitude toward the *Mysterium*. As is clear in the full title of Maestlin's book, *Epitome astronomiae: qva brevi explicatione omnia, tam ad Sphæricam quàm Theoricam eius partem pertinentia, ex ipsius scientiæ fontibus de-*

ducta . . . [*The Epitome of Astronomy, in which in a brief exposition all things pertaining both to its spherical and its theoretical parts are deduced from the sources of this science* [i.e. astronomy] *itself. . .*], like many at his time, he saw astronomy as having two distinct parts. One was the astronomical/theoretical part, which was not concerned with efficient or final causes. The other was physical. It addressed the material, formal, efficient, and final causes. Kepler's work clearly fell into the latter part. The title *Mysterium cosmographicum* is entirely consistent with Maestlin's use of the word *kosmos* when describing the formal cause of the world.

Kepler's account in chapter 22 of why a planet moves uniformly with respect to the equant was the only part of the *Mysterium* in which he introduced a cause into technical planetary theory. Maestlin was apparently concerned that Kepler was crossing the boundary between the physical and the more purely mathematical parts of astronomy. Thus Kepler's speculations, while appropriate in cosmography, threatened astronomy if extended too much. It is clear, however, that Maestlin was very much against the introduction of physical reasoning into astronomical theory. Physics, for Maestlin as well as nearly all other astronomers of the time, was a pursuit unrelated to astronomy.[32]

Kepler and Maestlin never resolved their differences over the proper relationship between physics and astronomical theory. Indeed, Maestlin's dim view of Kepler's unconventional research probably contributed to his refusal to answer Kepler's letters from late 1600 to early 1605, when Kepler was making the most important theoretical innovations for the *Astronomia nova*.[33] Even years after its publication, Maestlin chided his former student:

> I think that one should leave physical causes out of account, and should explain astronomical matters only according to astronomical method with the aid of astronomical, not physical, causes and hypotheses.[34]

Galileo Galilei

As noted above, Maestlin was generally very supportive of Kepler's work. And apparently Kepler's exposure to Maestlin at Tübingen left him with the impression either that most leading astronomers were also Copernicans (in the realist sense) or that they could be won over to this position. His naiveté in this regard was expressed best in a letter to an Italian reader of the *Mysterium*, who, Kepler wrote, "signs himself Galileaus Galileaus."[35] Kepler had blindly dispatched two copies of the *Mysterium* to Italy in the hope that they might find suitable recipients. Both of them ended up in Padua in the hands of Galileo Galilei, who was then a professor there.[36] Galileo wrote to Kepler in a letter of 4 August 1597 that while he had not had a chance to read the book, he expected great things because he, too, had been a Copernican for many years. Galileo even disclosed that:

> from this standpoint [i.e., Copernicanism] the causes of even many natural effects were discovered by me, which undoubtedly are inexplicable by means of the accepted hypothesis; I drew up many reasons and subversions of arguments against [Copernicanism], which I have however not yet dared to bring to light, terrified [as I am] by the fortune of our teacher Copernicus himself, who although he earned immortal fame among some, nevertheless among a vast number however (for such is the number of fools) appeared fit to be ridiculed and rejected. I would actually dare to bring forth my thoughts if there were more like you, but since there are not, I will refrain from this kind of business.[37]

Galileo's mention of investigating the causes of natural effects from a Copernican standpoint that could not be explained by the older hypotheses must have seemed to Kepler like the very philosophical approach he advocated in the *Mysterium*. And in contrast to Galileo's prescient caution, Kepler urged open espousal of Copernicanism. His perception of the obstacles to the reception of Copernicus is noteworthy. His isolation in Styria had taught him that it was neither the ignorant nor the moderately educated who were the greatest obstacles to Copernicus, for these individuals were content to follow the pronouncements of mathematicians. The major obstacle was instead ignorant mathematicians:

> For there are not only your Italians who cannot believe that they are being moved if they do not sense it, but also we here in Germany do not win the greatest favor with this doctrine. But there are reasons for which we should defend ourselves against these difficulties. First, I am separated from that vast multitude of men, and I do not take in the din of so many shouts at once. Besides, those who are near me are a crowd of men who, although they do not grasp these abstruse things, as they say, are nevertheless astonished, and do not ever think for themselves whether they want to believe or not. The moderately learned ones, the more knowledgeable they are, the more cautiously they involve themselves in these disputes of mathematicians. And furthermore they can be bewitched—I say this because I speak from experience—by the authority of those skilled in the mathematical arts, so that when they hear that we now have ephemerides constructed from Copernicus's hypotheses, [they believe that] everyone who now writes ephemerides follows Copernicus; and if it is demanded of them that they grant that it cannot be demonstrated except in mathematical principles, [they believe] the phenom-

ena cannot agree without the motion of the earth. For though these postulates or pronouncements are not credible in themselves, still they are to be granted by non-mathematicians; and although they are true, why should they not be passed off as irrefutables? Therefore only mathematicians remain, with whom this is done with greater effort. Since they have the same profession, they do not concede postulates without a demonstration. Of these, the more ignorant one is, the more difficulty he causes.[38]

Kepler advocated using the relative isolation of mathematicians to the Copernicans' advantage by means of a rhetorical letter-writing campaign in which the skeptical were to be won over by an appeal to the (apparent) consensus of the community of leading mathematicians:[39]

Nevertheless also here a remedy can be furnished: isolation. There is one mathematician in any place; where it is so, it is the best. Then if he has a sharer of ideas somewhere else, he should obtain a letter from him for this reason: when the letter has been exhibited (to which end yours is also of use to me) it can incite this idea in the minds of the learned, as if all professors of mathematics everywhere agreed.[40]

Finally, with his characteristically endearing honesty, Kepler ironically urged Galileo to carry through his defense of Copernicus:

Truly, how is deceit useful? Have confidence, Galileo, and step forward. If I guess correctly, few of Europe's principal mathematicians will want to distance themselves from us; so great a force is truth.[41]

Kepler closed his letter with a request to make some observations that would detect stellar parallax and provide proof of the earth's motion. Unaware of Tycho's unsuccessful attempts to measure stellar parallax, Kepler believed that, regardless of the outcome, such a measurement would be an unprecedented victory.[42]

Even though he would not hear from Galileo again for some time, Kepler certainly perceived him as a kindred spirit. He complied with Galileo's request for additional copies of the *Mysterium*, and with regard to Galileo's arguments for Copernicus he asked, "If doing so publicly is not acceptable, at least communicate anything that you have discovered for Copernicus's benefit by writing privately to me."[43] Galileo did not do so, but that did not keep Kepler from speculating what kind of arguments Galileo might have had in mind. In a letter to Herwart von Hohenburg—which in some sense reflects Kepler's rhetorical strategy of citing the consensus of mathematicians with regard to Copernicus—he wrote:

Now, with regard to your thinking that arguments for the motion of the earth can also be taken from reasons of the winds and the seas' motions, I too certainly have not a few thoughts about these matters. Recently when the

> Paduan mathematician Galileo testified in a letter to me that he had most correctly deduced the causes of very many natural things from Copernicus's hypotheses which others could not render from the conventional [hypotheses], although he did not relate any specifically, I suspected this [cause] of the tides.[44]

Kepler's guess that Galileo might be defending Copernicus on the basis of tides was discerning, especially in light of evidence that suggests that Galileo had first developed the tidal argument that he used in his *Dialogo . . . sopra i due massimi sistemi del mondo tolemaico, e copernicano* [*Dialogue concerning the Two Chief Systems of the World, Ptolemaic and Copernican*] (1632) in 1595, around the time he became a Copernican.[45] However, the phenomenon of tides also serves to illustrate the chasm that separated Kepler's and Galileo's views, for both of them later appealed in very different ways to this phenomenon as evidence in their defenses of heliocentrism. In the *Astronomia nova*, Kepler used the tides as evidence of the mutual attraction of the earth and the moon, writing that the moon's sphere of influence called forth the waters when overhead.[46] In the fourth day of the *Dialogo*, in contrast, Galileo explained the tides as the result of the changing compound speed of a spot on the earth's surface as its daily motion carried it with or against the earth's annual motion, compounded with a kind of resonance in the Mediterranean basin. Moreover, he publicly attacked Kepler's account of the tides as a belief in "occult properties":

> Thus everything that has been previously conjectured by others seems to me completely invalid. But among all the great men who have philosophized about this remarkable effect, I am more astonished at Kepler than at any other. Despite his open and acute mind, and though he has at his fingertips the motions attributed to the earth, he has nevertheless lent his ear and his assent to the moon's dominion over the waters, to occult properties, and to such puerilities.[47]

The rich tapestry of the *Mysterium*, which extended in increasing strides of speculation from the polyhedra through the astrological and numerological significance of the solids and their arrangement, surely generated the same disdain from Galileo.

The *Dialogo*, however, was published some thirty-five years after the *Mysterium*. Nevertheless, although Galileo was still developing his approach to science when he received the *Mysterium*, the evidence shows that his and Kepler's views were already rapidly diverging. By then, Galileo had been teaching for eight years. Although the notes he had first developed for teaching at Pisa were Aristotelian, at Padua he began to shed conventional philosophical tenets. Thus, for instance, although in his 1594 treatise on mechanics he regarded reasons and causes as essential to mechanics—in agreement with the pseudo-Aristotelian *Questions on Mechanics*—by 1602 Galileo had revised his view, abandoning causes.[48] Moreover, Galileo's engagement with Aristotelian mechanics was antagonistic, and indeed, one of the signs of his increasing scientific maturity was the growing realization that it did not need to be addressed, but abandoned.[49] The development of his clarifying process of abandoning Aristotelian preconceptions for an ideal geometrization of observed phenomena was underway.

Similarly, probably around the same time, Galileo added a preface on method to his "Treatise on the sphere, or cosmography" (c. 1586–1587).[50] Note that although Galileo was defining the same physical part of astronomy as Maestlin, his approach was starkly different:

> In the Treatise on the Sphere (which we shall more appropriately call Cosmography), as in the other sciences, the subject should be [first] pointed out, and then we should touch on the order and method to be followed in this [science].
>
> Accordingly, we say that the subject of Cosmography is the world, or we mean the universe, as indicated by the word [cosmography] itself, which only means "description of the world." But of everything that might be considered about the world, only a part belongs to the Cosmographer, which is to reflect on the number and order of the parts of this world and the shape, size, and distance of [each of] these, and especially about their motions, leaving to the Natural Philosophers consideration of the qualities of the said parts of the world.
>
> As to the method, the Cosmographer customarily proceeds in his reflections in four ways, the first of which embraces the appearances, or phenomena, and these are nothing but sensate observations we see every day, as for example the rising and setting of stars; the darkening now of the sun and again of the moon; the latter's showing herself now crescent, now at quarter, now full, and again completely dark; the moving of the planets with very different motions; and many other such appearances. In the second place there are hypotheses, and these are nothing but some suppositions relating to the structure of the celestial orbs such as correspond with the appearances, as it would be when, guided by what appears to us, we should assume the heavens to be spherical and to be moved circularly, sharing diverse motions, [and] the earth to be stable, situated at the center. Third there follow geometrical demonstrations with which, by means of some properties of circles and straight lines, the particular events that follow from the hypotheses are demonstrated. And finally, what has been demonstrated by lines being [then] calculated by arithmetic operations, [this] is reduced to tables from which without trouble we may later at our pleasure find the arrangement of the celestial bodies at any moment of time."[51]

Where Maestlin's definition had bristled with Aristotelian causes, Galileo's was nearly instrumentalist. He presented cosmography as being concerned primarily with appearances, "some suppositions," and geometrical demonstrations, leaving the rest to the natural philosophers.

Kepler's *Mysterium cosmographicum* must then have held little of interest to Galileo when he got around to reading it. For if Galileo was a Platonist,[52] Kepler was a Neoplatonist; for Galileo, nature could be idealized and mathematized, but mathematics in itself had no prior ontological meaning. Nonetheless, it is clear that Galileo did read the *Mysterium* and found at least one element of interest. As we might expect, it was none of Kepler's speculations but rather a table of data in chapter 20. When Kepler had begun to address the relation of the planets periods and distances, he had prepared a table to illustrate how the planets move more slowly the further they are from the sun. He presented the period each planet would have if it moved with its usual speed around the orbits of the planets inferior to it. Thus, for example, Saturn's period is 10759;12^{d}. At this speed around Jupiter's orbit, its period would be 6159^{d} (Jupiter's actual period is 4332;37^{d}), around Mars's orbit 1785^{d} (Mars's period is 687^{d}), and so on. Around 1602, Galileo apparently seized on these figures when thinking about the "moment" of a balance, and carried out some unsuccessful calculations using Kepler's data to determine whether the velocities of the planets and their distances might "balance" around some point.[53] If Galileo discussed these ideas with his students, then this is the most probable explanation for Edmund Bruce's otherwise bizarre accusation that Galileo was passing off Kepler's discoveries as his own.[54]

Galileo's fame as an astronomer rests on his telescopic observations; he was never much of an astronomical theorist. Nevertheless, it has been a matter of some speculation why he never mentioned Kepler's elliptical orbits after the publication of the *Astronomia nova*.[55] It is clear that Kepler's elliptical orbits were common knowledge within a few years of the publication of the *Astronomia nova*, as is shown in a letter from Federico Cesi to Galileo of 21 July 1612:

> I know, as you do, that many motions are concentric neither with the sun nor with the earth; some [are nonconcentric] with the earth; others [are noncentric] with the sun, and perhaps all [the motions are noncentric with the sun] if the orbit of the planets is elliptical, as Kepler claims.[56]

Various accounts have been offered as to why Galileo never mentioned Kepler's ellipses, ranging from aesthetic reasons[57] to the desire not to complicate unnecessarily a more purely philosophical consideration of planetary orbits.[58] But perhaps ultimately Galileo's relative lack of interest in astronomical theory is sufficient explanation.

Even if the details of Kepler's planetary theory did not appeal to Galileo, there is some intriguing, if inconclusive, evidence that Galileo drew upon Kepler's celestial physics in his defense of Copernicus against theological objections. The miracle in the book of Joshua (10:12–13), in which Joshua declared "Sun, stand thou still," occupied a prominent place in Galileo's attempt to reconcile Copernicanism with Scripture. It was this miracle that the Grand Duchess Christina specifically raised as an objection to Copernicus, and it was the only Scriptural passage Galileo directly addressed in his letter to Castelli and subsequently in the "Letter to the Grand Duchess Christina" (1615).[59] In addition to pointing out serious problems in a Ptolemaic interpretation of the miracle, Galileo sketched his answer to Castelli as follows:

> For I have discovered and conclusively demonstrated that the solar globe turns upon itself, completing an entire rotation in about one solar month, in exactly the same direction as all the other heavenly revolutions; moreover, it is very probable and reasonable that, as the chief instrument and minister of nature and almost the heart of the world, the sun gives not only light (as it obviously does) but also motions to all the planets that revolve around it; hence, if in conformity with Copernicus's position the diurnal motion is attributed to the earth, anyone can see that it sufficed stopping the sun to stop the whole system, and thus to lengthen the period of the diurnal illumination without altering in any way the rest of the mutual relationships of the planets; and this is exactly how the words of the sacred text sound.[60]

This conception of celestial motion, in which all motions—including the rotation of the earth on its axis—would cease if deprived of the sun's motive power, is in fact quite un-Galilean and is instead Keplerian. Indeed, Kepler had predicted the sun's rotation on the basis of his physical principles in the *Astronomia nova* before Galileo provided observational evidence of it in the *Letters on Sunspots* (1613). Even apart from the *Astronomia nova*, Kepler's views on the sun as the source of light and motive power were clearly articulated in the *Mysterium*, which we know Galileo read.

If Galileo was influenced by the *Mysterium* in this regard, why did he not credit Kepler? There is no conclusive answer to this question but at best some plausible circumstantial evidence. To begin with, we have already argued that Galileo was not sympathetic with Kepler's approach in general. Seizing on the motive force idea in the heat of the argument over interpretation of Scripture by no means entails the endorsement of Kepler's ideas. Indeed, in the context of Galileo's thought, making use of this idea of Kepler's is nearly inexplicable. Secondly, there is suggestive evidence that, upon reflection, Galileo did not want Kepler, a Lutheran, to be mentioned in this sensitive debate. For although in his private "Considerations on the Copernican Opinion" (1615), Galileo adduced in support of Copernicus a parade of figures who held to heliocentrism—Pythagoras, Philolaus, Plato, Heraclides, Aristarchus, Hicetas, Seleucus, and the moderns, Copernicus, Gilbert, Kepler ("a living illustrious philosopher and mathematician in the service of the former and current Emperor"), and Origanus—the latter three (non-Catholics) are nowhere mentioned in the "Letter to the Grand Duchess Christina."[61] Moreover, in the "Letter to the Grand Duchess," Galileo went out of his way to emphasize Copernicus's service to the Catholic church and the dedication of *De revolutionibus* to Pope Paul III. Finally, in his expanded exposition of the miracle in Joshua, although he "could quote the supporting statements of many serious writers" in support of the sun's motive power in the world, Galileo restricted himself to citing the Blessed Dionysius the Areopagite rather than Kepler,[62] who in the *Mysterium* and *Astronomia nova* presented by far the most developed treatment of this idea.

Conspicuous by its absence in the growing tension between Galileo and the Church from 1611 to 1616 was any mention of Kepler's reconciliation of Copernicanism with Scripture in the introduction to the *Astronomia nova*. As we have seen, Kepler's attempt was deemed so inappropriate by the Lutheran authorities that it was ordered removed from the *Mysterium cosmographicum*. Yet after its publication in the introduction to the *Astronomia nova*, Kepler was not, to my knowledge, ever mentioned in the debate over Copernicus and exegesis surrounding Galileo, nor was the *Astronomia nova* placed on the index, although Kepler's *Epitome astronomiae Copernicanae* was. Even if in the midst of his troubles with the church, Galileo did seek to distance himself from the Lutheran Kepler, nevertheless the two became linked in the eyes of seventeenth-century non-Italian readers, for the section of Kepler's introduction to the *Astronomia nova* devoted to the reconciliation of heliocentrism with Scripture was appended by Matthais Bernegger to his Latin translation of Galileo's *Dialogue*, the *Systema cosmicum . . . in quo quatuor dialogis de duobus maximis mundi systematibus, Ptolemaïco & Copernicano, rationibus utrinque propositis indefinite disseritur . . . Accessit Appendix gemina quâ SS. Scripturæ dicta cum terræ mobilitate conciliatio* (1635, 1641, 1663).[63] Kepler's words, though far less measured and sensitive than Galileo's magnificent "Letter to the Grand Duchess Christina," thus served to some extent to express what Galileo could never publish.

These events have brought us far beyond Kepler's perception of Galileo's response to the *Mysterium*. Let us remind ourselves that Galileo responded only vaguely in 1597 that he was a Copernican and looked forward to reading Kepler's work. From this brief note, Kepler wrongly inferred support for his own work. In fact, Galileo was little interested in astronomy at that time, and, apart from the nova of 1604, displayed little further interest in it, nor did he communicate further with Kepler until he had published his telescopic discoveries in 1610.[64] As Drake points out, perhaps the most remarkable aspect of their relationship during that period when they exchanged no correspondence, was Kepler's consistent good faith toward Galileo in the face of numerous attempts to undermine it.[65]

Johannes Praetorius

As noted above, late in 1597 Kepler established contact with Herwart von Hohenburg, partially in the interest of soliciting opinions about the *Mysterium*. Herwart complied with Kepler's request by sending a copy of the book to one of Germany's leading astronomers, Johannes Praetorius (1537–1616).[66] At that time, Praetorius was a professor of mathematics at the University of Altdorf. He had studied at Wittenberg, where he obtained a copy of *De revolutionibus* in 1560, and where he was most probably taught by Caspar Peucer. In 1562, he had gone to Nuremberg, where he spent about seven years constructing astronomical instruments, before traveling to Prague, Vienna, and Cracow, where he met Rheticus. In 1571, he returned to Wittenberg after receiving appeals from Caspar Peucer and the senate of the university to become a lecturer in mathematics. He taught there until he took the position at the newly founded University of Altdorf in 1576.

Praetorius was an astronomical conservative in the Wittenberg tradition. His position regarding astronomical instrumentalism versus physical reality in Copernicus was made clear in manuscripts written during the early 1590s, only a few years before the *Mysterium*. In a manuscript entitled *Hypotheses astronomicae* (1592), he wrote:

> Of all those who followed Ptolemy, Nicholas Copernicus alone deservedly stands forth. For he wished his astronomy to be eternal and that all observations, both past and future, might be in agreement [with his theory]. . . . Now, just as everyone approves the calculations of Copernicus (which are available to all through Erasmus Reinhold under the title *Prutenic Tables*), so everyone clearly abhors his hypotheses on account of the multiple motion of the earth. Here, in the present work, we shall not discuss at length the freedom of astronomers to form hypotheses, for elsewhere this matter is discussed more fully. Nevertheless, having omitted all ambiguities, we wish to say that inasmuch as we can properly approach astronomical hypotheses, and in

> those very hypotheses which are to be discussed here, we follow Ptolemy, in part, and Copernicus, in part. That is, if one retains the suppositions of Ptolemy, one achieves the same goal that Copernicus attained with his new constructions.[67]

As much as Praetorius abhorred Copernicus's moving earth, he admired his replacement of the equant with epicyclic constructions based on uniform circular motion.[68] He sought, in a manuscript titled *Theoricae Planetarum per homocentrepicyclos respondentes placitis Copernici* [*Theory of the Planets by Means of Homocentric Epicycles Consistent with the Tenets of Copernicus*] (1592), to further Copernicus's reforms by replacing simple eccentrics with equivalent epicyclic constructions. Praetorius even attempted to develop a geocentric Copernican cosmology in a manuscript titled *Compendiosa enarratio Hypothesium Nic. Copernici, Earundem insuper alia dispositio super Ptolemaica principia* [*A Succinct Exposition of the Hypotheses of Nicolaus Copernicus, as well as Another Arrangement of the Same on Ptolemaic Principles*] (1594). Like Kepler, he calculated the maximum and minimum distances of the planets based on a common earth-sun distance, but he became frustrated by the necessary overlapping of geocentric planetary spheres and eventually retreated to the Ptolemaic scheme.[69]

As a professional astronomer with a deep appreciation for Copernicus's planetary theories, as well as personal experience in trying to develop an acceptable Copernican cosmology, Praetorius was as qualified a critic as Herwart could have chosen for a verdict on Kepler's *Mysterium*. And initially, in a letter of 14/24 February 1598 acknowledging the receipt of Kepler's book, Praetorius's report was favorable.[70] But after having the opportunity to study the book in detail, Praetorius became scathing.

In a letter of 23 April 1598 (o.s.), Praetorius made it quite clear to Herwart that it was Kepler's overall approach that was misguided. Significantly, Praetorius singled out Kepler's comment in the preface that the arguments in his disputations were "physical or, if you prefer, metaphysical," showing that at least Praetorius understood this to refer to the approach of the *Mysterium* as well:

> Truly concerning this subtlety of it, I prefer others to pass judgment, who, not content with plain astronomical speculations, endeavor to pry into more secret principles of nature with physical or metaphysical reasons (as the author says), and are delighted by those same arguments. I approached the reading of this book with great eagerness, but to tell the truth (and I would not wish to speak in contempt or reprehension of anyone soever), the more I wanted to strain my nerves with a greater effort and to press on in reading, the weaker I became and the more frustrated of all hope. And should some-

> one ask for the reason, for my part I am not able to respond otherwise than that I do not understand the least part of these matters, and that I judge them to deviate not a little from the scope of astronomy or to pertain more to physics, and they cannot be of use to the astronomer in almost any way. I think the only science as such ought to be sought above all in the disciplines, that is, geometry and arithmetic (in the highest rank of the mathematical disciplines), and that the mathematician can be content with this science alone. And application, on the other hand, without which studies of the arts are undertaken without purpose, [ought] to be sought in the arts.[71]

For Praetorius, the *Mysterium* was metaphysics, or natural philosophy at best, and had little to do with mathematics or astronomy to begin with. He was deeply opposed to the idea that the distances of the planets should be derived a priori from the Platonic solids, rather than a posteriori from observations. For this reason, Kepler's suggestion in the *Mysterium* that the discrepancy between his distances and astronomical theory might be due to shortcomings in Copernicus and Reinhold incensed Praetorius:[72]

> Moreover, because astronomy, according to Aristotle, is more inclined to physics, and cannot be counted among the sciences per se, it is proper (I believe) that the astronomer should put his doctrines to use as follows: the phenomena perceived by the eyes and sense should be consistent with his hypotheses, as though the changes in motions were brought about by such a definite cause. But that speculation of regular bodies, what, I ask, does it bring the astronomer? Well (he says), it can be of use for defining the order or magnitude of the celestial bodies. In fact, the distances of the orbs are to be taken from other bases, from observations for example, and a posteriori. With these things determined, what does it matter whether or not they coincide with the regular bodies? If he truly thinks that after this kind of principle has been proposed, both calculations and calculation are to be bound to it, I think this is done in a preposterous order, unless he presents more certain things, with which he confirms that this doctrine of astronomy can be deduced and supported a priori or from principles, not with prior observations. . . . And he seems to want the phenomena themselves to be brought into agreement, so that the speculation of the bodies can stand.[73]

Given Praetorius's reaction to Kepler's introduction of physical (or natural philosophical) arguments into astronomy in the *Mysterium*, it would not be too much to suggest that Praetorius had Kepler in mind when he wrote in his manuscript *Planetarum theoriae inchoatae* (1605), "The astronomer who endeavors to discuss the truth of the positions of these or other bodies acts as a Physicist and not as an Astronomer—and, in my opinion, he arrives at nothing with certainty."[74]

Helisaeus Roeslin

Another person from whom Herwart von Hohenburg received an opinion about the *Mysterium* was Helisaeus Roeslin (1544–1616), personal physician to various counts Palatine and counts of Hanau–Lichtenberg.[75] Roeslin expressed his opinion about the book in letters to Herwart of 4/14 May and 20/30 July 1597. Though not remembered as a first-rate astronomer today, Roeslin was then recognized as a notable thinker.[76] His comments provide additional evidence of the skeptical reception of the *Mysterium* among Kepler's learned contemporaries.

It was clear to Roeslin that the book was meant as an argument for Copernicanism. He agreed that Kepler's critique of other astronomers was justifiable, but he rejected the necessity of Kepler's conclusions. In his initial assessment, he wrote:

> Kepler's invention is outstanding for confirmation of Copernicus's hypotheses. But whether it must then equally follow that specifically the cube gives the distance of the spheres of Saturn and Mars [*sic*], and that another of the five regular solids could not also give such [a result], I do not know. I leave this judgment to those skilled in geometry. Such a thing is beyond me, nor can I pursue it. And although there may be 5 such distances of 5 planets, with every body especially arranged for specific planets, still I will not be of Copernicus's opinion for that reason. To be sure, not unjustly do Kepler and Maestlin dispute and direct their argument against Ptolemy and the ancient hypotheses. And they are displeased by Tycho Brahe, that he gives too narrow a space to the three higher spheres of Saturn, Jupiter, and Mars. In fact, I also find fault with this in Tycho. For this reason I give those three spheres a large enough space from another right and equally necessarily-following basis. For this reason, Kepler's and Maestlin's writings do nothing against me and also do not refute my position. I would also have nothing more to do with that obscure physics, in which the earth is moved.
>
> How does Kepler want to account for that almost infinite distance between the sphere of Saturn and the fixed stars so the earth could be placed in some mobile sphere? With which geometrical figure does he want to account for such an infinite empty space? As I consider it, such an invention with the 5 regular geometrical bodies will thus just suit any business without risk. Not that there is a necessity of consequence there, or [that] for that reason its consequences must necessarily follow.[77]

Roeslin's criticism was manifestly rooted in his anti-Copernican position. His fundamental objection was to "that obscure physics, in which the earth is moved." He had himself accommodated Copernicus while maintaining a static earth in a geo-heliocentric cosmology published that same year, to which he refers in the passage above.[78] While expressing some doubt about the necessity of the polyhedral hypothesis, he was probably sympathetic to Kepler's attempt to derive the celestial proportions a priori, for in his own system he assigned the three superior planets orbits in the ratio of once, twice, and three times the diameter of the solar orbit. From these ratios he believed that God had revealed the height of heaven.[79] His second major objection to Kepler's scheme thus concerned its failure to account for the large space between Saturn and the sphere of the fixed stars. This criticism was not peculiar to Roeslin; Tycho also raised this problem.

About ten weeks after his initial assessment, Roeslin offered Herwart a slightly more developed critique. His respect for Tycho was evident, as was his conviction that Kepler stood no chance against Tycho. But Roeslin was also impressed by Kepler. By this time, he had remembered that he had corresponded with Kepler earlier with regard to some astrological matters, and offered an astrological account of Kepler's genius:[80]

> [He] is a first-rate wonderful genius, still very young, now only in his 26th year. For when he was born in the year 1571 on December 27 at 2:30, Mercury in the seventh house in Capricorn in quartile to Mars stimulated [him]. Which then means an exceedingly profound and precocious genius. . . . Now, so far as Kepler's writing is concerned, it is a miraculous invention, and seems that it is supposed to defend Copernicus's hypotheses.
>
> But it might not hold good. Tycho Brahe is a match for him. He will fully refute him, as he has previously refuted other arguments of Copernicus.[81]

Having ruminated on the polyhedral hypothesis, Roeslin thought up a way to usurp Kepler's idea of nested polyhedra in support of the geocentric position. There is no indication that he did any calculation to justify his arrangement, so perhaps he was not being serious. At any rate, recasting the polyhedra to support his own cosmology was entirely consistent with his opinion that Kepler's argument did not have any demonstrative necessity:

> Your Grace will see how he and I agree in our thoughts, that I do not believe anything other than that the same thing shall have revealed the divine will to him and to me in the same way. Kepler wants to defend his teacher Maestlin, who is wholly of the opinion of Copernicus. However, I will prove that Kepler's invention with the 5 regular bodies is not sufficient for defending the Copernican system of the world as follows. Because even according to Copernicus there is a sixth space between the sphere of Saturn and the eighth sphere of the fixed stars, and indeed an immense and almost infinite space, how does Kepler now want to defend this sixth space geometrically, since a sixth regular body is not left over? Maybe he wants to introduce the re-

> maining infinite irregular bodies to defend that immense and almost infinite and empty space. But it might not hold good. Rather, this invention of Kepler's defends much more my system of the world. So that the first regular body, the cube, gives me the size of the sphere of Saturn. The 2nd regular body, the tetrahedron, gives me the space of the [sphere][82] of Jupiter. The 3rd body, the dodecahedron, gives me the place of Mars down to the Sun. The 4th regular body, the icosahedron, gives the space from the Sun to the Moon, in which space Mercury and Venus traverse their course. The 5th regular body, the octahedron, gives me the size of the sphere of the Moon down to the uppermost air. I do not require a further geometrical demonstration, because I put the uppermost part of the sphere of Saturn to be contiguous to the eighth sphere [i.e. the fixed stars]. Thus Your Grace sees how his invention confirms my system far better than his. [And] thus that it is an argument applicable to either side. . . They are not demonstrative arguments. But Tycho Brahe will doubtless refute this for him. For he has a special talent, an easy style and without doubt a singular divine favor, so that one who simply loves truth must subject himself to him. I hope that Kepler will do the same, unless he is obstinate, which Mars's bad aspect to Mercury seems to signal. But the time will come, and it will be hard for him to be unresponsive to the goad.[83]

Both Roeslin's and Praetorius's criticisms were addressed to Herwart, and Kepler evidently was not informed of them in detail. The only information that Herwart passed on in response to Kepler's request for the judgment of others read, "Today I received an answer from a mathematician about your *Prodromus*. But that answer was nothing. He does not want to agree at all with the motion of the earth."[84] While Kepler did not learn of their criticism directly, their opinions nonetheless illustrate the widespread skepticism toward the *Mysterium*.

Furthermore, their opinions provide some indication of how the *Mysterium* was read. While Praetorius and Roeslin had different attitudes toward the appropriateness of a priori speculations in astronomy, neither was impressed by the polyhedral hypothesis. Nor did either even mention the motive power hypothesis, from which we can infer that the book was read much as one might expect; although Kepler had intended to present both arguments on more or less equal footings, the greater development of polyhedral hypothesis allowed it to overshadow the motive power hypothesis almost completely.

Tycho Brahe

Late in 1595, Kepler dispatched news of his discovery to the then imperial mathematician, Nicholas Reimers Ursus (1551–1600). A Styrian nobleman had urged Kepler to contact Ursus, and, as Kepler later wrote to Maestlin, he cultivated Ursus "because he is the imperial mathematician, capable of helping me and harming me, since his influence extends even into Styria."[85] He could not have foreseen that his letter would enmesh him in the eruption of a smoldering priority dispute between Ursus and Kepler's future patron, Tycho Brahe (1546–1601).[86]

Ursus did not respond to Kepler's letter of 15 November 1595 until he saw a listing for the *Mysterium* in the catalogue of the Frankfurt book fair in the spring of 1597. Now that Kepler was an established author, Ursus responded in a letter of 29 May 1597. With regard to Kepler's discovery, he wrote that he had not yet made up his mind and was in doubt. However, he related that he had by chance compared Kepler's distances with a drawing by Joost Bürgi[87] of the Copernican system, and "[b]etween his sketch and your conclusions, I for my part found not much difference (which astonished me greatly)."[88] Ursus went on to ask for a copy of the *Mysterium* that he might "more fully and more thoroughly learn and understand the opinion you have formed concerning this matter (the opinion which has just been made public) and whether you still want to maintain the same, or perhaps a slightly modified view."[89]

Now that Kepler had new standing as an astronomical author, Ursus realized that he could use Kepler's fawning introductory letter to his advantage in his dispute with Tycho Brahe. In his earlier letter, Kepler had written, "I love your hypotheses."[90] Although Kepler went on to say that he admired Copernicus more, for Ursus, the remark was enough to suggest Ursus's priority in formulating the Tychonic system. Consequently, he reprinted the letter in his own *De astronomicis hypothesibus* (1597), a thoroughly libelous attack on all aspects of Tycho's character. Oblivious to the impending maelstrom, Kepler also hoped to use Ursus to his advantage. He sent Ursus an extra copy of the *Mysterium* and asked that it be forwarded to Tycho. Despite two additional pleas for Ursus's judgment, Kepler never heard from him again.

About the time of his final letter to Ursus, Kepler decided to contact Tycho personally. In a letter of 13 December 1597, he asked for Tycho's judgment, noting that he could use the support of so noted an astronomer to give him courage in the face of the criticism he was encountering:

> In keeping with that intense love of truth which your reputation for greatness proclaims is likewise present in you, I beseech you to explain your view of this subject with that honesty and sympathy for which you are renowned, and to inform me in a brief letter. How happy I would be if Tycho would arrive at the same conclusion as Maestlin! With these two supporters I shall not hesitate courageously to withstand the opposition that has already been aroused on all sides. If, however, the outcome is that you decide to censure

> the feeble, inept, and childish remarks such as have escaped me in large numbers on account of my immaturity, would I not on the other hand prefer such blame to the approval, it might be, of the whole world?[91]

Kepler sent the letter to Denmark, not knowing that Tycho had left for Germany. It did not reach Tycho until March 1598.

Tycho replied to Kepler in a letter of 1 April 1598 (o.s.). His response, though critical, was encouraging. He wrote that he had seen the *Mysterium* and read it as much as his other occupations permitted. And he praised Kepler's ingenuity in attempting to derive the distances of the planets from the regular solids. Tycho was not unsympathetic in principle toward the idea of deriving the proportions of the celestial spheres from harmonic and numerological considerations, provided they were as accurate as possible. In his *Progymnasmata*, he commented favorably on Offusius's *De divina Astrorum facultate in larvatam astrologiam* (1570), in which Offusius had adopted the value of 576 earth diameters as the distance of the sun while pointing that this was the sum of the number of constituent triangles in the five regular polyhedra.[92] Though Tycho noted the general agreement between Kepler's hypothesis and the Copernican distances, he expressed little confidence that more could or should be expected:

> [M]any things in them seem to be consistent, notwithstanding that the Copernican proportions are by no means sufficient in all respect to the smallest degree, since these [i.e., the Copernican proportions] even by themselves deviate not a little from appearances. . . . I would thus not easily say whether it need be truly satisfied by them throughout.[93]

Kepler defended himself in a marginal note that reads, "Which I had anxiously justified in my little book itself."[94] Tycho did, however, hold out to Kepler the tantalizing possibility that more accurate determinations of the distances could be made with the observations he had been collecting:

> If more accurate measurements of each of the eccentricities for the individual planets, of the kind collected over many years I have at hand, were applied, they could furnish a more accurate test in these. But as I am most occupied in completing and publishing my astronomical works, which I had left to some degree unfinished in Denmark, the leisure is not now available to me for comparing them more accurately. Perhaps this will be done at a later time.[95]

Tycho did not fail to notice Kepler's physical explanation of the motion around the equant in chapter 22 of the *Mysterium*. This is not surprising, for Kepler had cited Tycho in that chapter in support of the immateriality of the spheres. Tycho had nothing to say about the idea of the planet-moving force per se, although in later years he objected vehemently to Kepler's "physical principles." Rather, he criticized technical aspects of Kepler's explanation of the equant. He took issue with Kepler's account of the equant on the basis of its dependence on the distance from the Ptolemaic equant to the center of the Copernican deferent being one-third of the whole Copernican eccentricity (the size of the epicycle is likewise one-third).[96] He explained that the actual proportions are different and vary in time for the different planets. Kepler's resolution of this problem would later turn out to be a potent confirmation of the *Mysterium*. Furthermore, Tycho noted that Copernicus's technique for deriving the orbital elements from three observations at opposition could not extract the apogees.[97] Only by means of a sustained series of observations taken throughout the orbit could the parameters be derived. Such a series of observations had just been completed, Tycho wrote, now that Saturn had returned to the point on its orbit where he first observed it. The basis of Tycho's argument was that nothing in the orbital parameters could be assumed a priori.[98]

He expressed this criticism of Kepler's a priori approach more plainly in a letter to Maestlin written three weeks later, on 21 April 1598 (o.s.). Taking issue with Maestlin's suggestion that Kepler's findings might be used to reform astronomical theory,[99] Tycho wrote:

> Yet in the same little book you apparently not only doubt the advancement of astronomy but also abandon the effort, although previously you had other hopes. Suppose that astronomy is going to be reconstructed in advance, as you suggest, by measurements based on those regular solids rather than on precise observations made afterwards. We shall surely wait too long, if not indeed forever, before anybody can furnish any such result. The determination of the size of these regular solids, if it were to be accepted throughout, ought to be based on and proven by earlier observations. In like manner, with the exception of the more general features, all the details cannot be obtained from this source with the required accuracy.[100]

Yet despite Tycho's insistence that the orbital parameters be more accurately derived from his abundant stock of observations, his attitude was more ambiguous with regard to Kepler's goal of searching for harmonies. Elsewhere in the letter of 1 April 1598, he criticized Kepler's hypothesis for not being complete *enough*. The symmetry is ruined, he wrote, because it does not give the distance to the sphere of the fixed stars.[101] Tycho knew better than anyone alive that the distance to the fixed stars was indeterminable because there was no measurable parallax. And indeed, this argument directly followed Tycho's appeal that Kepler adopt his geo-heliocentric cosmology. But Tycho did allow that:

> There is no doubt that everything in the universe is divinely mutually related and ordained by a certain harmony and proportion; in the same way as they are succinctly apprehended so much in numbers as in figures: just as it was foreseen to a certain degree by the Pythagoreans and the Platonics a long time ago. And so, direct the sinews of your genius to these things further, and if you invent in addition a consonance of every sort, which is not defective in any part and in which nothing is still missing, you will be a great Apollo to me.[102]

He added in a later letter that "harmony and analogy of arrangements will have to be sought ex posteriori, when motion and the circumstances of motion have been ascertained with precision, not however ex priori, as both you and Maestlin wished."[103]

Tycho's criticism of Kepler's astronomical method extended beyond the question of over-reliance on a priori methodology. It also included the proper means of constructing planetary theories, which was to be by means of compounds of uniform circular motion. Whether he was referring to chapter 22—the only chapter in which Kepler addressed the qualities of individual planetary motions—is ambiguous, for Tycho mentioned angles and oblongs, but it is clear from the earlier letter that Tycho had noticed the argument in that chapter. In the later letter, Brahe wrote:

> And although circular motions in the heaven might, by means of each one balancing curiously, be able to seem sometimes to produce certain angles and various figures and even oblongs by their compounded motion, nevertheless it would be done by accident, and reason shrinks from this supposition. For it is necessary the circuits of the stars be compounded wholly from circles, otherwise they would not perpetually return into themselves uniformly and equally, and they would frustrate [our efforts] perpetually, apart from the fact that the anomalies would be even less simple, and they would not even be suitable for study or use.[104]

The tenets expressed here were common also to Copernicus and suggestively similar in phrasing.[105] Regardless of whether Tycho was addressing Kepler's model for the equant, this passage clearly reveals Tycho's attitude toward astronomical theory. When he came to work with Tycho, Kepler would again encounter Tycho's opposition to constructing astronomical theories on physical principles.

In addition to purely astronomical matters, Tycho had two physical objections to Kepler's hypothesis. The first concerned the material of the spheres. Tycho had established in his *De mundi aetherei recentioribus phaenomenis* (1588) that the path of the comet of 1577 passed through the spheres of Mercury and Venus and that the spheres were therefore immaterial. Had he read the *Mysterium* carefully, he would have seen that Kepler cited him for precisely this reason. But Kepler's polyhedral hypothesis emphasized the planetary spheres, and even though he affirmed the immateriality of the spheres in chapters 16 and 22, Tycho chided him that, along with many others, he "determines there to be a certain reality in the celestial orbs."[106] Again, Kepler defended himself in a marginal note, writing that as far as he was concerned, Tycho was correct.

Tycho's second physical objection concerned the motion of the earth. It was a standard objection based on Aristotelian physics, in particular on the distinction between sublunar and supralunar matter:

> Nor will the earth be able to move in the annual motion . . . , howsoever the daily [motion] could plausibly be assigned to it in the interest of a smaller sphere, for such a heavy and dark body would simply be suited rather to rest than to be put in motion.[107]

Kepler recognized this for what it was. His marginal note read, "The sun's rest: he attacks the earth's motion physically, he has nothing that he opposes astronomically."[108]

Tycho did offer one significant astronomical finding in his second letter of 9 December 1599: the absence of stellar parallax. Kepler had hoped that observations of parallax would provide incontrovertible evidence of the earth's motion around the sun. He made observations himself, appealed to Galileo to do the same, and had described his efforts in detail to Herwart von Hohenburg.[109] Tycho reported that though he had carefully sought parallax, he had found nothing of the kind.[110]

Despite Tycho's detailed critique of the *Mysterium*, he was generally kind and supportive of Kepler's attempt. His demeanor and his invitations to Kepler to come to Prague to discuss astronomical matters with him were probably due to two complementary motives. First there was his redoubled search for talented assistants in the run up to completing the publication of his *Progymnasmata*. Second, there was the more compelling possibility of making use of Kepler in his feud with Ursus.[111] Together these factors may well have given Tycho a motive for sweetening his mien and mitigating his criticism.[112]

Conclusion

Despite the ways that Kepler might have misconstrued his contemporaries' comments about his work or been unaware of their most serious criticisms, the *Mysterium* was not wholly without genuine support. As noted above, many years after the publication of the *Mysterium*, Kepler cited the letters of support he had received from the professors of mathematics

Galileo, Ursus, Tycho, and Georg Limnaeus. It was the last of these from whom he received the most genuine outpouring of praise.

Georg Limnaeus (1554–1611) was a professor of mathematics at the University of Jena.[113] Kepler evidently wrote to him, sending a copy of the *Mysterium* along with a request for information about Tycho.[114] Limnaeus sent the requested information and thanks and praise for the *Mysterium* in a letter of 24 April 1598 (o.s.). His admiration for the work was manifestly rooted in his own enthusiasm for Platonism and the *prisca philosophia* of the ancients:

> Most illustrious sir,
>
> I was never so unsympathetic toward that most ancient philosophy of the Platonists that I thought it fit to be driven from the domain and borders of the republic of letters along with some petty philosophers of our time. Rather, I always thought there is greater wisdom in it, and that that most learned philosophy contained greater secrets, which could easily lead its pupils to the very penetration of nature, if leaders and teachers were trained in this kind of studies. But while I still longed for this, and indeed longed for it so very often, behold! The Frankfurt book fair brought us your *Prodromus cosmographicus* [i.e., the *Mysterium*], a book replete with the most ancient and true wisdom. Therefore I rejoiced for myself, having always awaited such a thing from the specialists; I rejoiced for students, especially for students of astronomy that a new path to knowledge of the heavens is now laid open; and yes indeed I rejoiced for the republic of letters that at long last the most ancient principle of philosophizing is practically revived. To state the matter clearly, I begin to expect the best possible things for me from that philosophy. For who is ignorant of that philosophy passed down as though by hand from the ancient Egyptian philosophers? If it was the Egyptians', then it was doubtless also the Patriarchs', from whom Joseph, the most authoritative author of Jewish history, declares they along with Berosus[115] received their philosophy. Therefore if it is praiseworthy to live by the laws and ordinances of the elders and not to change anything in them heedlessly, it will certainly also be praiseworthy to follow the studies of the first fathers and patriarchs, who are said to be of a most holy and ancient race, even if all the most ancient things were devised for the practice of spiritual and civic life, and even these things were devised, which claim for themselves the foremost place among those things. We certainly justly rejoice also for you, most illustrious sir, for your most worthy undertaking, and we beg that, however you had fruitfully set your course, you should continue more fruitfully in it, and complete it most fruitfully.[116]

Limnaeus did not make any specific comments about particular arguments in the *Mysterium*, nor did he offer any thoughts about heliocentrism. However, in much the same way that Praetorius criticized Kepler's general approach, Limnaeus supported it.

With the *Mysterium*, Kepler failed in many cases to breach the division in his readers' minds between cosmography and astronomical theory. Praetorius, Roeslin, Limnaeus, and Ursus had nothing at all to say about Kepler's motive power argument, and perhaps did not even perceive it. However, Praetorius, the most competent technical astronomer in this group, rejected the book outright—from an astronomical standpoint anyway. Those who did recognize that Kepler was attempting in part to make physical inroads into astronomical theory itself with his motive power hypothesis warned him off. Maestlin in particular clearly warned Kepler to abandon this line of argument on philosophical grounds. Tycho raised technical obstacles to Kepler's explanation of the equant.

Those who read the *Mysterium* as a cosmographical work and were intrigued by no means accepted it as a physical proof of heliocentrism. Indeed, a physical proof of heliocentrism was nearly a contradiction in terms. Tycho, Praetorius, Roeslin, and Ursus did not waver in their geocentrism. The only reader who agreed that Copernicus could be proven on physical grounds was Galileo. His support for Copernicus was vague enough that Kepler perceived it as support for himself. In fact, the only reader who really supported the utility of Kepler's a priori derivation of the sizes of the planets' spheres was Maestlin.

Kepler's direct awareness of the response to the *Mysterium* was scant. There is no evidence that Herwart informed him in any detail about the critiques he received from Praetorius and Roeslin. Limnaeus was wholehearted in his support. Ursus was intrigued but ultimately withheld judgment. And Tycho, while expressing some serious misgivings about what real agreement might be expected between Kepler's findings and accurate observations, was at least supportive, in principle, of cosmic harmonies. Kepler was then justified in inferring, as he did in the 1621 edition, that among "those who understood it," the book was a great success.

But despite his ignorance of the most serious critiques of the *Mysterium*, Kepler was well aware of the resistance the defense of Copernicus faced generally. This is evident already in his reproachful tone in the student disputation, where he admonished his audience to treat him more fairly. After the *Mysterium*, he sought Tycho's support, noting "the opposition that has already been aroused on all sides." And he learned, not surprisingly, from Herwart about a response to the *Mysterium* from a mathematician who wanted nothing to do with the motion of the earth. Kepler's awareness of the nature of the resistance to Copernicanism is expressed best in his response to Herwart's report.

After informing Kepler of the critical opinion of the *Mysterium*, Herwart expressed his willingness to accept the intermediate position of a geo-heliocentric cosmology with a rotating earth. Interestingly, Herwart suggested that good physical arguments could be made in favor of this position:

> Today I received an answer from a mathematician about your *Prodromus*. But that answer was nothing. He does not want to agree at all with the motion of the earth. I am surprised that someone does not treat this material overtly and resolve it, especially since the most splendid physical arguments are not wanting, with which it would be shown that motion is to be ascribed rather to the earth than to the fixed stars, especially from the reason of the winds, of the sea running through straits and the ebb and flow of the sea. Philipp Landsberg contemplates new hypotheses of the planets, in which he posits a fixed center of the earth, however he assigns the remaining motions to the circumference of the same prime mover, but as yet he has published nothing. But I do not see the reason why one could not also attribute [rotational] motion to the center of the earth. And the lunar anomaly could certainly be equally accounted for by means of the supposition of the motion of the earth's center. But I grant that the motion of the center of the earth, with which Copernicus moves the earth around the sun on the great circle [i.e. the ecliptic] seems to me truly to exceed all plausibility, etc.[117]

In his response of 26 March 1598, Kepler was quick to point out that the physical arguments to which Herwart referred were promised in the title of the *Mysterium*, in that it was presented as a forerunner to the cosmographical essays, and it was here that he gave the outline of these books,[118] as well as speculating about Galileo's physical arguments.[119] A major obstacle to continuing his plan, Kepler wrote, was that Copernicus had given no measure of the size of the sphere of the fixed stars, and he described to Herwart the program of parallax observations, which he hoped would both determine the size of this sphere and decisively demonstrate the truth of Copernicus's cosmology.

As for Landsberg's geo-heliocentric system, Kepler wrote that Ursus had already come up with the system, and geo-heliocentric systems in which the fixed stars, rather than the earth, rotated had also been proposed by Magini, Roeslin, and Tycho (who, Kepler pointed out, was apparently the first). But Kepler clearly viewed these systems as artifices, products of man's ingenuity in constructing hypotheses rather than true descriptions of the world. Kepler's belief in heliocentrism had a religious component which precluded such a compromise:

> In short, the hypotheses are not new, but a new form, that is, a mixture of the ancient hypotheses and Copernicus's new hypotheses. But I am of the opinion, since astronomers are the priests of God almighty from the standpoint of the book of nature, that it is proper to show that I praise the glory

> not of genius but principally of the Creator. One who is convinced of this does not easily bring anything else to light than what he himself believes, and he does not heedlessly change something in the hypotheses unless he hopes the phenomena can be demonstrated from them with more certainty. And he does not particularly desire to eclipse those great luminaries, Ptolemy, Copernicus, etc., with the glory of his invention. And thus even for me, who is fully convinced of the opinion of Copernicus, there is a scruple to put forth something else, either in praise of genius or for the favor of men, very many of whom are offended by the absurdity of the opinion. It is enough glory for me to be able, while Copernicus worships at the high altar, to guard the door of the temple with my invention.[120]

In dismissing these compromise positions, Kepler articulated his experience to this point in defending Copernicus and defined the field where the battle must be fought:

> There is not an astronomer who rates these new hypotheses a hair's breadth below the ancients. Copernicus's only quarrel, and his entire quarrel, is with natural philosophers, metaphysicians, and theologians.[121]

Kepler knew that the struggle was not astronomical but physical, metaphysical and theological. The *Mysterium* had been his first effort to open this front. In his subsequent efforts—most notably the *Astronomia nova*—he would refine his physical arguments. It is likely that criticism of the metaphysical, a priori approach in the *Mysterium* eventually played its part in the considerably more refined argument of the *Astronomia nova*.

PART 2

THE *ASTRONOMIA NOVA*

We have considered the genesis and early development of Kepler's notion of a physical astronomy from his student disputation through the reaction of astronomers to his *Mysterium cosmographicum.* We saw that physical astronomy was deeply rooted in Kepler's conviction that Copernicanism was a physically true representation of the world, and that this belief was itself a manifestation of Kepler's faith in the accessibility of the plan of God's creation to human intellect. The *Mysterium* was thus conceived as the first of a series of books devoted to proving the reality of Copernican heliocentrism (for Kepler had no particular devotion to the details of Copernican planetary theory) by means of arguments based on physical and metaphysical cause—arguments then alien to the traditional scope of astronomy.

The *Mysterium* was bound to fail. The contemporary philosophical consensus among astronomers regarding Copernicus was almost the opposite of Kepler's. To the majority of astronomers, manifest physical experience and theological authority precluded the motion of the earth. It was Copernicus's equant-less planetary theories—which were more orthodox to Aristotelian physics—that had won their support.

Although Kepler was enthusiastic to hear the opinions of astronomers regarding his book, he was probably initially unaware of their most serious criticism. However, his correspondence regarding the book had two fateful consequences. First, he came into contact with Tycho Brahe, the most prominent astronomer of the time. Tycho gave him his opinion of the book and, more importantly, held out the possibility of checking Kepler's celestial proportions against his accurate observations. Second, Kepler unwittingly became entangled in the fierce dispute between Tycho and Ursus, which threatened his good standing with Tycho. When his situation in Graz took a turn for the worse, Kepler set out to meet Tycho with an eye toward resolving the dispute and pursuing his research with the benefit of Tycho's observations.

In this part, we will consider the stage in the development and reception of Kepler's physical astronomy that began with his work as one of Tycho's assistants and continued through the publication of the *Astronomia nova*, Kepler's acknowledged masterwork. We will argue that this period saw not only the greatest technical development of Kepler's planetary theory but also his decisive realization of the resistance to his "new astronomy."

Kepler's awareness of the strong resistance his physical astronomy faced began with his work under Tycho's direction. Although he had sought out Tycho with the hope of an open exchange of information, Kepler found Tycho reticent and imperious. He began to work for Tycho with the clear intention of pursuing his own preconceived notions, which stemmed from the *Mysterium*. However, further development of the polyhedral hypothesis was impossible, in part because Tycho prevented it, and also because the individual planetary theories needed to be reduced before he could proceed. Kepler thus refocused his attention from the larger, system-wide arguments of the *Mysterium* to one of the most contentious, if unrecognized, points of his new astronomy—the use of physical considerations in astronomical theory. Tycho did not hesitate to express his displeasure.

With the turn of events surrounding Tycho's death, Kepler was swept into the office of imperial mathematician, and was left reasonably free to pursue his own direction. Kepler's release from Tycho's supervision corresponds almost exactly with the beginning of the most innovative and purely physical research on Mars's orbit.

But even after Tycho's death, his legacy threatened Kepler's independence. Tycho's son-in-law, Franz Tengnagel, managed to have responsibility for Kepler's most prestigious publication taken away from him, which forced Kepler to conceive the *Astronomia nova* as the vehicle for his most compelling physical arguments. Moreover, a conspiracy between Tengnagel and Tycho's longtime assistant, Longomontanus—the so-called Tychonics—threatened Kepler's intellectual freedom. As Kepler became increasingly worried about what he might be allowed to publish, Longomontanus attacked the *Mysterium*. If Kepler had remained unaware of the serious criticisms of the *Mysterium*, he knew them now, and they came from a man who had the ability to hinder him. Although his work had been more concerned with less metaphysical arguments for some time, I will argue that the threat of the Tychonics interference may well have influenced the less speculative nature of the *Astronomia nova*.

In the meantime, the physical considerations that led Kepler to an oval, and then an elliptical, orbit for Mars began to make the new astronomy infernally complicated. Even a sympathetic correspondent like David Fabricius, who was kept very well informed of Kepler's progress, became increasingly suspicious of Kepler's methodology. Time and again, he demanded specific demonstrations to win his assent. And Kepler provided them, both to Fabricius and to the public in the *Astronomia nova*. When Fabricius could no longer take Kepler's physical reasoning, he proposed a theory in the classical form, which apparently reproduced Kepler's results, and offered it for inclusion in the *Astronomia nova*. Kepler viciously attacked the theory and broke off the correspondence.

Kepler's specific problems with the Tychonics and Fabricius provide a hitherto unrecognized context for the rhetorical nature of the *Astronomia nova*. We can now better understand why he chose to present his findings in a pseudohistorical account. By describing in detail the considerations that led to his orbit for Mars, Kepler hoped to discredit traditional mathematical astronomy decisively and establish the absolute necessity of physical astronomy. At the same time, he would so enmesh physics and astronomy that neither the editorial meddling of the Tychonics nor the mathematical simplification proposed by Fabricius could disentangle them. The *Commentary on Mars*, as Kepler usually called it, would thus become a specific argument for the physical truth of heliocentrism, but still one that shared the motivation Kepler carried forth from his student days.

⌘ CHAPTER 5 ⌘

KEPLER AND TYCHO

Kepler ascribed to divine providence the series of events that intertwined his fate with that of Tycho Brahe.[1] The uncanny convenience of their encounter, which sees the accomplished but aging observer apparently passing the baton to the brilliant young theorist, tempts the historian, too, to take the action of fate seriously.[2] But we should take care lest the portent of this event distort its actual significance. Although cruel fates had apparently driven them together, Kepler and Tycho approached their encounter with distinct ideas of what they wanted from one another.

For Kepler, the political and religious situation in Styria had been progressively deteriorating as the forces of the Counter-Reformation began to gather strength. In late September 1598, the Protestant school where he taught had been closed and its teachers expelled from the land. Because of his status as district mathematician and perhaps the influence of some friends, an exception was made in Kepler's case, and he was allowed to return shortly thereafter. Nevertheless, as pressure continued to mount, it became increasingly clear that unless he were willing to convert to Catholicism, he would eventually be forced to leave.[3]

In the meantime, his involvement in the acrimonious dispute between Tycho and Ursus was beginning to weigh on Kepler. He had hoped as early as February 1599 to travel to Prague to confront Ursus, and then go on to Wittenberg to see Tycho.[4] When the opportunity finally arose in January 1600 to travel to Prague free of charge, Kepler gladly took leave of the oppressive atmosphere in Graz. Reconciliation with Tycho was foremost in Kepler's mind. It seems doubtful that he actually foresaw the possibility of being hired by Tycho. His explicit motive was rather the chance to obtain better observational data against which to check his work in the *Mysterium*. As he had written to Maestlin in a spirited response to Tycho's criticism:

> Actually, Tycho's reasons do not move me at all to doubt. . . . It will become clear [when the planets'] celestial narratives come to light. . . . On the other hand, I am impatient: come, Tycho, and do not delay. . . . I do not want to be frightened but to be taught. . . . I rate Tycho in this way: *he possesses riches in abundance, of which he does not make proper use*, in the manner of most of the wealthy. Therefore an effort must be made (which I too have

> done for my part, but with appropriate discretion) to wrest away that wealth of his, namely that we beg that he candidly publish his observations, and all of them.[5]

Even after meeting with Tycho and learning of the possibility of being hired temporarily, Kepler continued to assert that his primary motivation in seeking him out had been the desire to obtain better values for the eccentricities.[6]

For Tycho, on the other hand, Kepler's unexpected arrival was a boon. Tycho always had his eye open for good help, but when Kepler arrived, he was more interested than usual. At that time, Tycho was in the process of assembling a new staff to make a push toward completing his *Progymnasmata* and *De mundi*. His most senior assistant, Longomontanus, had returned only a few weeks earlier,[7] and Tycho was also trying to recruit Johannes Müller (fl. 1596–1601), mathematician to the elector of Brandenburg; Christoph Rothmann (fl. 1599–1608), mathematician to Wilhelm IV of Hesse; and David Fabricius (1564–1617), a preacher from East Frisia and later a close correspondent of Kepler's.[8] Later that year, Tycho would also contract work out to Melchior Jöstel and Johannes Oswald in Wittenberg, and at various times hire Ambrosius Rhodius, Matthias Seiffart, and Simon Marius.[9] Had he secured the participation of everyone he wanted, Tycho would have assembled a remarkable proportion of Germany's astronomical talent. In this context, Kepler was merely one of many potential assistants.

Kepler did have the distinction of having published an original and imaginative astronomical book, with whose major premise Tycho had at least some sympathy. Still, Kepler would have had the status of an assistant, and not even first assistant, as this role was still filled by Longomontanus.[10] Kepler's status was reflected in Tycho's first attempts to secure imperial funding for him. Tycho initially attempted to have the emperor write a letter summoning Kepler to work on the *Rudolphine Tables* for a year or two, for which he would be paid a stipend on top of his continuing salary as district mathematician in Styria.[11] This was the same arrangement that Tycho was then trying to make with Johannes Müller.[12] Like Kepler, Müller arrived to work for Tycho at Benatky castle early in 1600, but he left again in the spring of the following year. In fact, Müller also had seniority over Kepler by virtue of his previous work for Tycho at Wandsburg Castle in northern Germany, Tycho's first home in exile. At Wandsburg, Tycho had put Müller in charge of astronomical matters and had him direct the activities of the staff.[13] There is little reason to believe that Tycho would have had any reason for preferring Kepler to Müller at Benatky. Tycho attempted to make more permanent arrangements with Kepler only during Kepler's second visit to Prague, by which time both

Kepler and Tycho were more desperate: Kepler's position in Graz had been effectively canceled, and Longomontanus had finally left Tycho for good.

Kepler's most unique qualification had little to do with astronomy. Because Ursus had reprinted Kepler's obsequious letter in his scandalously libelous *De hypothesibus astronomicis tractatus* (1597), Kepler had become an unwitting pawn in the war between Tycho and Ursus.[14] From the beginning of their relationship, Tycho had been plotting to make use of Kepler's testimony in his relentless effort to exact revenge. In his first letter, Tycho asked Kepler to deposit with him a statement of his opinion regarding Ursus's behavior, doubtlessly with the intention of adding this to his documentation of the dispute.[15] And within a month of Kepler's arrival at Tycho's residence outside Prague, Tycho solicited, through his future son-in-law, Franz Tengnagel, a written statement of Kepler's understanding of the technical issues of the priority dispute.[16]

Having Kepler on hand became useful to Tycho. Should the case against Ursus proceed, Kepler could be brought personally before the court as a witness. Also, since Kepler shied away from the dispute, thinking the whole matter should simply be dropped, it would be useful to have him on hand so that Tycho could stiffen his resolve should he falter.[17] Taking into account Kepler's eminent qualifications as well, the logical conclusion was to add him to the staff.

Kepler's Research under Tycho

Once Kepler arrived at Benatky in early February 1600, Tycho was faced with the question of what to do with him. Even had he harbored no doubts about Kepler's trustworthiness, Tycho well knew the value of his own observations and would not have allowed Kepler the kind of open access he desired. As it was, Tycho did have his doubts about whether Kepler was in league with Ursus, on top of a general distrust of his collaborators that had been considerably intensified by his dealings with Paul Wittich some years previously.[18] After a congenial and productive exchange of ideas with Wittich on Hven in 1580, Tycho became enraged when he learned that Wittich later passed on his secrets at the court of his rival Wilhelm IV, Landgrave of Hesse-Kassel. After this searing learning experience, Tycho ruefully vowed never to be so free again in sharing his inventions.[19]

Tycho made the fateful decision to risk giving Kepler access to the observations of one planet, Mars. Undoubtedly Tycho felt that in this way he could limit the damage caused by a possible breach of confidentiality on Kepler's part while getting some productive work out of him. For

Kepler, the limitation barred him from the cosmological data that had brought him to Tycho in the first place. His frustration with being prevented from pursuing his primary goal, Tycho's mistrust, and the decision to unleash him on Mars are all reflected in the familiar scene Kepler described in a letter to Herwart:

> Among the most important reasons for visiting Tycho was also that I learn from him more accurate proportions of the eccentricities, with which I might test both my *Mysterium* and the aforementioned *Harmony* [*of the World*]. For these a priori speculations ought not to conflict with manifest experience, but ought to be reconciled with it. But Tycho did not confer the multitude of them upon me, except in as much as in an aside while dining he incidentally mentioned now the apogee of one [planet], now the nodes of another. But when he saw that I had a daring mind, he decided perhaps the best way to deal with me was to give me my way with the observations of a single planet, namely Mars. I spent my time doing this, and was not concerned with observations of other planets. I hoped every day for a solution of the theory of Mars, after which [I thought] I would also have the other [observations].[20]

When Kepler arrived, Longomontanus was still occupied with the theory of Mars, and Tycho undoubtedly felt more comfortable having Kepler work under Longomontanus's supervision.[21] When it became clear that Kepler would be staying for a few months, Tycho tried to allay his suspicions about Kepler's making off with any of his discoveries by formalizing a confidentiality agreement. On 5 April 1600, he had Kepler sign a sworn statement pledging to "keep everything that the well-informed Herr Tycho has communicated to me, or shall hereafter communicate to me, in whatever manner, top secret."[22]

Apart from security concerns, another motivation for having Kepler work under Longomontanus's supervision was familiarizing him with local practice. Tycho had long-established routines and methods for his assistants, and Kepler would have been expected to learn the ropes. An illustration of what this entailed can be found already on the first two pages of Kepler's extant Mars manuscripts.[23] There are few words on these pages, but the calculations tell the embarrassing tale of Kepler's initial day on the job. His first calculation involved the analysis of the difference that would ensue from referring Mars's orbit to the true sun rather than the mean sun (see below). In order to do this, the first step entailed solving a plane triangle, two of whose sides represented the total eccentricities of Mars's and the earth's orbits, with the included angle being the difference between the orientations of their lines of apsides. Kepler's first error involved scaling the earth's eccentricity relative to that of Mars. It was a common practice since the time of Ptolemy to use a standard common value for the size of the deferent in the theories of the planets. For Tycho, that value was 100,000 parts. Kepler obtained the eccentricities of the earth (3,590 parts) and of Mars (16,380 parts) and the size of the earth's orbit relative to Mars (66,056 parts, where Mars retained the value of 100,000). He now should have multiplied 3,590 by 66,056/100,000 to obtain the scaled earth eccentricity of 2,371. Instead, he divided to get 5,435—a transparent, elementary error. He then proceeded to solve the triangle in a reasonably laborious manner, dropping a perpendicular and solving the smaller triangle as a step toward solving the whole triangle. His first attempt used a value of 23°45′ for the difference between earth's and Mars's lines of apsides. It is not clear where he got this value, nor why the following attempt used exactly twice this value. After finishing the first solution, he crossed it out and squeezed in another solution of the triangle using the same method, but changing the angle to 47°30′ (more or less correct) and a slightly adjusted value of the earth's eccentricity of 5,439. Turning the page over, he proceeded to do the calculation again using the same method, this time using the original erroneous value for the earth's eccentricity (5,435) and the angle 47°45′.

We can only imagine that, at this moment, Longomontanus took a look at what Kepler was doing, for in the left-hand margin of the second page, the whole problem is solved in short order. First, the value for the Martian eccentricity is corrected to 20,160. Longomontanus had been using a Copernican version of the theory of Mars, in which the total eccentricity is distributed between the eccentric (16,380 parts) and the epicyclet (3,780 parts), and Kepler had to take both parts into account.[24] In addition, a more precise value of 3,584 for the earth's eccentricity appeared (this was Tycho's final value for this parameter). Similarly, the precise value of 47°59′16″ for the difference of the lines of apsides is also used. Second, the scaling of the eccentricities is corrected. The earth's orbit is left at 100,000 parts, and the size of Mars's eccentricity is scaled by multiplying by 1.525, the size of its orbit relative to the earth's. Finally, there is a completely different method for solving the triangle with far fewer intermediate steps. Rather than an intuitive working through of the problem geometrically, the method is a recipe of the form:

$$a = \frac{1}{2}\,(\,180^\circ - c\,) - \arctan\left(\frac{\tfrac{1}{2}\,(A + B) - B}{\tfrac{1}{2}\,(A + B)}\;\tan\frac{1}{2}\,(180^\circ - c\,)\right)$$

where A is the larger side, B is the smaller side, c is the included angle, and the answer a is the angle opposite the shorter side. Where did this method suddenly come from? This exact solution is given in Dogma IV of Tycho's *Triangulorum planorum et sphaericorum praxis arithmetica*, "Given the two sides and included angle of an oblique triangle, to find the angles and the remaining side," in a supplement titled "A shorter solution of this dogma without a dropped perpendicular."[25] The *Praxis* was a manuscript handbook for Tycho's students and associates that simply gave recipes for solving plane and spherical triangles. The sudden appearance of the method in Kepler's notebook is certainly evidence of his training in Tycho's establishment, and a valuable reminder that, even though he would grow to become a mathematician of great talent and inventiveness, at this time he was just an assistant who would need to be trained like any other.

Being allowed access only to the Mars observations put a considerable constraint on Kepler's research, but despite Tycho's restrictions, Kepler was nevertheless able to pursue an existing line of inquiry. As Kepler mentioned repeatedly, his primary goal was to get on with the proof of Copernicus's heliocentrism that he had begun in the *Mysterium*. He wrote in the *Astronomia nova* that he "entered upon the work [on the theory of Mars] girded with the preconceived opinions expressed in my *Mysterium cosmographicum*."[26] Tycho's secretiveness with his data prevented Kepler from proceeding with the *Mysterium's* primary finding—the polyhedral hypothesis—but despite being restricted to the data from a single planet, he could pursue other findings.

The most important of these was the transfer of the planets' orbits from the mean to the true sun. This change bore on both main arguments of the *Mysterium*. First, Kepler had concluded that the sizes of the orbs in the polyhedral hypothesis were best calculated with respect to the true sun.[27] Second, the explanation of the equant based on the planet-moving force in chapter 22 depended on the planets' orbits being referred to the source of their motion, the true sun. The thrust of chapter 22 of the *Mysterium* had been that the planet-moving force coming forth from the sun causes the planets to move nonuniformly around their orbits in the same way that is modeled by an equant (or by Copernicus's eccentric epicyclet). The equant had become the mathematical representation of Kepler's physical principles. Thus, it was important both that Kepler show that the orbits must be referred to the true sun and also that their nonuniformity of motion corresponded to that represented by an equant.

The chapter on the equant had also produced some of the more unsatisfying qualifications in the *Mysterium*. Kepler had noted in chapter 22 that in the Copernican theories of Venus and Mercury, a planet's speed was not referred to the distance to the sun but to the earth's motion; its motion around the equant was determined not by its own but by the earth's position relative to the sun.[28] In addition, he noted that the earth's eccentricity was not bisected; that is, it did not have an equant at all.[29] From Hipparchus through Tycho, the theory of the sun (or earth) had always consisted of the planet moving uniformly around the center of a simple eccentric. These inconsistencies were doubly dissatisfying because, in addition to the fact that Kepler's argument explaining the action of the equant was not applicable to those planets, they highlighted a distinction in the form of theories for the planets that clashed with Kepler's belief that the theories of all of the planets ought to be of the same form.

Thus, although Kepler could not proceed to a verification of the polyhedral hypothesis, he could still try to resolve these problems with the mean sun. His interest in pursuing this line of inquiry is clear in his description in chapter 7 of the *Astronomia nova* of how he had come to work on the theory of Mars. He wrote that

> when I found out during the first week that, like Ptolemy and Copernicus, [Brahe] made use of the sun's mean motion, while the apparent motion would be more in accord with my little book (as is clear from the book itself), I begged the master to allow me to make use of the observations in my own manner.[30]

Kepler first had to demonstrate that changing from the mean to the true sun could achieve results at least as good as those that had already been attained. The theory for Mars developed by Longomontanus based on the mean sun already produced heliocentric longitudes adequate to within 2′, and there was some discussion of whether basing the theory on the true sun would be a worthwhile change:

> At the beginning there was great controversy between us as to whether it were possible to set up another sort of hypothesis that would express to a hair's breadth so many positions of the planet, and whether it were possible for the former hypothesis [i.e., Longomontanus's Mars theory] to be false despite its having accomplished this so far over the entire circuit of the zodiac.[31]

Kepler thus began the demonstration that appears on the first page of his working manuscripts for the *Astronomia nova*, which seems substantially identical to chapters 5 and 6 of that work.[32]

His analysis in chapter 5 of the *Astronomia nova* showed that the error in heliocentric longitude introduced by choosing the mean sun rather than the true sun would be almost insensible.[33] Conversely, changing to the true sun could produce nearly the same results as the theory already developed by Longomontanus, and Kepler could thus be expected to produce a theory based on the true sun that would account for the heliocentric longitudes as well as that already being used. The major obstacle that Longomontanus faced was not in heliocentric longitudes. It arose when he attempted to calculate the position of Mars when the earth was not directly between the sun and Mars, i.e., out of opposition. Because his theory was constructed from observations of Mars at opposition, it was relatively insensitive to the earth's position with regard to the true or mean sun in that situation. Longomontanus had run into trouble when trying to apply the theory to situations where the position of the earth becomes significant, which constitute the vast majority of observations. Kepler was able to show that, in this situation, the choice of the true or mean sun was significant. The thrust of chapter 6 of the *Astronomia nova* was that, whereas the error in referring the orbit to the mean sun was negligible at opposition, the error outside of opposition could be as large as one degree, an enormous error even by pre-Tychonic standards.

Whereas Kepler's first step in his investigation of the theory of Mars corresponded to the first step presented in the *Astronomia nova*, the steps that followed do not proceed in its logical sequence. In the largest scale, the *Astronomia nova* would be divided into five parts. The first concerned the shift from the mean to the true sun, which corresponds in fact to the first work Kepler did under Tycho. In the second part, Kepler constructed a theory of Mars "in imitation of the ancients."[34] The result was what Kepler called the "vicarious hypothesis," a theory that, although it gave accurate values for heliocentric longitudes, failed to give accurate positions of Mars out of opposition. When Mars is at opposition, the earth is by definition somewhere along the line linking the sun and Mars, and Mars's geocentric longitude equals its heliocentric longitude. Outside opposition, the position of the earth becomes a significant factor in Mars's position. Therefore, if the vicarious hypothesis could produce adequate heliocentric longitudes at opposition but not elsewhere, the fault lay in the theory of the earth. In the *Astronomia nova*, the failure of the vicarious hypothesis thus logically led to part 3, in which Kepler examined and reformulated the earth's orbit. He then returned to the orbit of Mars in the fourth part, applied his physical principles, and, after a few fits and starts, arrived at the ellipse.

Although the working manuscripts for the *Astronomia nova* confirm that Kepler's first effort was to set about moving the orbit from the mean to the true sun, the account of his researches that he gave in the *Astronomia nova* is not a true chronology. Indeed, Kepler never made this claim. The structure of the work was avowedly rhetorical,[35] for reasons we will discuss more completely below. Kepler's actual progress was far more complex and can be assessed only from the manuscripts and periodic progress reports in letters from the period.

It is significant, therefore, that the manuscripts indicate that before Kepler did any of the laborious work on the vicarious hypothesis, he devised his triangulation to investigate the earth's orbit,[36] which he did not present until part 3 of the *Astronomia nova*. This is significant because it reflects Kepler's interest in the dynamics of the earth's orbit. There was no other compelling reason to reformulate the theory of the earth. The solar theory was one of the few planetary theories Tycho had already

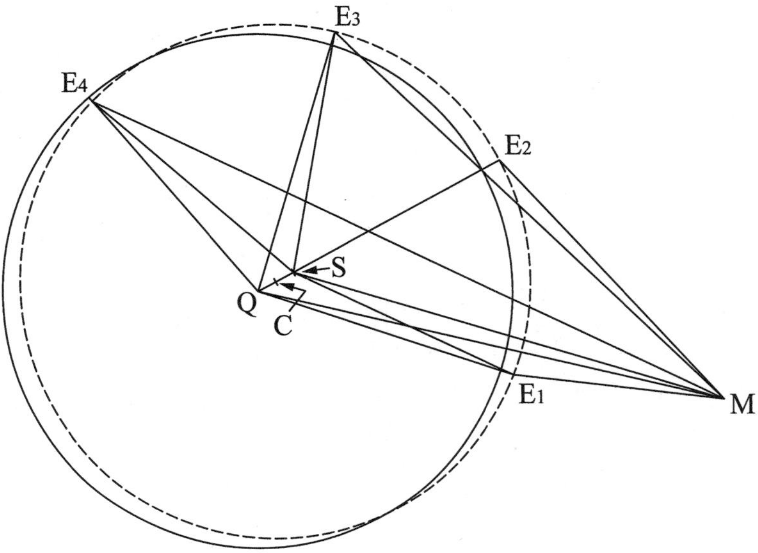

Figure 5.1 Kepler's triangulation from observations of Mars to the earth's orbit

Mars (M) is observed at integral number of 687-day intervals from different positions of the earth (E_1, E_2, E_3, E_4), after which it can be assumed to have returned to the same position in space. The construction is used in two different ways in the *Astronomia nova*. In chapter 24, Q is the mean sun. Accordingly, the angles of the positions around Q correspond to the time between observations. The angles E_1M, E_2M, etc. are from Tycho's observations. The direction QM is known from the mean sun using Tycho's Mars theory. Any three locations of the earth are sufficient to determine a circle (dotted), whose center C is not coincident with the center of mean motion. In chapter 26, Kepler repeats the derivation using the angles around S. His vicarious hypothesis gives the heliocentric longitude of Mars SM. Tycho's solar theory gives the orientations of the lines SE_1, SE_2, etc. The observations again yielded E_1M, E_2M, etc. The result was similar. Since the triangulation in the Mars notebook precedes the vicarious hypothesis, the latter version could not have been carried out early in Kepler's work for Tycho.

highly developed, and he thought it accurate to within 1′ of arc.[37] However, although he had redetermined the parameters, Tycho had maintained the simple eccentric form of the theory. Kepler's dynamical view of planetary theories from the *Mysterium* demanded that the earth should move more quickly when close to the sun and more slowly when far from the sun in the way that is modeled by an equant.

Kepler's triangulation was an ingenious technique that made use of the large number of Mars observations at his disposal to investigate the earth's orbit (figure 5.1). By selecting four observations of Mars separated

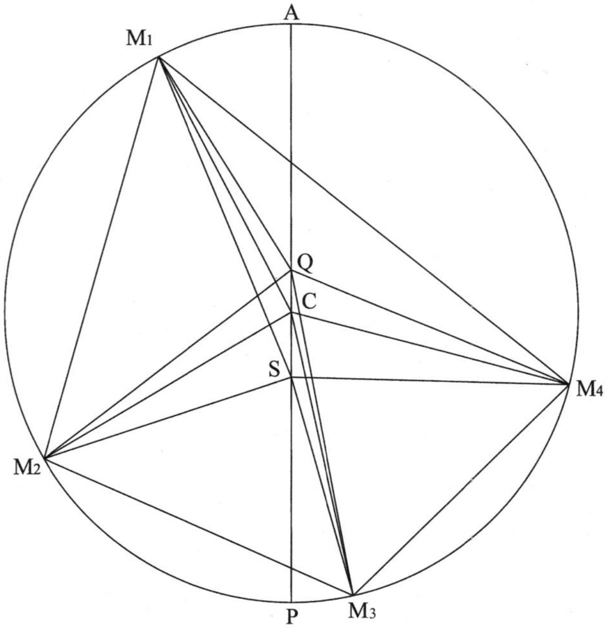

Figure 5.2 The derivation of the vicarious hypothesis
The lines SM_1, SM_2, etc. represent observations of Mars at opposition. Since the earth lies between the sun and Mars at opposition, its position along the SM lines is irrelevant. Q is the center of mean motion or equant point. The time between observations gives the angles between QM_1, QM_2, etc., given an orientation of the apsidal line AP and an epoch. The four triangles SQM_1, SQM_2, SQM_3, and SQM_4, with common side SQ can be solved to give four points, which overdetermine a circle. Kepler approached the solution iteratively in the absence of a direct solution, first adjusting the orientation of the apsidal line until all points fell on one circle, and then adjusting the date until the center of that circle C fell on SQ.

by integral periods of 687 days, he was able to observe Mars at the same place in space from different positions of the earth, that is, to observe four different positions of the earth from Mars. This allowed him to determine that the center of the earth's orbit was not coincident with the center of mean motion, which indicated that an equant was necessary.

During this first period of work, Kepler also undertook his first attempt at the vicarious theory. The method he employed was, again, very elegant. Kepler had a genius for setting up derivations in the right way to include or exclude just enough information. In the triangulation to the earth's location, by selecting observations of Mars when it was in the same place in its orbit, he had been able to remove Mars's motion from consideration. In the same way, in the work on the vicarious theory, he used four observations of Mars at opposition, when the earth is between the sun and Mars. These "acronychal" observations (meaning "rising at nightfall," which happens at opposition) gave Mars's heliocentric longitude independent of the orbit of the earth. In itself, this was not too unusual; the same kind of derivation enabled Longomontanus's Mars theory to provide accurate heliocentric longitudes. Kepler's unprecedented step was to allow the eccentricity of the equant to vary independently of the eccentricity of the orbit. Ptolemy had assumed the total eccentricity to be precisely bisected, which made these partial eccentricities equal. In Kepler's derivation (see figure 5.2), the observations provided the direction of Mars from the sun. The dates between observations gave the angles around the equant, given a direction for the apsidal line and a date to align it. The two angles and the common sun-equant side of the Mars-sun-equant triangles gave four positions in space that overdetermined a circle.[38] This procedure does not admit of a direct geometrical solution. Kepler's recourse was to approach the solution iteratively. Initially, he guessed the orientation and date of the apsidal line. He then required that the four points fall on a circle. If they did not, he adjusted the orientation of the apsidal line. Finally, he required that the center of the circle fall on the sun-equant line. If it did not, he adjusted the aphelion date.

There is one aspect of Kepler's work on the vicarious theory that is apparently difficult to reconcile with his pursuit of his physical theory. The theory of the equant in chapter 22 of the *Mysterium* relied on the fact that the equant and the sun are equally spaced from the center of the orbit. Only in this configuration does it hold geometrically that—at least around apsides—the speed of the planet varies inversely with its distance from the seat of the motive source. Yet the vicarious theory allowed the equant and the sun to be unequally spaced from the center in a ratio determined by the best fit with observations. The freedom that Kepler allowed in the vicarious theory thus apparently represents a retreat from the presuppositions of the motive force hypothesis from the *Mysterium*. Such second thoughts are confirmed in a section of the Mars manuscripts Kepler wrote around the beginning of 1602, when he reexamined the motive force hypothesis with some rigor. We will examine this section of the manuscripts, titled "Consideratio hypotheseos circa aequantem," more completely below;[39] now we will consider only what it reveals about the decision to allow the two components of Mars's eccentricity—the eccentricity of the equant and the eccentricity of the eccentric—to vary independently in the vicarious hypothesis. Since this document has not been widely referenced, I will quote it at length.

Kepler began this document with a recollection, a clear general statement of why he abandoned Copernicus's eccentric-epicyclet in favor of a return to the equant. As we have seen in part 1, this line of reasoning was certainly already in place at the time Kepler was writing the *Mysterium*:

> Copernicus recast the Ptolemaic equant into a small circle, in such a way that he composed the motion from 2 circles, with the result that it is not as simple as it is in Ptolemy. Consequently there is something wanting in the motion or the revolution, so it is not perfectly circular as it is in Ptolemy. But it makes no difference that it differs a very little from a circle. Is the perfection now undone if it differs a hair's breadth with regard to geometrical beauty? Add to this that Copernicus's means of calculating is difficult. Finally, it is not amenable to geometrical contemplation. For these reasons I have taken up again the Ptolemaic path which Copernicus rejected, since a natural cause has been found as a result of which a planet proceeds non-uniformly with regard to its center. For in the first place there are no orbs; Copernicus therefore ought not to be disturbed that the same orb is contrived proceeding now slowly, now quickly. Then, that the planet proceeds slowly or quickly according to its approach or recess from the sun is physical.[40]

He then turned to the argument from chapters 20 and 21 of the *Mysterium* that links the ratios of the sizes of the orbits with the motive force hypothesis. Continuing, he wrote:

> The planets compared among themselves two at a time confirm it. For the longer [a planet's] radius, the longer is the delay in an equal space also. Furthermore, there is this difference, that two by two the slowness is not wholly proportional to the circuit (otherwise, if as the radius of Mars's [circuit] to Jupiter's is 1 to 3, so the period of Mars to the period of Jupiter would also be in an equal interval, [and] the periodic revolution would be different). But in fact [the slowness] is fitted to harmonic ratios by design. In one and the same [circuit], the slowness or delay is entirely proportional to the distance. For what else could the reason relating the eccentricity of the equant to the eccentricity of the eccentric be?[41]

This represents Kepler's thought beyond the *Mysterium*. In the final section of the passage we see that he has realized the difference between applying the motive force hypothesis to a single orbit in order to cause the nonuniform motion around the equant and applying the hypothesis to pairs of orbits to determine the ratio of their distances. He had decided around 1599 that the planets are spaced according to harmonic ratios.[42]

Perhaps it was the appreciation that he had used the motive force hypothesis in two different senses in the *Mysterium* that caused him to reevaluate it. In any case, he now realized the error he had made in the *Mysterium* when he had added, rather than multiplied, the two different effects that caused the planets to move more slowly with their distance from the sun:

> However that may be, a consideration might be inserted as to whether I was correctly engaged in the *Mysterium* in the year '96. And indeed, it seems that I was not. For I said a double ratio of the distances is observed in the motions; it is not true (even if the hypothesis is conceded), except where the distances are in a double proportion. Rather, the proportion needed to be doubled, that is, brought into itself [i.e., squared], or the root of the composite proportion of the periodic motions needed to be taken in place of the simple proportion of the distances or slownesses, with the added proviso that the motion at mean longitudes is not precisely moderate.[43]
>
> As this proportion occurs in the periodic motions of Saturn and Jupiter: 10759;12::4332;37, $\sqrt{645552}$ = 803.46;[44] $\sqrt{259957}$ = 509.86 (:803.46) = 0.63458.
>
> Thus Mars has 686;59, $\sqrt{41219}$ = 203.04 (:509.86) = 0.39823,
> and the earth has 365;15, $\sqrt{21915}$ = 148.04 (:203.04) = 0.72911.
> Venus has 224;42, $\sqrt{13482}$ = 116.11 (:148.04) = 0.78432.
> Mercury has 87;58, $\sqrt{5278}$ = 72.65 (:116.11) = 0.62570.
>
> In the case of Saturn-Jupiter and Jupiter-Mars, it would be more correct than the solids allow. It agrees in the case of Mars-Earth, Earth-Venus, and Venus-Mercury. And in fact, the difference between the arithmetic mean which I employed in the year '95 and the geometric mean which I employ now is small. However, the means in Copernicus are all larger. If I should inquire into the root's root: $\sqrt{80346}$ = 283½; $\sqrt{50986}$ = 255½, then the terms will converge even closer.[45]

Kepler's immediate resort to checking the fourth root indicates his displeasure with what has become of his argument. However, it is precisely this displeasure that is significant.

It is notable in itself that Kepler's continuing interest in the research program from the *Mysterium* included the aspect of the motive force hypothesis from chapters 20 and 21 that attempted to correlate the planets' periods and distances overall. But it is most startling to discover that this also turned out to have important implications for the form of the vicarious hypothesis. Kepler's dissatisfaction with the ratios of the periods and distances of the planets compared among themselves also led him to abandon his explanation of the action of a Ptolemaic equant. If the speed of the planets around their orbits did not agree with the motive force hypothesis on a systemwide scale, why should he maintain his belief that the motive force hypothesis explained the relation between a planet's speed and distance around its own orbit? Kepler's motive force explanation of the equant had depended on the total eccentricity being precisely bisected, as Ptolemy had done. His disillusionment with the motive force argument freed him to consider whether a better fit with observations might be achieved by allowing the two components of the eccentricity (the distance from the equant to the center and the distance of the sun to the center) to be in some ratio other than one-to-one:

> After these things regarding the causes I came to this idea that the slownesses are not proportional to the motions, but are taken from harmonic music. But if it is the case that it is not consistent between the planets two by two, [it would] certainly not be present between the apogeal and perigeal motions of the same planet. In accordance with this basis of the undertaking, I sought the proportion of each of the eccentricities in Mars in such a way that it could be of whatever size.[46]

The reason for the unprecedented freedom that Kepler allowed in the ratio of the eccentricities in the vicarious hypothesis thus stands revealed: consideration of the other aspect of the motive force hypothesis so discouraged him that he abandoned his explanation of the equant based on motive power for a time, ignoring his physical interpretation of Ptolemy's bisection of the eccentricity.

This account is to some extent confirmed in the *Astronomia nova*, chapter 16, where Kepler explicitly recalled chapter 22 of the *Mysterium* and his ongoing concern for the issues in that book that drove the research of the *Astronomia nova*:

> Now I myself could also have taken the bisection of the eccentricity as certainly established, and with better reason than Ptolemy, because in ch. 22 of my *Mysterium* I had brought forward a physical cause for the bisection. Indeed, it was for this very reason that I had come to Tycho, that I might use his observations to inquire further into my opinions expressed in that book. I of course did this without prejudice, and continue to do so. And if I survive to see astronomy achieve its purity and perfection, so that a verdict can be given in the case which I have brought before her tribunal in that book, I promise the reader I shall retract that book and confirm what is seen as true, faithfully revealing the remainder which has turned out not to be so.[47]

(In this interesting pledge, we see the first reference to the amended 1621 edition of the *Mysterium*.) However, the account Kepler offered in the *Astronomia nova* differs from that in the "Consideratio . . ." in the notable omission of any consideration of the architectonic use of the motive power hypothesis from *Mysterium* chapters 20 and 21. In the *Astronomia nova*, Kepler wrote only:

> And so, since I thought then that this was altogether too much to assume, and also saw it pointedly called into question by Copernicus when he argued for a change in Mars's eccentricity on the basis of his figures which were not

> in accord with this bisection of the eccentricity, I envisaged a method which would lead me to a knowledge of the ratio of the two eccentricities (which, as I said, in not indubitably 2:1). . . . And so in the year 1600, having acquired knowledge of this art, I came to Tycho, and was happy to learn that he too did not assume this ratio, but made an investigation as his figures indicate.[48]

This line of reasoning was perfectly valid, given the precedence in Copernicus and Brahe for not precisely bisecting the eccentricity of Mars.[49] And indeed, Kepler noted in the "Consideratio . . . " that "Acronychal observations of Mars made the eccentricity of the eccentric greater than half [the eccentricity] of the equant, which even Tycho and Copernicus said."[50]

The accounts that Kepler offered in the *Astronomia nova* and the "Consideratio . . ." are not contradictory. In fact, we might use the account offered in the *Astronomia nova* to date Kepler's reconsideration of the motive force hypothesis to 1600 or before, given that he says he came to Tycho with the idea of the possibility of an unequal bisection already in his mind. However, why did Kepler make no mention of the architectonic aspect of the motive force hypothesis? It is possible that this was because it was simply not germane to the issue of the bisection of the eccentricity. On the other hand, it may also have been a conscious effort during the composition of the *Astronomia nova* to minimize the importance of the architectonic principles of the *Mysterium* in the face of Longomontanus's severe criticism of that work.[51]

Finally, we should note that although Kepler abandoned the precise bisection of Mars's orbit and thus the detailed explanation of the equant from the *Mysterium*, he did not abandon the idea of the equant being a representation of the action of the planet-moving force. As we will see below, his idea of how the equant might model the force became more complex. However, the equant itself remained an important manifestation of Kepler's physical principles, and thus his subsequent discovery that the theory of the earth needed an equant was still an important vindication of his approach.

Kepler made his first attempt at the vicarious hypothesis in the spring of 1600. After obtaining a solution, he tested it against four other observations of Mars at opposition. The error was as great as 6′. Even more distressing was the fact that it was 8′ in error for one of the observations he had used to derive it! "I do not know where the error lies," he wrote, just before this work was broken off by his return to Styria in June 1600.[52]

Despite the troubling failure of the first version of the vicarious hypothesis, Kepler had successes to report in a letter to Herwart of 12 July 1600 written after his return to Graz. Note that he reported his results with explicit reference to his continuing research based on the *Mysterium*:

> [Mars] confirmed also my *Mysterium* in a remarkable way in two points: in fact, I had referred the eccentricities of the planets to the true Sun, and I was very much afraid that Tycho Brahe would refer them to the mean Sun, as Copernicus had done. Now, Mars obstinately refused any point other than the centre of the body of the Sun. Tycho Brahe was satisfied with my attempts; for, he told me, that he himself had long been of the same opinion, but had wanted to avoid making the complicated calculations, and therefore wished to know the opinions of others besides.[53]

Tycho's reaction to the work on transferring from the mean to the true sun was apparently ambivalent: it was something he had been considering yet wanting to avoid.

Kepler's other major result concerned the results of his triangulation to the earth's orbit. It had revealed that the center of the earth's mean motion and the center of its orbit did not coincide, that is, that a simple eccentric theory would not suffice and that the theory of the earth would require an equant. This was more directly related to his physical theory of the equant in chapter 22:

> Next, although I lamented what I pointed out at the end of my *Mysterium*—that an equant had been denied to the sun (or earth) alone—now Mars has also most plainly provided evidence that the relation [*ratio*] of an equant is in the Sun [i.e., in the theory of the earth]. And it is this which Tycho had written to me [in his letter of 1 April 1598] in the guise of an enigma (as he is sometimes wont to do) about the variable size of the annual orb, which makes a difference of 1° 40′ in [the case of] Mars.[54]

Kepler referred here to the fact that a simple eccentric theory for the sun (that is, the earth) is not sufficient for the theories of other planets. The epicycle in a Ptolemaic theory or a simple eccentric referred to the mean sun in a Copernican or Tychonic theory would function well, provided that the motion around the mean center were uniform. Kepler had discovered that the center of the orbit and the center of mean motion were not coincident, and that Tycho's enigma was caused by the nonuniform motion of the earth, which would be reflected in "the variable size of the annual orb." Put another way, an equant in effect accentuates the nonuniformity of motion present in an eccentric without further increasing the eccentricity. A simple eccentric, without an equant, exaggerates the change in the earth-sun distance, which had to be compensated by a corresponding change in the size of the annual orb in Tycho's Mars theory. In 1598, Tycho had brought up this "enigma" as an argument against Kepler's physical theory of the equant, on the grounds that the varying eccentricity was too complicated to explain in Kepler's way.[55] Now, at the same time, Kepler was able to introduce the equant he wanted into the theory of the earth and resolve the enigma for the three superior planets. He was undoubtedly delighted. As for Tycho's reaction to this, Kepler wrote to Herwart only, "I am in doubt as to whether I should therefore now expect Tycho to emend his theories of the planets."[56] Among other things, Tycho was concerned for his work on the solar theory. He had already gone so far as to have the section of the *Progymnasmata* that dealt with the theory of the sun printed up.[57] In the *Astronomia nova*, Kepler wrote, "Brahe feared that if I bisected the sun's eccentricity I would vitiate his equations of the sun's motion."[58] To make matters worse, Kepler's results were radiating out and affecting the theories of all the other planets, for each planet would in principle suffer from the same effect. We can only conjecture that Tycho resisted the idea of starting everything over again on the basis of Kepler's Mars research.

At least one thing is clear about Tycho's reaction to the introduction of an equant into the theory of the earth (which was represented by him in his highly developed solar theory): If Kepler argued for it on the basis of his planet-moving forces and physical causes, Tycho would have objected vehemently. The evidence for such a reaction is in a letter Kepler wrote to Archduke Ferdinand at just about the same time as the letter above to Herwart. The letter was undoubtedly a last-ditch attempt to salvage his position in Styria by preparing a special gift for the archduke in the form of a letter on the lunar theory. Although it concerned lunar theory, it is one of Kepler's most significant early writings on the application of physical principles to planetary theory, and we can extract from it a picture of Kepler's and Tycho's disagreement over Kepler's use of physical principles in his investigation of Mars. Kepler wrote:

> Following in Copernicus's footsteps, [Tycho] greatly abhors the non-uniformity of the movements which Ptolemy introduced. Consequently, in order to represent all the kinds of deviations of the moon by means of circular motions uniform around their particular centers, he employs very many small circles, by the accumulation of which contemplation of the system is rendered unintelligible. It seems to me, *although Tycho opposes this very much*, that simplicity is more consonant with nature. In order to obtain it, help [to explain] the Ptolemaic non-uniformity of the motions must be sought from the innermost secrets of nature. Thus in contrast to Copernicus and Tycho, I say that nature is exceedingly consonant, that the more a planet gives up its quickness of motion, the further it is from its center, because observations of the ancients before Ptolemy and of Copernicus and now Tycho in our time consistently show that the motor virtue is dispersed from the center into the capacity of the circumference. And thus the further the virtue departs the more it

> is attenuated, since it has been dispersed into a larger orb. Therefore less motion is impressed into the planet when by means of another element of motion it is moved further from the seat of virtue.[59]

Later in the letter, he wrote of Tycho's use of epicycle upon epicycle in the lunar theory, "To me, on the other hand, this seems absurd in natural philosophy: for which reason I resort to my natural causes and non-uniform motions, with Tycho objecting greatly."[60] And later, when Kepler concluded that the motion of the moon is physically nonuniform, he mentioned that "Tycho objects vehemently" to the idea.[61]

Tycho objected to Kepler's physical principles and his consequent insistence that the nonuniformity of motion was physically real. Like many of his contemporaries, Tycho found the reformulation of the equant into uniform circular motions to be the most compelling aspect of Copernicus's work. As he had explained in a letter to Kepler, only compounds of uniform circular motion were suitable for planetary theories.[62] It is difficult to assess the degree to which their philosophical differences might have influenced Kepler's work during the time he worked under Tycho. However, Longomontanus followed Tycho's dictum. When Kepler arrived at Benatky, Longomontanus was using a double epicycle version of Copernicus's eccentric epicyclet, which would have been eminently acceptable to Tycho. Kepler's ready translation from Copernican into Ptolemaic form, i.e., with an eccentric and an equant, was certainly just as easily made by Tycho in reverse. Thus Tycho may have had no objection to Kepler using the equant as a working hypothesis, which could easily be translated into an acceptable epicyclic form when it was finished.

In his letter to Herwart of 12 July 1600 written from Styria, Kepler reported on his work on the vicarious theory, but with regard to the flawed solution he wrote only, "I have not come to the conclusion of any model these four months," with the marginal note, "In fact I came to one, but it was useless for it was false due to one of the assumptions."[63] He included a lengthy postscript with the letter outlining his four acronychal method in the hope that Herwart might pass it along to the French mathematician François Viète (1540–1603) for solution. Kepler was familiar with Viète's simpler method of determining the apsidal line and the eccentricity from three observations. The manuscripts show that he subsequently used it to unmask the bad datum in the first vicarious theory by examining the rough results from Viète's method for the four permutations of three observations from the original four he had used. Only one permutation was close to what Kepler expected, so the culprit was clearly the observation common to the other three permutations.[64]

Kepler probably did not accomplish much more on the theory of Mars in Graz during the summer of 1600. To begin with, he took some time to compose a calculation of the solar eclipse on 10 July for Archduke Ferdinand (this was the topic of the letter discussed above), again undoubtedly with the intention of seeking some beneficial influence. He wrote to Maestlin that since leaving Benatky, "In the mean time I have been occupied entirely with calculating and observing the eclipse of the sun."[65] In addition, by the end of July, the situation for Protestants in Graz began to deteriorate. From 31 July to 2 August, the citizens of Graz were assembled for an examination of their faith. Protestants who did not pledge to convert were ordered to leave within forty-five days after paying an exit tax of 10 percent of their assets. Kepler was among them.

The new, more oppressive measures caused Kepler to be quickly dismissed from his position as district mathematician, though with half a year's severance pay and a letter of recommendation. His hoped-for arrangement with Tycho was thus ruined, for it had been based on the assumption that he would retain his salary as district mathematician while working temporarily with Tycho. Meanwhile, Protestants were forbidden to lease property and thus were forced to sell. Real estate values collapsed, devaluing most of Kepler's wife's holdings. Even though half the exit tax in Kepler's case was eventually rescinded, he had to make arrangements to leave. He appealed desperately to Maestlin for a position at Tübingen or at least for some advice.[66] Not hearing from Maestlin, Kepler finally left Graz on 30 September 1600 with his family and belongings in tow, trusting his fate to Tycho in Prague.[67]

August and September were obviously tumultuous months for Kepler. It would be perfectly understandable if he did not work on the Mars theory at all during that time. This is roughly confirmed in an account of his activities that Kepler sent to Longomontanus in 1605:

> I spent the greatest part from February to May 1600 primarily hoping and thinking. You know, of course, that I wanted to finish all of the mysteries [of the Martian orbit] within eight days on a bet. However I succeeded only in so much as I saved the eccentric inequality tolerably (except that one of the fundamental observations was assumed 20′ in error) and discovered the sun's eccentricity had to be reduced. From June to October I was abroad [in Styria] and moved my family.[68]

Upon his return to Prague, Kepler fell ill, which restricted his activities. In addition, now that Kepler was utterly dependent on Tycho and Longomontanus had left, Tycho apparently took a more active role in directing his activities:

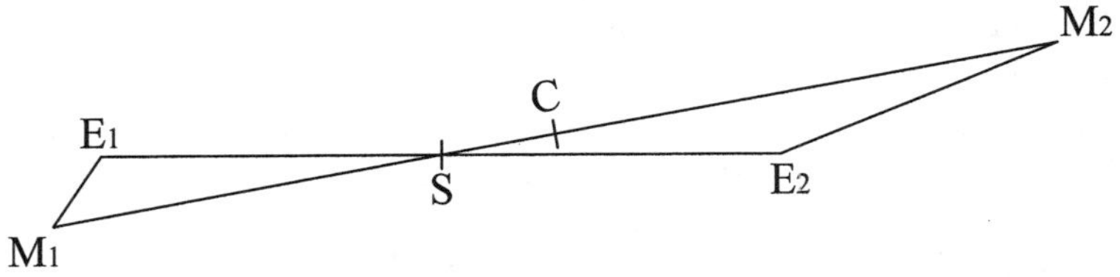

Figure 5.3 Kepler's use of Martian latitudes

In chapter 19 of the *Astronomia nova*, Kepler describes a method of estimating the eccentricity of Mars's orbit using latitudes. E_1E_2 is the plane of the earth's orbit, M_1M_2 the plane of Mars's orbit. The inclination of Mars's orbit yields the angles $\angle E_1SM_1 = \angle E_2SM_2$. The angles $\angle SE_1M_1$ and $\angle SE_2M_2$ are the complements of Mars's observed latitude. The distances E_1S and E_2S are obtained from Tycho's solar theory. Solving for M_1S and M_2S yields the eccentricity of the sun from the center C, SC. The method is error-prone but yields a solar eccentricity close to half of Mars's total eccentricity, rather than the division yielded by the vicarious hypothesis.

> A quartan fever had a hold on me from October 1600 to August 1601. In the meantime at Tycho's bidding I wrote against Ursus, and supported his other studies according to his judgment and my strength. Although Tycho was indignant, I speculated about Venus, Mercury, and the Moon—profitably for the former, for the moon clearly in vain.[69] I speculated also about Mars. With the false fundamental observation corrected, I corrected the first inequality.[70]

Thus, sometime between October 1600 and April 1601, when he returned again to Styria, Kepler resumed his Mars work. Having found his error in the first attempt at the vicarious hypothesis, he redid the calculations and came up with a second version, which he considered satisfactory.[71]

Perhaps made wary by his experience with the solar theory, where Tycho had obtained satisfactory heliocentric longitudes despite errors in the distances, Kepler turned to Martian latitudes to confirm his findings. He had been aware of the potential of latitudes to provide information about the relative distance of the earth and Mars from the sun since early in composition of his research notebook.[72] In principle, the method involved considering the latitude of Mars at opposition (where the vicarious hypothesis had already proved satisfactory) (see figure 5.3). Given the relative distances of the earth and Mars and the inclination of their orbits, Kepler would expect a certain latitude. If the observed latitude were greater, then the distance between the earth and Mars at opposition would be smaller than expected. Likewise, a smaller than expected latitude would indicate that the distance between the two would be larger. Kepler immediately discovered that the distances computed from the vicarious hypothesis did not correspond to those observed in the latitudes. As Gingerich noted:

> Here, perhaps for the first time, the harsh conflict between what he was later to call the "vicarious hypothesis" and physical reality impressed itself into Kepler's consciousness. (It is interesting to note that Kepler became aware of the celebrated 8′ discrepancy only at a much more sophisticated point in his researches.)[73]

Kepler recognized that the latitudes were sensitive to the distance not only of Mars from the sun (as given by the vicarious hypothesis) but also of the earth from the sun. He also decided at this time to re-reduce Tycho's opposition observations in light of his shift from the mean to the true sun.[74] Both of these considerations required that he return again to the triangulation from constant positions of Mars to the earth's orbit. With a new reduction of the same opposition observations, Kepler derived a third version of the vicarious hypothesis, but it too contained an error in one of the fundamental observations.[75] At this point, Kepler abandoned the work with disgust. It was now April 1601, and he returned to Styria to attend to some leftover business.

We have another detailed account of the progress of Kepler's research from the summer of 1601 in a letter to the prominent Italian astronomer Giovanni Antonio Magini (1555–1617). Kepler did not know Magini, but he knew of him through his letters to Tycho, some of which Kepler read or heard recited. His recent experience with the third version of the vicarious hypothesis had renewed his frustration with his indeterminate, iterative procedure, and he hoped that Magini would know a direct means of making the calculation.[76] First, however, Kepler provided an outline of what he had found thus far. Having had to pledge an oath of secrecy to Tycho, Kepler was beginning to appreciate the mores of the astronomical community, so he gave the same pledge to Magini, along with an account of his research, to establish his good faith.[77]

Although he knew Magini to favor some sort of geo-heliocentrism,[78] Kepler did not hesitate to identify himself as a Copernican. "Having obtained Tycho's observations," he wrote, "I have been examining the Copernican hypotheses for a whole year now, principally in the case of Mars."[79] His continuing interest in pursuing the lines of inquiry established in the *Mysterium cosmographicum* is clear. He explained to Magini that his reason for seeking out Tycho and remaining with him was his own investigation of the "harmony of the world," and he explained his progress with explicit reference to the *Mysterium*, a copy of which he hoped Magini had received.[80] He began with the *Mysterium*'s table of planetary distances with respect to the true sun calculated by Maestlin, noting that his work with Mars confirmed his use of the true sun. He then mentioned the qualifications in chapter 22 regarding the lack of an equant in the solar theory. His work with Mars had shown, he wrote, that the sun (or earth) did not extend as far in apogee as a simple eccentricity would require, but instead fell closer to the center by 1/3 of its eccentricity.[81] Kepler thus concluded that an equant is needed for the theory of the earth. As we saw above, Tycho was concerned that the use of an equant would vitiate his own solar theory. By this time, Kepler was able to conclude that the maximum discrepancy introduced by an equant would be 1′ 10″ in the anomaly at 135°.[82]

Kepler now moved on to some newer results. He mentioned in the chronology given to Longomontanus that during this period, "although Tycho was indignant," he speculated about Venus and Mercury. We can date this work more precisely because Kepler mentioned in the letter to Magini that his thoughts about Venus and Mercury occurred to him when he was away from Prague on a matter of inheritance and without his books, that is, in late April or May 1601.[83] His letter to Magini reveals that this speculation also regarded the qualifications in chapter 22 of the *Mysterium*. He had been bothered since writing the *Mysterium* that the motions of both Venus and Mercury are referred to the earth. It turns out to be a consequence of Copernicus's transformation of the Ptolemaic theories for the inferior planets into heliocentric form that their equations of center ended up depending on the motion of the earth with respect to their apsidal lines, and at the same time, it became necessary to regulate their distances from the mean sun with respect to the motion of the earth.[84] In the summer of 1601, Kepler realized that "these supposed inequalities of the inferior planets are nothing other than parallax caused by the earth's motion either in its approach to or withdrawal from Venus's orbit, [which was] not fully understood before now."[85] Being without his books, Kepler was unable to do any calculations to confirm his insight. He suggested to Magini that he do some calculations for the orbit of Venus based on an orbit for the earth with a bisected eccentricity and an equant. Nevertheless, he was confident enough to proclaim, "I say I have called these new inequalities distinctly into question."[86]

Kepler's insight into the cause of the peculiar form of the Venus and Mercury theories confirmed the result from his Mars work that the theory of the earth ought to have an equant. At the same time, it resolved the last of the leftover problems from the *Mysterium* by suggesting that there were no exceptions to the equant form of planetary theory, with which Kepler had modeled the planet-moving force. Thus the annoying qualifications in chapter 22 became potent demonstrations of the truth of his conviction that the theories of the planets ought to be the same:

> If you grasp these things with your mind's singular insight, most clever Magini, you will certainly decide with me that the form of all seven theories that concern the stars' real motions are clearly the same and most simple, for whichever you like goes about a most constant and exact circle over one revolution, [moving] more slowly when higher and more quickly when lower, that is, close to the sun, and this is not by way of fancy but the truth. In fact, Tycho even employed an equant in the case of the moon. From this elegance and simplicity, that is, from the perfection of the celestial motions, you easily perceive how much strength should be added to Copernicus.
>
> For although Tycho imitates and represents Copernicus as closely as possible while keeping the earth immobile in the center, nevertheless it cannot be avoided that he constantly and differently twists the paths along which the planets are wheeled round through the clearest aether (which I myself easily believe) into non-uniform spirals.
>
> But the calculation does not therefore become easier. By no means; the more simple this form of astronomy is to comprehend, the more laborious it is to compute and intricate to discover.[87]

Kepler's last comment marks his first realization that, even though the astronomy he sought to create was harmonious and conceptually far simpler, it would actually be more difficult for the astronomer. This would eventually prove to be one of the greatest obstacles to the adoption his work.

Having lamented the difficulty of making the calculations, Kepler turned to the problem with which he wanted Magini's help. He knew from his work with the triangulation to the earth's orbit that it was not a simple eccentric, and that the difference in distance that this introduced was reflected in Mars's orbit. He assumed that the orbit of the earth was therefore of the same general form he had assumed for the orbit of Mars when he began constructing the vicarious hypothesis. They were both perfect circles with the sun and an equant located opposite each other on the apsidal line, but with independent proportions of the total eccentricity. The problem of determining the parameters for both orbits independently of one another must have seemed increasingly intractable, for Kepler proposed a bold simultaneous solution of the two orbits. He did not have a clear idea of how to solve the problem, but proposed it to Magini in the hope that he would be able to provide the solution.

Kepler assumed that some of the elements of the earth's orbit were sufficiently known. He assumed as given the direction of the earth's line of apsides, the time of aphelion, and the value of the whole eccentricity. This left as an unknown only the location of the center of the orbit on the line separating the sun and the equant, that is, the proportion by which the whole eccentricity was divided by the center of the orbit. On the other hand, he assumed nothing about the orbit of Mars except that it was a circle. This left unknown (1) the size of Mars's orbit relative to the orbit of the earth, (2) the orientation of Mars's apsidal line, (3) the amount of Mars's whole eccentricity, and (4) the proportion by which the whole eccentricity was divided by the center of the circle. Kepler proposed a hybrid of the two methods by which he had been investigating the orbits of the earth and Mars. The observations he proposed were three acronychal observations of Mars, which gave its true heliocentric longitude. Along with each of these three acronychals, Kepler suggested two observations separated from the time of the acronychal by whole periods of 687 days, for a total of nine observations. These latter observations allow one, given the orbit of the earth, to triangulate to the location of Mars in space for each of the three acronychals. These three points in space define the circle of Mars's orbit. The acronychals provide the location of the sun on Mars's line of apsides, and the dates of the acronychals provide the angles around the equant and thus its location. Inversely, given an orbit for Mars, one can triangulate in reverse to the earth's orbit. The acronychals again provide the location of the sun, and the dates provide the angles around the equant. Given the information Kepler assumes about the earth's orbit, one should be able in principle to solve for the earth's orbit. However, Kepler did not know how to solve the two problems simultaneously. He appealed to Magini for a solution, in the unrealistic hope that he could provide it.

Kepler's appeal to Magini reveals his understanding of how intertwined the problems of the orbits of Mars and the earth were. In contrast, his account of his research in the *Astronomia nova* had a more logical and systematic structure: he began with Mars, achieved an interim solution for its orbit, then turned to the earth, solved its orbit, and given this solution, returned to Mars's orbit. Neither Kepler nor Magini could achieve the bold simultaneous solution of the orbits that Kepler proposed, so it is perhaps not surprising that he did not bring it up in the *Astronomia nova*. However, Kepler's unanswered appeal to Magini does illustrate one aspect of how the account in the *Astronomia nova* has been sanitized into an expository text. He knew before he arrived at a satisfactory version of the vicarious theory that the theory of the earth needed an equant. Throughout his working notebooks, Kepler moved back and forth between the two problems. Furthermore, Kepler moved back and forth between trying to solve the problems by means of planetary longitudes and then by means of latitudes. (He was aware of the benefit of the independence of these two approaches; although he did not trust the latitudes because of the small angles involved and the large amount of error, he saw them as a valuable check for his solutions from longitude.)

A true chronological account of his research in the *Astronomia nova* would have served only to confuse the reader. Instead, Kepler would construct a didactic account, which separated the problems of Mars's orbit and the earth's orbit, as he would lead the reader from one to the other. Such an approach did make sense, but its sequence introduced a false logical necessity, for it suggested that Kepler was driven to a re-examination of the earth's orbit by the failure of the vicarious hypothesis. Since Kepler introduced his physical astronomy in the *Astronomia nova* only when treating the motion of the earth, the structure of his argument implied that he was driven to physical considerations only when a theory in the classical form with unprecedented flexibility failed. As we will see below, this was a major premise of the argument of the *Astronomia nova*.

Kepler's interest in the latitudes was also reflected in the letter to Magini, for after asking for the simultaneous solution from longitudes, he proposed the same thing by means of latitudes. His description of the method was confused, for what "I see in my mind, I cannot yet express in words."[88] The method involved three observations of Mars's latitude in opposition to the true sun, along with a direction for the line of nodes and a value for the inclination between the orbits. The observations would yield, according to Kepler, "the eccentricity of the planet confused with the earth's eccentricity and each aphelion confused into the same common aphelion in an intermediate place,"[89] that is, rather than separate eccentricities and locations of aphelion for each planet, the result would be a sort of average of the quantities for the two planets. From this confusion, Kepler hoped to find a way to extract the individual eccentricities, which would serve as a check on the method from longitudes.

Finally, Kepler asked Magini's help on two more things. The first was a simplification of the iterative method Kepler had been using to determine the vicarious hypothesis, the same problem that he had asked Herwart to forward to Viète. Second, Kepler went into a lengthy query of how tables could be composed, given the interrelation of the earth's and Mars's orbits. Kepler's concern about this complicated problem reveals again his awareness of the difficulty his kind of astronomy would introduce. Kepler had heard that Magini excelled at shortcuts, and he asked that Magini consider the problem and send him his thoughts.[90] Although Magini evidently did not answer this first letter from Kepler, the query preceded a proposal by Kepler some years later that they collaborate on ephemerides.[91]

Kepler did not intend to stay long in Graz during the summer of 1601—this is probably why he did not have his books with him—but his visit dragged on until the beginning of September.[92] When he returned to Prague, his findings regarding the theory of the earth had evidently sufficiently troubled Tycho that he set Kepler to re-examining the solar eccentricity. Kepler consequently concentrated on the triangulation to the earth's orbit, and did not return to the vicarious theory until some time later.[93] He and Tycho were engaged in this work until Tycho's death on 24 October 1601. On 26 October, Kepler was appointed his successor as imperial mathematician and the care of Tycho's instruments and unfinished studies was put into his hands.[94]

Tycho's death brought an end to Kepler's work as his assistant. This was a liberating turn of events for Kepler. Although we have shown that his explicit intention was to continue the work he had begun in the *Mysterium* and that the nature of the work he did on Mars was clearly related to this goal, during the time he worked for Tycho he had had to keep his light under a bushel. He knew that Tycho objected vehemently to his use of physical causes in astronomy and this had apparently constrained him. In his letter to Magini, Kepler did not refer at all to physical causes, although the revelation that the form of all planetary theories was simple and the same certainly held great physical significance for him. But in a letter to Maestlin written after Tycho's death and in passages in his notebook that date from early 1602 we see a startling new emphasis on the physics of the situation.[95]

Even if Kepler found his work under Tycho odious, it was not in his character to denigrate Tycho or his accomplishments publically. However, it is clear that he felt both that Tycho was past his prime and that he lacked the vision to exploit his vast stock of observations. Kepler expressed his feelings about his relationship with Tycho and his own research when discussing his state of affairs in a letter to Maestlin written shortly after Tycho's death. After explaining his succession as imperial mathematician and the uncertainty about his salary, he wrote:

> I am full of hope. For if God has taken care of me for the better, namely in the matter of exercising my intellect, He will by all means also take care of my expenses. And if astronomy is likewise in God's care, as it is devout to believe, I hope I will now accomplish something in it, since I see how fatefully God has joined me to Tycho, such that a way was not open to tear myself away from him [even] in the most oppressive circumstances. What Tycho accomplished he accomplished before the year '97. From that time his affairs took a turn for the worse, and, distracted by his enormous worries, he himself grew childish. . . . Most excellent of the whole Tychonic opus are the observations, in the same number of full books as the years he presided over this work. Beside these, there is the *Progymnasmata* (in which there are the fixed stars, and the motion of the sun and moon to this day). These exude pure ambrosia. . . . Those things that pertain to the moon were completed in these last years principally through the work of a certain Dane Christian Severin

> Longomontanus, [although] with Tycho holding the tiller. These do not exhibit the divine quality that there is in the solar theory. [Tycho] was going to write another book on comets, and commented most diligently and masterfully on all the planets, but for the most part in the Ptolemaic fashion, with the things that needed changing changed, as Copernicus also did. You see how God dispenses His gifts, such that we cannot all do everything. Tycho did as Hipparchus did: it is in the foundations of the edifice that he expended the greatest effort. We cannot all do everything. Hipparchus needed a Ptolemy, who added on the 5 remaining planets.[96]

The juxtaposition of Tycho's decline with his life's accomplishment—which Kepler placed almost entirely in the observations and the solar theory—and the mention of Tycho's planetary work as being merely a continuation of Ptolemy in the spirit of Copernicus show that Kepler expected Tycho to produce nothing more. The mention of the oppressive circumstances in which Kepler had been trapped, along with that of Tycho's direction of Longomontanus's work on the lunar theory, suggests a liberation with Tycho's death.

But despite the circumstances of his work under Tycho, Kepler did his best to honor him in the analogy of the relationship between Hipparchus and Ptolemy. It was in the foundations of the edifice that Tycho had expended the greatest effort. It fell to Kepler to be the architect of the structure that rose upon them. He would exploit this architectural analogy several times over the rest of his life. In its most public form, it was displayed graphically on the frontispiece of the *Rudolphine Tables*, where Kepler was unobtrusively depicted in a side panel on the base of the temple of Urania laboring by candlelight on the design of the structure that soared above him.

The *Apologia pro Tychone contra Ursum*

We have seen how Kepler was able to continue to pursue his research program by redirecting his attention exclusively to the planet-moving force hypothesis and the way that it played out in the orbits of Mars and the earth. Much of the last year of his work with Tycho was devoted (at Tycho's insistence) to a philosophical inquiry on the nature of astronomical hypotheses. In this too, Kepler was able to harness an assignment by Tycho for his own ends. In the *Apologia pro Tychone contra Ursum*, Kepler composed his most focused treatment of the status of astronomical hypotheses and of their necessary physical commitments.

One of Tycho's primary motives for enlisting Kepler as an assistant had been the possibility of pressing him into service in his legal proceedings against Nicholas Reimers Ursus. After Kepler's return from Styria in the fall of 1600, this pursuit came to occupy more and more of his time. Kepler wrote miserably to Maestlin on 8 February 1601, "On account of this illness of mine I am doing nothing but writing against Ursus."[97] At this time, Kepler had become utterly dependent on Tycho, and writing against Ursus was virtually a condition of his employment.[98] Tycho may have refocused his attention on the dispute following Ursus's death on 15 August 1600. Unable to haul a dead man before a court, Tycho conceived a plan to have Ursus's book condemned and to publish the proceedings of the trial. Along with this, Tycho planned to include a section on mathematics and astronomical hypotheses, which undoubtedly would have been the fruit of Kepler's effort, his *Apologia pro Tychone contra Ursum* [*A Defense of Tycho against Ursus*].[99]

Kepler did not have the stomach to participate in Tycho's personal vendetta against Ursus. In his opinion, the matter was better left alone.[100] Indeed, as he had made clear to Tycho, "I was not willing to deliver my judgment about the insults before a holder of public office."[101] Kepler avoided involving himself in the vicious personal aspects of Ursus's slander, and instead limited himself to analyzing Ursus's pronouncements on the nature of astronomical hypotheses. Thus he sidestepped Tycho's personal complaint, and in his quasi-legal defense, as Jardine puts it, "Kepler treats astronomical hypotheses as his client."[102]

There was more to Kepler's approach in the *Apologia* than distaste with involving himself in Tycho's dispute. Since he was a sincere Copernican, Kepler did not ultimately agree with Tycho. A major thrust of Ursus's attack had been to ridicule the idea that astronomical hypotheses have any claim to represent reality. Since Kepler and Tycho differed over what they believed this reality to be, Kepler was to some extent restricted to defending a common locus of their beliefs.[103] However, even though he wrote at Tycho's behest and suppressed his Copernicanism, Kepler's philosophical stance in the *Apologia* should not be discounted. His argument is a more developed version of the philosophical argument presented in the *Mysterium*.[104] And to some extent, Kepler can be seen as defending the *Mysterium* itself, for Ursus had published Kepler's letter in the course of ridiculing the idea that the proportions of the celestial spheres could be derived a priori.

The fact that Kepler was defending Tycho against Ursus's attacks rather than writing a treatise devoted to his own epistemology and methodology also shapes the work to a certain extent, for in refuting Ursus's arguments, he is forced to address those arguments specifically. However, this specificity does not affect the work much. Kepler cogently extracts the essential skeptical arguments from Ursus's rambling invective, and these turn out to be rather conventional skeptical arguments for the time.[105] The *Apologia* thus represents a fine summary of Kepler's philosophy of astronomy.

Personal insults aside, the main thrust of Ursus's attack was that Tycho's insistence on priority in the invention of the geo-heliocentric system was foolish, for astronomical hypotheses were so inconsequential that their authorship would convey but little honor. In his effort to disparage astronomical hypotheses, Ursus cast against them whatever aspersions came to mind. Thus on one level, he mounted an etymological attack by equating *hypotheses* with *aitemata* in the sense of "illegitimate or false postulates."[106] The core of Kepler's *Apologia*, however, was his refutation of what Jardine has labeled the "argument from evidential insufficiency."[107] There were several related facets of this argument, at least one of which was already well known to Kepler. It was the argument, which he had first addressed in the *Mysterium*, that true conclusions sometimes follow from false premises.[108] Taken to a more agnostic extreme, it denied the possibility of deciding between two observationally equivalent hypotheses, such as an eccentric and the geometrically equivalent arrangement of an epicycle and deferent. Ursus went so far as to taunt Tycho that he could spin off an endless series of distinct yet equivalent hypotheses.[109] This argument could be extended to the choice between geocentric and heliocentric systems, and Kepler's reply is interesting in this regard, although he could not pursue it too far because he was defending Tycho.

Kepler laid the groundwork for his general refutation in his specific reply to the etymological argument. He did so by carefully defining his terms. By the plural "astronomical hypotheses," he wrote, we "designate a certain totality of the views of some notable practitioner, from which totality he demonstrates the entire basis of the heavenly motions. All the premises, both physical and geometrical, that are adopted in the entire work undertaken by that astronomer are included in that totality."[110] With this definition, Kepler accomplished a great deal of his argument. It had always been his assertion that the physical content of astronomical hypotheses was essential. By including in his definition of astronomical hypotheses both the physical and geometrical premises involved, he preempted the skeptical stance inherent in the dissociation of astronomy and physics on which the "Wittenberg interpretation" of Copernicus was based. According to Kepler's definition, one cannot speak of following the hypotheses of some practitioner while ignoring the physical content of his theory. Accordingly, he pointed out, "As in every discipline, so in astronomy also, the things which we teach the reader by drawing conclusions we teach altogether seriously, not in jest. So we hold whatever there is in our conclusions to have been established as true."[111]

By insisting on the commitment of the astronomer to the truth of his conclusions and widening the scope of that commitment to include his physical premises, Kepler prepared himself to answer Ursus's skeptical challenge on the larger, cosmological scale. At the same time, he dismissed the importance of geometrically equivalent hypotheses on the smaller scale. He could not deny that these exist—it had been known since the time of Apollonius that an eccentric can be replaced by an equivalent form of epicycle and deferent. To answer Ursus, Kepler drew a distinction between an "astronomical hypothesis" and a "geometrical hypothesis," narrowing the definition of the singular "astronomical hypothesis" to refer only to the general motion being modeled. "Thus," Kepler explained,

> if some astronomer says the path of the moon describes an oval shape, it is an astronomical hypothesis. But when he shows by what circles a drawing of this sort of oval can be constructed, he uses geometrical hypotheses. And when Ptolemy said that the motions of the planets slow down at the apogee and are speeded up at the perigee, he set up an astronomical hypothesis; but when he introduced the equant, he did so as a geometer for the sake of calculation, so that it could be worked out how much that motion of the planet slows down at any given moment in time.[112]

This argument was the fruit of the realization to which Kepler had come while writing the *Mysterium* that Ptolemy's equant and Copernicus's eccentric epicycle both model the same kind of motion. He had pointed out then that, although Tycho had shown the spheres not to exist, a "whirling spirit" could cause the same phenomena and that "even without the spheres the same results will follow."[113] It was by banishing epicycles and eccentrics to the realm of geometry that Kepler claimed to have created an astronomy without "fictitious hypotheses," and jokingly suggested to Maestlin that he might now claim the chair that Ramus had promised to anyone capable of constructing an astronomy without hypotheses.[114]

According to this second set of definitions on the level of the individual planetary theory (rather than the sum of a practitioner's ideas), there were three things in astronomy: geometrical hypotheses, astronomical hypotheses, and the apparent motions of the stars themselves. And there were two distinct tasks for the astronomer: the task of setting up astronomical hypotheses from which apparent motions follow, and the task of setting up geometrical hypotheses of whatever kind from which the astronomical hypotheses follow.[115] Thus, for instance, the astronomer might note the phenomenon that the sun spends more time in one part of the zodiac and less in the opposite part. From this, he would form the astronomical hypothesis that the sun's motion is on a circle not concentric with the earth. And from this, he would proceed to a geometrical means of modeling this motion, either a simple eccentric or an equivalent epicycle and deferent. By means of these definitions, Kepler made the existence of geometrically equivalent hypotheses irrelevant to the status of the astronomical hypotheses (plural) of some notable practitioner, which he defined more broadly earlier to include both physical and geometrical considerations.

In addition to dismissing the significance of geometrically equivalent hypotheses, Kepler denied the danger of true conclusions following from false premises. This was in fact a rare possibility, because even if some false hypothesis does perchance produce true conclusions, continual application of the hypothesis will eventually reveal its falsity:

> And just as in the proverb liars are cautioned to remember what they have said, so here false hypotheses, which together yield the truth once by chance, do not in the course of a demonstration in which they have been combined with many others retain this habit of yielding the truth, but betray themselves. Thus in the end it happens that because of the linking of syllogisms in demonstrations, given one mistake an infinite number follow.[116]

Kepler also used the opportunity of explaining how false hypotheses reveal themselves to reinforce the importance of the astronomer's commitment to the physical aspects of his hypotheses. It was because of this cascade of errors that astronomers were careful not to admit anything into their hypotheses that is the slightest bit doubtful, "thus almost all notable authors to date assess [their hypotheses] both on geometrical and physical grounds and want them confirmed in all respects."[117]

The importance of the physical aspects of astronomical hypotheses was most prominent in Kepler's more general treatment of equivalent hypotheses, that is, in his treatment of equivalence on the cosmological scale. In this context, he did not deny the possibility of different hypotheses yielding the same predictions, but he did deny that hypotheses that are not simply identical are indistinguishable. Physical considerations played the essential role in countering Ursus's skepticism. Kepler exploited his more encompassing definition of astronomical hypotheses to distinguish even between those that yield identical results, for, as he defined the term, there was necessarily a physical consideration that could be used to distinguish two astronomical hypotheses. In the case of Ptolemy and Copernicus, Kepler denied their equivalence on the basis of Copernicus's broader *explananda*. Rather than accounting only for the phenomena, Copernicus also explained the *causes* of the phenomena. Kepler described this difference in terms of physics:

> And though some disparate astronomical hypotheses may provide exactly the same results in astronomy . . . nevertheless there is often a difference between the conclusions because of some physical consideration.[118]

Kepler did not define what he meant by physics, but his emphasis on causes recalled and contradicted the ancient distinction between physics and astronomy.[119] In the broadest sense, Kepler's physical astronomy was astronomy that explained not only the phenomena but the causes of those phenomena. This was most obviously expressed in the title of the *Astronomia nova*, which read in part *Astronomia nova* ΑΙΤΙΟΛΟΓΗΤΟΣ, *seu physica coelestis*, that is, *New Astronomy* Based upon Causes, *or Celestial Physics* (emphasis added). In the *Astronomia nova*, the emphasis was on the cause of the motions or the motive force hypothesis. Ultimately, physical considerations included the cause of the arrangement of the world itself. It was in this sense that Kepler considered all the arguments in the *Mysterium* to be physical arguments for heliocentrism, and it was in this way that Kepler was able to see his astronomical vocation as religious devotion.

Limited as he was in the *Apologia* to defending Tycho's views, Kepler did not make such extreme claims. However, the most noteworthy aspect of the *Apologia* is Kepler's complete reliance on physical causes as the means to differentiate and judge hypotheses. In this way, his defense of Tycho was a defense of his own, unique interpretation of the role of physics in astronomy. He thus spoke beyond the immediate context when he replied to Ursus's assertion that astronomers should be interested only in prediction by saying:

> For even though what he mentions is the primary task of an astronomer, the astronomer ought not to be excluded from the community of philosophers who inquire into the nature of things. One who predicts as accurately as possible the movements and positions of the stars performs the task of the astronomer well. But one who, in addition to this, also employs true opinions about the form of the universe performs it better and is held worthy of greater praise. The former, indeed, draws conclusions that are true as far as what is observed is concerned; the latter not only does justice in his conclusions to what is seen, but also . . . in drawing conclusions embraces the inmost form of nature.[120]

Before going on, we should consider the implications of Kepler's views in the *Apologia* for the argument of the *Astronomia nova*. In the *Astronomia nova*, Kepler went to some length to present his argument in all three major cosmological forms: the Ptolemaic (geocentric), Copernican (heliocentric), and Tychonic (geo-heliocentric). And he acknowledged that these hypotheses "are for practical purposes equivalent to a hair's breadth, and produce the same results."[121] Establishing the truth of Copernicanism therefore hinged on celestial physics and what Kepler could establish about the causes of the motions. This was the foundation of the structure of the work, for Kepler could not simply lay out his findings, trusting astronomers, who he knew resisted his inclusion of physics in astronomy, to pick the true one. Instead, he bade the reader to follow his account of how he came to his findings to establish *rhetorically* that "[no] other [approach] could succeed than the one founded upon the motions' physical causes themselves, which I establish in this work."[122]

⌘ CHAPTER 6 ⌘

KEPLER'S WORK AFTER TYCHO'S DEATH

There is perhaps no better illustration of the liberating effect of Tycho's death for Kepler's research than the significant change in emphasis we see shortly thereafter. Kepler was generally explicit about the influence of the work he had done in the *Mysterium* on the direction of the work he undertook under Tycho, and while Tycho was alive, we see that Kepler described his successes in terms of the research program he had begun in that book.[1] But very shortly after Tycho's death, Kepler paused to assess his work to date and then embarked on a new line of research. His previous research had employed physical reasoning, but in a secondary capacity, as explanatory principles for more or less conventional mathematical astronomy involving circular orbits and uniform motions. Kepler's new line of research was revolutionary, for he decisively abandoned the traditional forms of classical planetary theory for a completely new astronomy in which consideration of the physical cause of planetary motion was used not only to describe but also to derive it.

The first mention of a "physical hypothesis" in the Mars manuscripts apparently dates from late November 1601, very shortly after Tycho's death.[2] A more detailed and deliberate reassessment of Kepler's work is found in the section titled "Consideratio hypotheseos circa aequantem," which was composed shortly afterward, around the beginning of December.[3] Kepler began with the material discussed above, in which he reviewed his decision to use an equant instead of Copernicus's eccentric epicyclet on the grounds that Copernicus's theory was not amenable to geometrical contemplation, and "since a natural cause has been found as a result of which a planet proceeds non-uniformly with regard to its center."[4] He then reviewed the use of the motive force hypothesis in the *Mysterium* to derive the relation between the periods and distances of the planets and the error in that procedure. With this application of the motive force hypothesis called into doubt, he abandoned it and proceeded to derive a model for Mars in which the two eccentricities could be in any proportion. The result was the observationally determined proportions

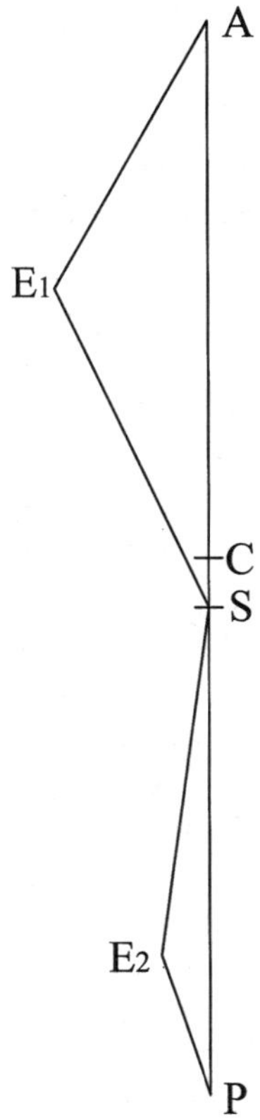

Figure 6.1 Another test of the division of Mars's eccentricity

Mars is observed near aphelion A and perihelion P, at which time earth is at E_1 and E_2. The angles $\angle SE_1A$, and $\angle SE_2P$ are Mars's observed position. The angles $\angle ASE_1$ and $\angle PSE_2$ are known from the difference of Mars's longitude as taken from the vicarious hypothesis and the earth's longitude from Tycho's solar theory. The distances E_1S and E_2S are likewise from Tycho's solar theory. The distances SA and SP can be calculated. C is the midpoint of AP. The eccentricity of the sun CS is shown to be smaller than the solar eccentricity from the vicarious hypothesis, though the exact value is too uncertain to be trusted.

of the vicarious hypothesis, in which the eccentricity of the eccentric was greater than half of the eccentricity of the equant.

As noted above, Kepler was not particularly surprised by this finding, because it was consistent with both Copernicus's and Tycho's inexact bisection of Mars's eccentricity. Nevertheless, ever cautious, he devised a method of checking the division of the eccentricity using annual parallaxes. The vicarious hypothesis provided the orientation of the line of apsides and the epoch, so Kepler could select from Tycho's data observations of Mars when it was very nearly at aphelion and perihelion. They were not opposition observations, and consequently the earth was off to one side of the line connecting Mars and the sun (the line of apsides, in this case). Kepler could use the parallax caused by the position of the earth (the annual orb, hence "annual parallax") to triangulate to the position of the sun on Mars's line of apsides, assuming an accurate solar theory.[5] (See figure 6.1.) The method did not yield precise results, but it showed the eccentricity of the eccentric, that is, the sun's distance from the center of the orbit, to be around half of the total eccentricity. The annual parallaxes thus indicated that the eccentricity to be indistinguishably close to being bisected. "From this," Kepler wrote,

> a new speculation came to me that the eccentricity of the equant does not remain the same throughout the whole circuit, and that this is the reason why a greater height of the eccentric should be produced from acronychal observations, since the eccentricity of the equant was assumed to remain the same in all four acronychal positions. Perhaps half of the equant can be the height of the center of the eccentric, since the parallaxes show it, whose evidence in more important in this business.[6]

With this speculation, Kepler attempted to reconcile his contradictory findings. His parallax method told him nothing about the location of the Mars's equant, but it provided firm evidence that the eccentricity of the eccentric seemed to be half the total eccentricity. Perhaps the eccentricity of the equant varied. In the derivation of the vicarious hypothesis, he had assumed it was constant. Perhaps this assumption was at fault.

A variable eccentricity was not an unprecedented innovation in astronomical theory. Complicated arrangements that had the effect of varying the eccentricity had long been used in the theories of the moon and Mercury. And, had Kepler wanted to arrange a simple periodic up and-down motion of equant along the line of apsides, the mechanism now known as a Tusi couple was available to him through Copernicus.[7] But the use of such a contrivance would have been utterly alien to Kepler's approach to astronomical theory, as his reintroduction of the equant as a substitute for Copernicus's eccentric epicyclet has shown. Furthermore, Kepler proceeded immediately to a cogent physical justification for a variable equant. Since he worked from the assumption that the motion of a planet was inversely proportional to distance, and the distances changed more slowly around apsides than around quadrants, he concluded that the eccentricity of the equant, which would accentuate the planet's speed, must be smaller around apsides. Thus Kepler for the first time began rigorously to examine his physical hypothesis of the equant away from apsides:

> Furthermore, no matter how nearly true as possible it might be that the height of the eccentric is half [the height] of the equant, still neither Copernicus's, as was said before, nor Ptolemy's hypothesis of the matter is satisfactory to the extent that from these principles the point of the equant does not remain the same (since we assumed the motion [to be] proportional to the distances, as was said above). For the distances change very little around apsides and much in the middle distances, because there [i.e., around apsides] lines from the sun strike the circumference of the circle nearly perpendicularly . . . , but here [i.e., in the middle distances, the lines strike] obliquely. Therefore, if the

motion, or time of delay, which the planet makes in the position of apogee is to the delay which it makes in the position of middle distance as the longer length is to the mean, and if whatever delays you please are as their lengths, then very little time is added to the planet in apogee, very little is taken away at perigee, and much more around middle distances. Therefore the equant point is low in the one case [i.e., around apsides] and high in the other [i.e. around middle distances].[8]

A moving equant also explained to Kepler how it was that in the vicarious hypothesis he had found the eccentricity of the equant to be smaller than the eccentricity of the sun, since most of the acronychal observations he had used to derive it were nearer apsides than middle distances.

The provisional hypothesis of a moving equant made sense in a number of ways. It reconciled the contrary results of Kepler's two methods of determining the eccentricity, and at the same time provided a reason for the disparity. Even if the idea of an equant of variable eccentricity seems unwieldy, it only underscores the fact that Kepler placed no faith in the equant's physical existence. It remained simply a mathematical representation of the distance law. In fact, his use of an equant was actually becoming more rigorous, for Kepler was now seriously considering the action of the distance law all around the orbit. Perhaps the complexity of using an equant of variable eccentricity even pushed Kepler along the road to the area law, for he immediately turned to the question of "how a table of equations is to be calculated."[9]

His solution marks one of the most fundamental turning points of the new astronomy:

Although there are infinite points in a circle, we can express it accurately enough if we say a circle has sixty times 360 points. Let us imagine just as many radii from the center and just as many from the sun in that same [circle] in turn. Then, therefore, as the sum of all the radii is to the periodic time of the planet, so is the sum of distances collected in one sum in some array to the time which elapses while the planet makes all those distances. In this way both the simple [i.e., mean] anomaly and the perfectly equated [i.e., true] anomaly are to be built up from the whole degrees and minutes of the semi-equated [i.e., eccentric] anomaly. This [i.e., the eccentric anomaly] first, from this the distances, and from them the simple anomaly, which subtracted from the equated anomaly reveals the whole equation, if it is necessary. However I believe in a table the perfectly equated anomaly juxtaposed to the simple anomaly suffices.[10]

In other words, divide the circle from the center into points one minute apart,[11] determine the distance from the sun to each point on the periphery, add up all the distances to find how much time corresponds to how much distance (based on the proposition that the total distance will represent the total period), and then construct a table correlating the position from the sun (equated anomaly) to the time the planet takes to get there from apogee (simple or mean anomaly). In addition to being a significant step toward the area law, this technique also presaged the unique form of planetary tables that Kepler was forced to introduce after the *Astronomia nova*, although he probably did not fully appreciate the intractable problem he was getting into. The essence of the problem that became clear later was that there was simply no direct way to calculate the planet's position exactly from its date, as had always been possible before Kepler, and never was after.

Only a few pages later in the Mars notebook Kepler investigated a method to calculate the sun-planet distances along the lines of the work in chapter 29 in the *Astronomia nova*.[12] In the following chapter of the *Astronomia nova*, a table of earth-sun distances for each degree around the earth's orbit, which might have been supposed to be the result, was then laid out. The information was presented in the *Astronomia nova* as background for Kepler's account of his attempt to model the motion of the earth using the distance law. But he pointed out that, although the distances in the table would suffice for this purpose, they corresponded in fact to a slightly oval orbit.[13] To further confuse the issue, the values for the "mean anomaly" and "equated anomaly" in the table did not correspond to those angles, but were instead the eccentric anomaly (the equated anomaly plus the optical equation) and another angle to be introduced later, which consists of the equated anomaly minus the optical equation. Kepler helpfully pointed out that there was no table for the actual equated anomaly because it is the mean between the two angles in the table.[14] Clearly, there was a great deal of confusion here. What Kepler did in the *Astronomia nova* was to make a table that acted as a bridge between his theories, and that would retain its usefulness after the first, a perfectly circular orbit, had been disproved. This is an excellent illustration that the *Astronomia nova* was not written as a faithful account of the research, but was crafted to instruct and lead the reader through Kepler's various attempts.

A less fragmentary report of the state of Kepler's research about this time was given in a letter to Maestlin of 10/20 December 1601.[15] Unfortunately, we cannot gauge Maestlin's reaction because, for a reason obscure even to Kepler, Maestlin had then begun a long silence during which he refused to respond to his former student's letters.[16]

The most striking aspect of the letter to Maestlin is the sudden reappearance of physical speculations in Kepler's correspondence. Kepler's sustained interest in physical considerations while Tycho was alive is manifest in his work during this period, and it is probably too much to maintain that Tycho kept him so cowed that he would not mention them in private to his beloved teacher. However, earlier that year when he had written to Magini, he had not mentioned physical considerations. Kepler's greater frankness with Maestlin can certainly be attributed partly to his close relationship with his former teacher. Even so, I do not think it is too much to suggest that Kepler felt some sense of relief at again being master of his own research program and was consequently freer in the discussion of his work.

A noteworthy feature of this summary of his results is Kepler's realization of the great simplification he wrought with the change to the true sun and the changes in the theory of the earth. As we have seen, he had discovered that those features of Mars's orbit that had been thought to be peculiar to it are in fact merely exaggerated in the case of Mars: "The reason why it was believed that its motions were more uncertain is not peculiar to it, but is common to all the planets, although most evident in Mars."[17] He then went on to explain how the problem of Mars's variable eccentricity had been resolved by the bisection of the earth's eccentricity and the introduction of an equant in the theory of the earth.[18]

Another great simplification of Mars's orbit was in the theory of latitudes. Kepler had discovered that both Ptolemaic and Copernican theories of latitude—Copernicus had slavishly followed Ptolemy in this regard—were seriously misconceived. Contrary to their schemes of variable inclination, Kepler found the inclination of the plane of Mars's orbit to be constant. (Although the use of the mean sun rather than the true sun did introduce a variation in the inclination, this was a far smaller effect and in a different sense than the error Ptolemy originally introduced due to bad observations.[19]) Kepler's finding regarding the theory of latitude was presented in chapters 13 and 14 of the *Astronomia nova*.

In addition to the simplification in latitude, assigning the cause of most of the remaining peculiarities of Mars's theory to the theory of the earth revealed, "Thus the theory of Mars would be as simple as can be, consisting of single circle with an individual period."[20] At the same time, it had become clear that "[t]he theory of the sun or earth would clearly be similar to it, for it too is based upon an equant."[21] Although, as we saw, Kepler had been concerned to allow maximum flexibility in both theories with regard to the division of the total eccentricity, the evidence (from the annual parallaxes, for instance) kept pointing toward an exact bisection: "In both cases, the proof and exigency of numbers obliged us to divide the compound eccentricity into two, as did Ptolemy."[22] When Kepler checked a simple bisection of the earth's eccentricity against Tycho's solar theory, he found no sensible difference and so concluded, "Thus the whole Tychonic restitution of the sun stands."[23]

This marvelous simplification in the mounting evidence that in both cases the eccentricity of the equant was simply bisected led him back to the motive force hypothesis:

> Speculation supervenes when these things in the cases of the two planets are considered, and it finds the cause of the equant to be merely physical, although amenable to geometrical measurements. For as any distance is to another, so is the delay [*mora*] of the planet at the point of that distance to the delay at the point of the other distance.[24]

This is a concise summary of the distance law, as we saw in the manuscripts from this period. In this letter to Maestlin, Kepler gives no statement of the area law, nor does he mention the idea of a moving equant. He may not yet have come up with these ideas (the "Consideratio" and the letter to Maestlin are so close in time that their precise sequence is not clear), but it is more likely that he did not see fit to go into these mathematical expedients when discussing the general principle. The distance law was still his primary physical concept.

The realization that the theories of Mars and the earth were essentially identical in structure led on to consideration of the others. "I have already been consumed with hope that the remaining planets will be reduced to the same laws, and even Mercury itself"[25] (the theory of Mercury had always been the most enigmatic of the planetary theories). Kepler disclosed that the orbits of the inferior planets are also subject to the same physical explanation as the other orbits:

> Actually inasmuch as those orbs which carry these two around the sun are eccentric to the sun, they also participate in the same physical cause, in that the more eccentric they are the slower they become in their aphelion. . . . All these things have not yet been proven from observations, but the greatest hopes of finding them are furnished in the example of those things which have already been established.[26]

Kepler's hope of reducing the theories of the inferior planets to the same physical principles is evidence not only of the importance he invested in the simplicity and harmony of having all planetary theories in the same form, but it also serves to remind us that Kepler was not at this time engaged in a commentary on Mars. The largest of his newly assigned tasks was the *Rudolphine Tables*, which would require theories for all of the planets. Because Kepler had more experience with Mars, it was in the front of his mind, but he did not ignore the theories of Venus, Mercury, or even the moon, as is also apparent in repeated references to these other planets in the *Astronomia nova*. There is further evidence that Kepler was looking forward primarily to the *Rudolphine Tables* later in the letter to

Maestlin, when he speculates about what form of planetary tables can most efficiently take into account the various corrections and speaks of the six different tables for the six planets, excluding the moon.[27]

The last decisive development in Kepler's reassessment of his work in the time shortly after Tycho's death was also the final fundamental step in the development of the *Astronomia nova*. It was the definite conclusion that the orbit of Mars was not simply a circle but some kind of oval. His first fragmentary thoughts on this have recently been published.[28] The Mars notebooks show that Kepler's investigation into the physical cause of Mars's motion—the distance law and the attendant area formulation—was central to this realization.

Kepler's substantial departure into a new derivation of Mars's orbit based squarely on consideration of the physics involved dates from early 1602.[29] Kepler began writing in the middle of the page and his writing grew more compact as he approached the bottom, as though he was suddenly releasing a number of considerations that he had had in his mind for some time.[30] It appears that, frustrated with the vicarious hypothesis and newly freed to pursue his own research program, Kepler jotted down some thoughts on the physics of the situation, made a list of his findings thus far, and recognizing their implication took off in a new direction.

The list began with the same matters he had addressed when investigating the distance law. First, he knew that the vicarious hypothesis provided accurate heliocentric longitudes. On the other hand, the parallax from different points on the earth's orbit showed that the eccentricity should be bisected, rather than divided in the ratio of roughly 3:5, as the vicarious hypothesis did. Kepler also introduced considerations of a "physical hypothesis." It is not clear exactly what he was referring to, although it seems that he was referring to the use of the distance or area law on a circular orbit. That this physical hypothesis was consistent with the parallax observations along the line of apsides was obvious to Kepler, since he had long thought of the distance law as being equivalent to the use of equant at aphelion and perihelion. But comparing the vicarious hypothesis with the physical hypothesis showed that the physical hypothesis did not provide accurate longitudes away from apsides.

Kepler's use in this context of the phrase "physical [hypothesis], especially adapted" suggests the conjecture that he may have been investigating different versions of the hypothesis.[31] The words *peculiariter aptata*, however, can also be translated as "applied in a particular manner," and might then refer to the method of adding up all the distances to estimate the cumulative effect of the planet-moving force or to the use of the area law, which Kepler still found suspect. The qualification would then be due not to multiple versions of the hypothesis but to Kepler's dissatisfaction with the technique. If Kepler had gone to the length of calculating out a table of Mars-sun distances similar to the table of earth-sun distances in the *Astronomia nova*, chapters 30 and 31, and calculated the motion of Mars based on such a table, it would be of interest, because no such work was described in the *Astronomia nova*.

Finally, Kepler added to his list the further consideration that calculated distances differ from those from the physical hypothesis away from aphelion and perihelion, meaning that the distances from the sun to Mars are shorter around middle distances. From the present state of the research on the Mars notebook, it is not clear what or how many calculated distances he was referring to, but this finding was presented in chapters 41, 42, and 44 of the *Astronomia nova*, where he brought these distances forward (with some very small variations) from triangulations from multiple positions of the earth to the same point on Mars's orbit presented in chapters 26, 27, and 28. Reusing these painstakingly calculated distances would be an understandable saving in labor.

With these points enumerated, Kepler considered the physics of the orbit. He began with the still doubtful—in his mind—supposition that the area of the triangle defined by the sun, the center of the orbit, and the planet measures the physical equation. (This is discussed in chapter 40 of the *Astronomia nova*.) Compared to the Ptolemaic hypothesis, the physical hypothesis made the planet too fast at aphelion and perihelion and too slow at quadrants. This was the crux of the realization that the orbit was an oval, for if the distances were somehow shorter at middle distances, Mars would move more quickly there and consequently comparatively more slowly at apsides:

> Therefore near apogee and perigee Mars is slower than the physical [hypothesis] permits, because it traverses less of the path than the physical [hypothesis] shows. The opposite happens about the middle longitudes. This is entirely consistent with the fact that about the middle longitudes Mars approaches nearer than about the apsides. For thence it happens that the area of Mars's oval circle is heaped up more about the line of the apogee—and therefore the same happens to the elapsed time.[32]

Having come to the conclusion that Mars's circle was squashed in toward the sides, Kepler proceeded to analyze the physical implications in terms of the sun's and Mars's faculties. The sun, Kepler reasoned, distributes a natural motion according to the distance and measured in time. The planet, on the other hand, makes its own contribution to its motion by varying its distance from the sun. This concept had long been in Kepler's mind. He had proposed in the *Mysterium* that the motion of a planet on an eccentric results from a spirit coming forth from the sun, which pushes the planets around, along with "an individual spirit in whichever planet, by whose set of oars [*remigio*] the star ascends in its orbit."[33] (The analogy of a planet with a steering oar was significantly developed in the *Astronomia nova*, chapters 38, 39, and 57, although Kepler did not use this analogy in this section of the Mars manuscripts.) For Kepler, concentrating on a philosophical consideration of the planet's faculty, the problem was that in order for the planet to accomplish a nearly circular orbit, it would have to regulate its distance with respect to space rather than time, since the time along each part of the orbit was determined by the sun's faculty. In order that "the offices of the Sun and the planet remain distinct," he attributed to the planet an animate, geometrical faculty, by which it could either know its location or perceive it in the relation to the observed locations of the sun and stars.[34]

Kepler then returned briefly to the form of the orbit, noting that a symmetrical oval would produce the necessary adjustments at octants. He then introduced the fact that the orbit is "narrower below," on the perihelion side, and suggested that the planet ascended more near apogee, which would cause it to slow and thus produce a broader oval above. These remarks are somewhat cryptic, but very suggestive in the light of the egg-shaped orbits that Kepler first employed in the *Astronomia nova* when attempting to model the physical theory.

He now abruptly switched back to the planet's faculty and to an epicyclic representation of the orbit—presumably as an aid for investigating how the planet might bring about the orbit. But he quickly dismissed it as absurd, for the same reason that he dismissed epicycles in the introductory material in the *Astronomia nova*, namely that Mars and the sun could not regard the imaginary point at the center of the epicycle in their actions. The same reason forced him to abandon momentarily the idea that Mars might attempt to travel in a circle centered on the eccentric: "But can it be said that Mars intends a circle? How is it established everywhere? Could it be [established] about some point of the orb? False!"[35] In these and a few more sentences, he laid down the groundwork for chapter 39 of the *Astronomia nova*.

However, although Mars could not regard an empty point around which to travel on a circle, it could regard the sun, and Kepler now considered whether it might be endowed with knowledge of the distances it would have to achieve in order to traverse a circular orbit (which Mars could compare with its actual distance from the sun, perceived by means of the sun's angular size):

> Therefore, the distances from the Sun are established, or computed, through the eccentric anomalies [i.e., the distances are measured from the center of the oval], but are picked out through the simple anomalies [i.e., the distances

> the planet has on its list are uniformly spaced with respect to the equant]. It has this sort of acquaintance with its eccentric and with the periodic time by which the apogee moves forward. For the rest, it is not anxious about whither it is carried.[36]

The reason that the distances would have to be uniformly distributed with respect to the equant or the simple anomaly was apparently to avoid the problem of the planet having to act nonuniformly around the top and bottom halves of its orbit (i.e., near apogee and perigee, respectively). The sun's uniform motion then caused it to spend more or less time according to its distance.

Kepler summarized:

> We shall explain the matter thus.
>
> The Sun tries to accomplish a circle. The planet does just the same, but each hinders the other. The Sun was going to move [the planet] uniformly in a circle, since it [i.e., the sun] does not have the force for pulling or for propelling, but simply for agitating. Therefore, it ought to be circular about the Sun. But the planet too tries to accomplish a circle, so as to nod towards and away from the Sun. In this way it hinders the Sun, so as to move unequally in equal times, for in departing from the Sun it is in a weaker power. And now the planet, at rest and free from the Sun's motion, will set out equal circumferences in equal times, and will approach the Sun uniformly in a circle. (Just as previously the Sun, free from the planet's motion, restored the planet to its position uniformly in a perfect circle.)[37]

Thus, the compound of the planet's uniform motion on the epicycle and the nonuniform motion induced by the sun at different distances produced the oval orbit.

The discussion of an oval orbit broke off shortly after and the Mars notebook itself continued for only a few more pages. But it is clear that in the fruitful few months after Tycho's death, Kepler's research into the physics of Mars's orbit took a highly significant turn. He began to put his physical convictions into use and derived from them the shape of Mars's orbit. If his powers of analysis failed him on occasion, this was to be expected in this strange new realm. In hindsight, Kepler described the sequence of thoughts we have just examined (which correspond to the model introduced in chapters 45 and 46 of the *Astronomia nova*) in this way:

> When I was first informed in this manner by Brahe's most certain observations that the orbit of the planet is not exactly circular but is deficient at the sides, I judged that I also knew the natural cause of the deflection from its footprints. . . . But then what they say in the proverb, 'A hasty dog bears blind pups,' happened to me. . . . If I had embarked upon this path a little

more thoughtfully, I might have immediately arrived at the truth of the matter. But since I was blind from desire, I did not pay attention to each and every part of chapter 39 [i.e., where he concluded that the planet regulating its motion with respect to an epicycle would be absurd], staying instead with the first thought to offer itself—a wonderfully probable one, owing to the uniformity of the epicyclic motion—and thus entered into new labyrinths, from which we will have to extract ourselves.[38]

⌘ CHAPTER 7 ⌘

THE TYCHONICS

While working under Tycho, Kepler was able more or less to continue with the program of physical astronomy that he had begun in the *Mysterium cosmographicum*. Tycho's secretiveness with his data may have prevented Kepler from pursuing the investigation of his system of nested polyhedra, but even so, that meant only that he had concentrated his effort instead on developing his idea of the planet-moving force, and he had had a string of successes with the theories of Mars, the earth, and even Venus and Mercury that had only strengthened and confirmed his speculations. With Tycho dead, and Kepler appointed imperial mathematician in his stead, the way was apparently open for Kepler to proceed as he wished. Under these circumstances, the *Astronomia nova* would probably never have been published at all. The very existence of that work and its unusual rhetorical structure are artifacts of Kepler's bitter struggle with Tycho's heirs to exploit his legacy.

It was now spring or early summer 1602. Although he had had to dance attendance at court for two whole months to confirm his appointment as imperial mathematician, Kepler was finally paid for the first time on 9 March 1602.[1] As Kepler had hoped, "God had looked after [his] expenses," but the philosophical freedom he experienced as a consequence of Tycho's death proved to be fleeting, due to the interference of Tycho's legal and scientific heirs, to whom Kepler referred broadly as "the Tychonics." Foremost among these was the Westphalian nobleman Franz Gansneb Tengnagel van Kamp. As a former assistant of Tycho's and his son-in-law by virtue of his marriage to Tycho's daughter Elisabeth, Tengnagel assumed a leading role in the family's dealings with Kepler.[2]

The conflict between Kepler and Tycho's heirs arose out of the disposition of Tycho's astronomical legacy. After his death, responsibility for his instruments and unfinished publications—most importantly the *Rudolphine Tables*—was assigned to Kepler as the new imperial mathematician.[3] The choice of Kepler for this assignment was obvious but at the same time fortuitous. Though he had been associated with the *Rudolphine Tables* from their inception, his rise to the status of Tycho's chief assistant was recent and due in large measure to the departure of other members of Tycho's household. The new Uraniborg[4] that Tycho had hoped to build in Bohemia had never quite achieved the critical mass that Tycho needed to press on with his life's work, the enormous multivolume *Astronomiae instauratae progymnasmata*. And in the final year of his life, Tycho saw his staff of assistants dwindle. His most talented and trusted assistant, Christian Severin Longomontanus, who finally resolved to find a permanent academic position independent of his longtime patron, had left Tycho's service for good in August 1600 to return to Denmark.[5] Tycho's other top lieutenant, Tengnagel, though not much of an asset astronomically, had married Tycho's daughter Elisabeth in June 1601, and the couple departed for a year-long stay in Deventer, taking another assistant, Johannes Eriksen, with them.[6] Johannes Müller, whom Tycho arguably valued more than Kepler and who had worked continuously for Tycho since March 1600, finally departed at the end of May 1601, after more than a year of waiting for imperial patronage that was never forthcoming.

It seems to have been the consistent failure to obtain additional imperial funding with which to pay assistants that led Tycho to conceive of the *Rudolphine Tables* in the first place. Tycho had long proposed that Müller and Kepler be summoned by the emperor from their existing appointments to assist him in his work, for which they would be paid an imperial stipend on top of their continuing salaries, but he had never been able to obtain additional funds from the emperor to pay these collaborators. By September 1601, when Kepler returned to Prague from Styria, Tycho had a plan. He arranged an audience with the emperor at which he proposed a new set of astronomical tables, which he asked permission to name the *Rudolphine Tables*.[7] It was a tantalizing proposition. Previous patrons of astronomical tables, King Alfonso X of Castile and Duke Albrecht of Prussia, had been immortalized by their association with the *Alfonsine Tables* and the *Prutenic Tables*. Tycho was proposing what were justifiably expected to be the greatest astronomical tables to date, which would likewise immortalize Rudolf II. Kepler recognized this when drawing up a list of Tycho's unfinished projects shortly after his death. Of the *Rudolphine Tables*, he wrote:

> This work is of the highest importance, in the respect that Emperor Rudolf justifiably ought to prize it most highly, because more people remember King Alfonso today on account of the *Alfonsine Tables*, however faulty they might be, than on account of his wars and exploits, of which memory is lost in time.[8]

But there was a catch. At the same meeting, Tycho introduced Kepler as his assistant in the project, thereby apparently opening the way for an imperial appointment and salary for Kepler.[9] Whether the paperwork for this new appointment was even in place at the time of Tycho's death less than two months later is unclear, but the ardent desire of the emperor to have the project completed undoubtedly contributed to the unusual alacrity with which Kepler was appointed imperial mathematician after Tycho's death.

The choice of Kepler to take responsibility for Tycho's unfinished publications was therefore straightforward. He had been associated with the *Rudolphine Tables* since they were first proposed, and, by virtue of the attrition in Tycho's staff, he had risen to the status of primary assistant. Ownership of Tycho's instruments and observations, on the other hand, passed by right to his heirs, from whom Emperor Rudolf II offered to purchase them. Reckoning what it had cost Tycho to make them, the heirs asked at first for 100,000 florins, an astronomical sum.[10] The emperor countered with an offer of 20,000 florins, which was still a princely sum, sufficient to purchase half a dozen country estates in Bohemia or to pay Kepler's old salary as district mathematician in Styria for a century. He prudently specified that this enormous outlay should come from the government's rather than his own personal treasury.[11] As was usual, the imperial treasury did not have sufficient funds to cover such a lavish expenditure, so it was specified that until the debt was paid, the heirs would be paid 6 percent interest retroactive to the date of Tycho's death.[12]

Kepler had experienced the ponderous action of the imperial court and treasury, and perhaps foresaw interminable delays before he would legally receive Tycho's observations. In any case, interpreting his mandate as imperial mathematician broadly, he seized control of the observations, ostensibly in the interest of their preservation. He was certainly the most qualified person around to take charge of this priceless resource. Moreover, he rightly considered possession of the observations to be necessary for fulfilling his new role of completing Tycho's posthumous publications, most especially the *Rudolphine Tables*. However, he recognized the questionable legitimacy of this preemptive seizure, as he explained some years later in a letter to Christopher Heydon:

> I do not deny that upon Tycho's death, because of either the absence or insufficient expertise of the heirs, I boldy and perhaps arrogantly took charge of the observations he left behind, against the will of the heirs, but nevertheless according to the not obscure command of the emperor. Since he had entrusted the care of the instruments to me, I interpreted the mandate broadly and took especially the observations to care for.[13]

Open conflict between Kepler and Tycho's heirs over possession of Tycho's observations did not break out until Tengnagel returned to Prague in the summer of 1602. By then, some months had passed since Tycho's death, and his heirs had received none of the 20,000 florins promised by the emperor. Experience in the imperial capital had shown that if one were to get anything from the imperial treasury, one had to get it quickly, so the family put into motion a series of steps designed to put pressure on the emperor. The first was to retake possession of Tycho's instruments in order to demonstrate that the family would not hesitate to keep the priceless curiosities for themselves or sell them to another buyer if payment were not made.[14] It was probably Tengnagel who recognized especially the value of the observations, essential as they were to the emperor's valued *Rudolphine Tables* project. That summer, he demanded the return of the heirs' rightful property; Kepler refused, and Tengnagel brought suit against him.[15] Tengnagel's imperious manner and Kepler's confessed "delight in being insolent [*insultandi libidine*]" soured their relations from the outset.[16]

For their part, in their haste to capitalize on Tycho's legacy, his heirs antagonized Kepler as well. In the autumn of 1602, impatient to profit, the heirs took control of the publication of Tycho's unfinished works, usurping Kepler's responsibility. They hastily completed the printing of volume 1 of Tycho's *Progymnasmata*, incorporating an appendix Kepler had written without attribution as well as some of Kepler's private notes.[17] The project was rushed through in such a slapdash manner that Kepler indignantly complained to Herwart and sent him a list of errata. "In this most miserable state of the book," he fumed, "which the author carried for twenty years, the heirs, hurrying to market, finally aborted it."[18]

By October, the heirs had still not been paid more than a fraction of what was due them, and Tengnagel apparently concluded that he was more likely to capitalize on the observations if he was directly involved in the production of the *Rudolphine Tables*. His intrusion distressed Kepler all the more because Tengnagel involved himself under the pretense of philosophical differences and with the accusation that Kepler was dishonoring Tycho by not carrying out Tycho's work the way he had wanted it done. As Kepler explained the situation in a letter to Herwart von Hohenburg:

> Tengnagel is obtruding himself in anticipation of the *Rudolphine Tables*. Perhaps his affairs in Westphalia are such that he has need of this support. . . . Nevertheless I am concerned by the precedent. In order to excuse his purpose, or to veil his true reason—poverty—he pretends to the courtiers that he is concerned with the honor of his parent or father-in-law, that he fears that the studies that have been undertaken are being carried out very clearly against [Tycho's] intention, and that he knows how [Tycho] wanted it to be done. These things have the appearance of truth, because both previously and while Tycho lived and was conscious of it, I followed Copernicus's hypotheses. And thus, as far as I can see, there will be either something to be defended contrary

> to what I believe in philosophy or something that disagrees with Tengnagel. The former I have never done and will not do in the future. The latter I would do in such a way that I speak of disagreement, not of hatred. Whether he would also act the same, I do not know.[19]

Tengnagel would have been justified in upholding the Tychonic system against Kepler's Copernicanism. But in addition, he was aware of Kepler's innovations in referring to the mean sun and bisecting the earth's eccentricity, both of which contradicted Tycho's solar theory. David Fabricius had written to Kepler earlier that year that he had received a visit in East Frisia from the "Tychonians" Tengnagel and Eriksen, from whom he had learned of Kepler's findings.[20] If Fabricius shared his views with the two, as seems reasonable to suppose, they would have learned that he, too, was critical of Kepler's innovations.[21] As will be shown below in reference to lunar theory, Tengnagel knew enough about Tycho's work to detect and prevent large-scale changes in Tychonic theory.

In October 1602, or shortly afterward, the Tychonics prevailed in their suit against Kepler. Without doubt impatient to realize some profit from the *Rudolphine Tables*, the heirs accused Kepler of sloth and saw to it that Johannes Pistorius, the emperor's father confessor, was assigned the task of overseeing his work.[22] It was apparently also at this time that Tengnagel secured for himself the commission to complete Tycho's work on the *Rudolphine Tables*.[23] With Tengnagel having been given responsibility for the *Rudolphine Tables*, Kepler was now ordered to name the works that he would complete as imperial mathematician.[24]

It was Kepler's awkward position of having to justify his employment as imperial mathematician without reference to the *Rudolphine Tables* that was responsible for his decision to publish his Mars work in the form of a commentary, and thus to publish the *Astronomia nova* at all. By 7 October 1602, he had already recognized the importance of the Mars work for overhauling the theories of all the planets. As he wrote to Herwart:

> In short, I have beheld the sun in the theory of Mars as though in a mirror, in that I see how and to what extent it affects all planets. Truly, from Mars I take the example for treating the others. And thus I hope presently for all the best for every part of astronomy.[25]

In other words, through his investigation of Mars, Kepler had learned how the motion of the earth was involved in its theory and, necessarily, in all the other planets' theories as well. He anticipated that once he had finally solved Mars's theory, he would use that knowledge as a kind of template for generating the other planets' theories, presumably for use in the *Rudolphine Tables*. But it was only when he was compelled to name the works he would undertake as imperial mathematician unrelated to the *Rudolphine Tables* that he resolved to publish these findings in the form of a commentary devoted almost entirely to Mars. Thus in his next letter to Herwart of 12 November 1602 he wrote:

> [B]ecause I have had my diligence called into doubt, I have assumed the obligation for two works. The one to be ready for Easter [1603] will be *Commentaries on the Theory of Mars* (or whatever else the name might be), or the *Key to a Universal Astronomy*. . . . The other, to be finished off within 8 weeks, is the *Optical Part of Astronomy*.[26]

Ironically, grasping among his half-finished projects for something he could complete quickly, Kepler came up with two of the most significant scientific works of the seventeenth century. The latter, *Astronomia pars optica* (1604), became a foundational work in seventeenth-century optical theory. The former, of course, became the *Astronomia nova*.

It is of great importance to note that by late 1602, Kepler already saw the work on Mars as the "key to a universal astronomy." This fact underscores the importance of the discoveries he had made by this time. More importantly, it shows that Kepler did not perceive the *Astronomia nova* as a book devoted exclusively to the elucidation of Mars's orbit. As is clear from its eventual title, it proclaimed a *New astronomy based on causes or celestial physics* treated by means of *commentaries on the motions of the star Mars* (emphasis mine).

What had led Kepler to conclude he held the "key to a universal astronomy"? It was not the orbit of Mars. He knew at this time that the orbit of Mars was some kind of oval, but this was a problem that still needed working out. More fundamental had been the discovery that the earth's eccentricity needed to be halved and that its motion was governed by an equant. This and attendant discoveries about the orbits of Mars, Venus, and Mercury showed that the orbits of all planets were essentially the same. That they all turned out to be circles with bisected eccentricities (disregarding Mars's oval for the moment) was a startling confirmation of Kepler's conclusion in chapter 22 of the *Mysterium* that the equant was a mathematical model for the motion caused by the planet-moving force issuing forth from the sun. This discovery of the bisected form of the earth's orbit led into the investigation of the physical cause of all the planets' motions and became the foundation of the new astronomy, as is manifest in the title of part 3 of the *Astronomia nova*, "Investigation of the second inequality, that is, of the motions of the sun or earth, or the key to a deeper astronomy, wherein there is much on the physical causes of the motions."[27] It was here that one of his initial titles for the book, "The Key to a Universal Astronomy," was echoed, indicating the great significance of this section of the work.

Ironically, the net effect of the heirs' efforts to take control of the *Rudolphine Tables* and hurry their production along was that progress on the project ground to a halt, whereupon a new round of squabbles with Kepler erupted. The heirs were finally granted a down payment of 4,000 florins toward the money owed them in September 1603.[28] In the meantime, having been relieved of his responsibility for the *Rudolphine Tables*, Kepler busied himself with the other projects he had promised the emperor. In addition to the manuscript of the *Astronomia pars optica* (1604), he also prepared a Mars ephemeris and a transformation of Tycho's lunar tables to present to the emperor as a New Year's gift in January 1604. After that, he returned to the work on the commentaries on Mars. That Kepler was continuing his astronomical work with the benefit of Tycho's observations while not contributing to the production of the *Rudolphine Tables*—all the while profiting on the heirs rightful property by collecting honoraria—was annoyance enough. But it may have been the final straw when Kepler attempted to extort one-quarter of Tengnagel's 1,000-florin salary in exchange for his assistance on the *Rudolphine Tables*. Who had possession of the observations at this time is not clear, but it is clear that Tengnagel wanted to prevent Kepler from having access to them. Around February 1604, Kepler wrote to Fabricius:

> I cannot be certain of the situation of the Tychonics, because Tengnagel has sent me away. He is like the dog in the manger, who neither eats the hay himself nor allows others to do so. Every year he receives a thousand florins. Now he wants me to support his salary with my discoveries. I was willing, if he handed a quarter of his thousand to me, to appear before the emperor with all my things in our common names. But because he wants to keep this thousand to himself, I cannot also pledge myself to the emperor for this thousand, and I am compelled to look after my own salary myself, which I have also done by handing over the *Optics* [i.e., the *Astronomia pars optica*], a Mars ephemeris, and a transformation of the lunar tables on the first of January [1604]. Seeing this, he devised the pretext that I am overthrowing Tycho's teaching and that he did not wish to arm me with the observations. But the truer cause is this: he wants to hinder me so that he has time to work out something. For he openly declares that he hopes to put forth something. But he is very inconsistent in doing this, for immediately after he interjects that this is not his profession.
>
> I swear to you on holy faith, that I am doing nothing against him or his salary. I am doing this only so that I can have the observations which I desire, and secondly so that by recounting the truth I can defend myself against criminal rumors, wherever they are pointed out to me. They received a fifth of 20,000 florins. They hope for the rest. Tycho took a wife noble in character but meager in fortune.[29]

To be fair, Kepler probably saw his scheme to secure some compensation from Tengnagel as entirely justified. Initially, when Kepler was still responsible for the *Rudolphine Tables* and the terms of his employment as imperial mathematician were being negotiated, he had been asked to specify his terms. On the advice of friends with courtly experience, he had left the amount up to the emperor, although he had suggested half of Tycho's salary of 3,000 florins, promising that if the emperor saw fit to grant him Tycho's entire salary, he would then be well set up to hire assistants and pay for the counsel of other learned men.[30] Despite his suggestion, he received a salary of 500 florins.[31] From Kepler's "bid," it is clear that he understood that the salary granted by the emperor would include expenses related to the tables. With Tengnagel having usurped responsibility for the *Rudolphine Tables* and receiving 1,000 florins, Kepler certainly understood that sum to include expenses, to some of which he would surely have been entitled.

Tengnagel's pretext for barring Kepler from the observations was that Kepler was in effect debasing the family's rightful resource. They felt that by publishing his own work on the basis of Tycho's observations, Kepler was raising his prestige at the expense of Tycho's. The heirs were continually on guard to maintain the appearance that Kepler was merely elucidating Tycho's unfinished work rather than working independently on the basis of Tycho's observations. Even some twenty years later, when the *Rudolphine Tables* finally appeared, they insisted that the title page be reprinted to excise the comment that Kepler had "corrected [*emendavit*]" Tycho's observations.[32] Similarly, in 1604, what Kepler meant as a "transformation" [*transformatio*] of Tycho's lunar tables was taken by the heirs to be a "refutation" [*refutatio*], and Tengnagel forced him to rewrite the preface to remove the suggestion that he was refuting Tycho's lunar theory (see below). Publishing the *Rudolphine Tables* would have allowed the family to realize their return on Tycho's assets, but, as Kepler saw it at least, Tengnagel had neither the time nor the ability to pull it off.

The solution to the problem of Kepler debasing the family's resource in the meantime would be a legally binding agreement that would put Tengnagel in the position to control Kepler's publications until such time as the *Rudolphine Tables* were finished. Such an agreement followed in the wake of further bickering between the parties in the spring of 1604. Finally, the emperor assigned Pistorius to mediate the dispute.[33] The resulting two agreements reached on 4 July 1604, one each signed by Tengnagel and Kepler, came to be known as the Contract of 1604.[34] According to the terms of the agreements, Tengnagel conceded to allow Kepler free access to the observations. In return, he would be allowed to make use of any demonstrations or tables that Kepler prepared on the basis of the observations. Although he would be allowed to include them in the *Ru-*

dolphine Tables, Tengnagel was prohibited from altering Kepler's work in any way without his consent and was required to identify Kepler as the author of any such work. These conditions seem to have been aimed largely at the problems that plagued the printing of the *Progymnasmata*, both the failure to give Kepler credit for the sections he wrote and the annoying errors that arose from incompetent editing, which would have damaged Kepler's reputation had he gotten credit. For his part, in return for the use of Tycho's observations, Kepler agreed not to publish anything based on them without Tengnagel's advice and consent, and he specifically granted Tengnagel the right to stipulate changes. The agreement was a considerable burden to take on; Kepler consistently called it unfair. But he did have an out: the agreement was only valid until the publication of the *Rudolphine Tables*, after which Kepler would explicitly regain his freedom to publish what he wished. Based on an oral promise Tengnagel gave to the emperor around the same time, they would be published within four years.[35]

Elements of the Contract of 1604 point to the different motivation of the parties. Tengnagel's contract conceded to give Kepler credit for his work and not to alter it—a concession he was prepared to make to have the theoretical work done by Kepler, who was appointed at the same time as the *mathematicus* for the project—but then explicitly stated, "But I, as the son-in-law of the primary author, Lord Tycho Brahe, heir of his studies, and owner of the observations (which are the basis [of the work]), reserve for my judgment the form, dedication, and editing of the entire book."[36] The emphasis on form over content, and particularly the explicit concern with the dedication, reveal Tengnagel's intention to position himself to receive any and all of the resulting patronage. Kepler's contract reveals his concern for something of a different nature. He pledged, "I will set forth in public nothing of those things which pertain to the work *The Rudolphine Tables* and the restoration of astronomy, derived and demonstrated from those same observations and discoveries of the said Lord Brahe, or things which are built up by me from them, without advising or against Tengnagel's will, until the said tables are perfected and published, *leaving my philosophical freedom intact afterwards*" (emphasis mine).[37] Kepler was rather unconcerned with personal gain.[38] He focused more on pushing forward his program of physical astronomy, for which he would require his "philosophical freedom." As he explained to Christopher Heydon the following year:

> A great dispute [has arisen] between the heirs and me over [my possession of the observations], with me regarding them as a public interest, and they as their private property on account of the emperor's promise, which is only slowly being paid off. Still, the conflict was a shame, to which the false accusa-

tion of robbery clung, and so [I entered] into an unfair agreement that I would not publish any of my work based on them without the heirs' consent. And in the end, with this I brought it about that the bulk of the observations were left with me by the heirs' own free will. And thus according to this I am not claiming for myself the patronage of Tycho's night studies, but I am happily returning it to the shoulders of the noble sir, Franz Gansneb, called Tengnagel, from Westphalia, Tycho Brahe's son-in-law, who since he is one of the Councilors of Bohemian Appeals to the emperor, not reluctantly leaves the better part of the work to me. What each of us will accomplish, time will tell. Consequently, whereas I promised *Commentaries on the Motions of Mars* [i.e., the *Astronomia nova*]—having done so before this agreement of ours—now because of [the agreement], the situation has been brought about that unless [Tengnagel] gives his assent, and the emperor grants the [printing] expenses, they [the *Commentaries*] cannot be published. Still, something should be said about them for your sake. For what you expect of them—extremely correct positions of the star Mars—I scarcely care about compared with other things which I have attempted here and pursued by the grace of God.[39]

In this passage, Kepler makes a seamless transition from speaking about the dispute with the Tychonics, to the constraint the contract put on what would become the *Astronomia nova*, to an account of the revolutionary findings he wanted to publish. These findings would lead in turn to a physical argument in favor of Copernicanism. Tengnagel's right of censorship, however, threatened to stand in Kepler's way, for which reason, Kepler explicitly retained his right to his philosophical freedom once the *Rudolphine Tables* were published.

This jockeying over the right to exploit Tycho's observations had important implications for the composition of the *Astronomia nova*. Tengnagel perceived Kepler's work as a potential threat to Tycho's honor and, more tangibly, to the cash value of Tycho's legacy. It was in his interest to prevent Kepler from overstepping Tycho, from publicly disagreeing with Tycho, indeed, from carrying out recognizably original work. The very nature of Kepler's research, on the other hand, with its emphasis on a physical "proof" of the truth of heliocentrism, was highly unusual and original. And—to be fair to Tengnagel—the derivation of planetary theory from the physical analysis of the cause of the planets' motion had been something to which Tycho had objected vehemently. The situation surrounding the publication of the *Astronomia nova*, in which, first, he had been cornered into conceiving the work at all, and then, Tengnagel had contrived to obtain the right of censorship over any works based on Tycho's observations, therefore created an atmosphere in which Kepler had to be very careful about how his work was presented. In essence, in order to be allowed to publish at all, Kepler had to plan to present himself as blameless for overthrowing Tycho's legacy. It is from this that the apparently aimless narrative of the *Astronomia nova* ultimately derived.

Because Kepler had made astronomy a kind of religious avocation, there was far more at stake for him than merely the publication of astronomical results. In this respect, the *Astronomia nova* sprang from the same source as the *Mysterium cosmographicum*: the physical proof of heliocentrism was directed toward showing God's glory. Consequently, Kepler's "philosophical freedom" continued to preoccupy him through the end of 1604, especially when he became sick and depressed, and feared that he might die without having made his findings public. Perhaps morbidly reasoning that he could not be obligated to seek Tengnagel's permission to publish after his own death, he reached beyond his immediate imperial milieu to his Protestant "paternal academy" with a scheme to publish his work posthumously if necessary. In a letter of 12 December 1604 to the University of Tübingen, he explained that he was engaged in a commentary on the motion of Mars based on Tycho's observations, on which he had been working for five years. He did not know if he would live to resolve the remaining problems, but he had already found "so many and such things that astronomy can simply be regarded as new."[40] This mention is the first clue to the book's eventual title, *Astronomia nova*. Kepler continued:

> But I do not yet see an opportunity to publish, for I have come to an agreement with the noble lord Franz Tengnagel, his holy imperial majesty's counselor of Bohemian appeals, in his capacity as the late Tycho Brahe's son-in-law, that he will share the Brahean observations with me henceforth (for he keeps possession of them until the heirs receive another sixteen thousand florins from the emperor) but I am obliged to publish nothing without his consent before he himself completes the *Rudolphine Tables*, for which he is slowly preparing himself. And there is the danger that, just as there has repeatedly been contention with him up to now, so it will be also in the future, while he looks out for his father-in-law's honor and I for my philosophical freedom. In this case, faith in my discoveries compels me to take the precaution of depositing a copy of the *Commentaries* with my paternal academy, which would attend to their publication if it should befall me to die.[41]

It must have been but little comfort to Kepler when the university, apparently understanding the implications of his proposal and cautious about involving itself, expressed its sincere hope that he would not die, but agreed to publish what he had written only "in so far as this can be done without offense to his holy imperial majesty."[42]

After the noncooperation of his paternal university thwarted Kepler's plan for the unregulated posthumous publication of his work, he conceived of another plan to circulate his findings clandestinely. Within two weeks of the demurral from the University of Tübingen, Kepler wrote to Herwart von Hohenburg proposing the circulation of the book in manuscript:

> If he [Tengnagel] is going to be difficult, I will circulate handwritten copies (lest I should abandon the work because of death), if some outlay is arranged for the copies. But this is also to be kept secret, lest he should perhaps want to set himself against this as well.[43]

Nonetheless, Kepler must have started feeling better and more optimistic, for apart from Herwart's subsequent offer to pay for two manuscript copies,[44] there is no further discussion of the plan.

Longomontanus and the Tychonics' Influence on the *Astronomia nova*

Tengnagel's usurpation of responsibility for the *Rudolphine Tables* compelled Kepler to conceive of the *Astronomia nova* as a kind of preliminary announcement of his results treated in terms of the theory of Mars. In addition, Tengnagel's legal right to censor Kepler's writings most probably refocused Kepler's attention on the unpopularity of his new astronomy, and was thus partly responsible for the form that the *Astronomia nova* took. Criticism of Kepler's ideas was not confined to the Tychonics, as we will see in the following chapter, but they posed the most serious threat because of the power Tengnagel held over Kepler's publications. Indeed, it was in association with his difficulties with the Tychonics that Kepler's rhetorical strategy of justifying his work by appeal to its historical development first appeared. This occurred first around the beginning of 1605 in Kepler's response to a strongly worded letter of criticism from Longomontanus, the most able astronomer among the Tychonics.

Christian Severin Longomontanus [Christian Sørensen Longberg] (1562–1647) had been Tycho's astronomical heir apparent. Had he remained in Prague until Tycho's death, he rather than Kepler most probably would have inherited responsibility for Tycho's remaining work.[45] Coming from an extremely humble background, Longomontanus had entered Tycho's service on Hven in 1589 after working his way through the University of Copenhagen. He became Tycho's most highly esteemed assistant, doing much of the work on Tycho's lunar theory. After the closing of Uraniborg in 1597, Tycho and Longomontanus parted ways for a time, but Tycho felt increasingly in need of his seasoned talent. Tycho was able to recall him during the summer of 1598 while he was at Wandsburg (near Rostock) for a few more months, and finally got him to come to join him at Prague, at the beginning of 1600, about a month before Kepler arrived. In Prague, Longomontanus clearly assumed the role of senior assistant.[46] Shortly after Kepler arrived, Tycho assigned Kepler to relieve Longomontanus, who had been working on the theory of Mars, and moved Longomontanus on to a further revision of the lunar theory. After wrapping this up, Longomontanus finally left Tycho's service for good around 4 August 1600.[47]

The enmity between Kepler and Longomontanus arose partly from this division of labor. Longomontanus was clearly annoyed whenever Kepler strayed into the lunar theory during the course of his speculations. He expressed his disapproval of this for the first time in a letter of 3 August 1600, written when he was just preparing to leave Tycho for the last time.[48] As we saw above, while Kepler was away in Styria during the summer of 1600, he had turned his attention to the problem of the lunar theory and was approaching it on the basis of the physical cause of the moon's motion, much to Tycho's disapproval. Kepler must have shared some of his thoughts with Longomontanus in a now-lost letter of 20 June 1600, for in his response, Longomontanus enumerated a number of questions. His tone was that of a—perhaps somewhat exasperated—veteran advising an overenthusiastic beginner.

In this initial letter, Longomontanus ran through a number of contentious points in Kepler's lunar theory, pointing out his errors and deficiencies, ending with a statement that underlines their philosophical differences:

> Sixth and last, you should know I want nothing more willingly than that physical reasons be intertwined in the hypotheses. For this reason, if it should move me during a period of leisure—having kept this letter of yours for myself so long already—I will leave no stone unturned to understand and explore this universal [physical reason?]. For otherwise, as you know, we scoop up water with a sieve. For this reason I beg you all the more to issue forth your singular speculations or meditations at the proper time for astronomy, when I will sincerely say they will be useful. You should see to it that you do not set something up before you have a general idea of all the phenomena of each star reduced to a standard. And do not put forward as an excuse for [such an idea] either insufficiencies of the observations or the impossibility of having them.[49]

(Note Longomontanus's apparent taunting at Kepler's then-restricted access to Tycho's observations.) Although Longomontanus expressed his approval of addressing physical questions, he proposed doing this post facto, after the planetary theory had already been constructed. This was the standard approach of the mathematical astronomer. Kepler's triumph, as he would try to make clear in the *Astronomia nova*, lay in doing precisely the opposite.

Kepler's disagreement with Longomontanus flared up again as part of his conflict with the Tychonics. Part of the context of Longomontanus's second letter to Kepler is missing, because neither of Kepler's first letters to Longomontanus survives. It is clear, however, that Longomontanus held a grudge against Kepler dating back to the summer of 1600, and that the lunar theory was again the proximate cause of their clash. He had not forgotten Kepler's previous intrusion into the lunar theory nor his own attempt to dissuade him from this course. He renewed his complaint after he heard through Tengnagel about the transformation of lunar tables that Kepler had submitted to the emperor in 1603.[50] He apparently became infuriated when the note that Kepler sent him in February 1604 contained some "uncivil and false accusation" rather than the apology he had been awaiting.[51] Still, he maintained a veneer of calm, and sought first to get Kepler out of the lunar theory, and second to convince him to abandon his unorthodox research program.

As before, Longomontanus expressed sympathy with Kepler's attempt to reconcile physics and astronomy, although he also pointed out that Tycho would not have approved of Kepler's conception of this:

> I regard things that are more correct and more consonant to physics with pleasure, although, constrained by observations, I was partly unable to satisfy our desire in this respect, or to extend the work beyond the astronomer's duty. But in truth, those things that you judge to be very manifest in astronomical calculation as far as you are concerned will certainly scarcely accord with the mind of the renowned Lord Tycho.[52]

After reviewing some problems in the lunar theory, and challenging Kepler to tell how he would make common tables for all the planets even though the true phenomena had still not revealed themselves to him (had Kepler told him of his hopes for a unified and simplified astronomy?), he wrote:

> Therefore convince the ignorant, and desist from telling the intelligent absurd things any further. But let us not proceed farther into your Augean stable. Do you remember, my friend Kepler, that back then at Benatky the work was divided between us, namely that you attended to Mars while I attended to the moon under the direction of the great Tycho? Is perhaps either your lack of confidence or despair of Rheticus's experience[53] now the cause for your taking up the moon again (forgive me for knowing nothing that you have produced up to now in that respect)? The facts simply speak of it. But, I pray, remain where you are now for a little while, and regard with judicial eyes in all honesty with how much sweat we endured whatever obstacles when we set out on the path before you. For indeed, after the moon's mean motions had been corrected and taken all the way back to Ptolemy; after the inade-

quate equation of time had been endured and the reason of the true luminaries' syzygies had been found through the mean [motions]; after a simple equant had been applied to the lunar prosthaphaereses, and the anomaly of the latitude and nodes coming between syzygies and quadratures had likewise been sought, found, and attended to in the same way; and finally after having defined the ends of the separation and proportioned the semidiameters of the little circles, at last through a long and most difficult effort before all these things were known—since they had been considered by no one previously—we firmly established the hypothesis of the moon to conform to the ingenious collection of very many observations, certainly not to your proportions or motor virtues (however I certainly never shrank back from that beauty wherever the possibility was given). Rather, the phenomena of the moon themselves at whatever time must be drawn forth backward and forward from the same hypothesis using geometrical demonstrations. In [our hypothesis] they are duly balanced with observations to within scarcely 3 minutes, as we defied you ever to do. 28 eclipses, in addition to countless other observations with which our lunar hypothesis is in agreement, bear witness to this. Lunar phenomena observed afterwards and yet to be observed show likewise. These and perhaps all other things that were discovered and worked out by Tycho during his restoration of astronomy for our eternal benefit, you, my dear Kepler, although submerged in shit in the Augean stable of old, do not scruple to equal. And you promise your labor in cleansing them anew and even triumph, as if we should recognize you as Hercules reborn. But certainly no one does, and prefers you to such a man [i.e., Tycho], unless when all of it has been cleared away, he understands that you have substituted more appropriate things in the heaven and in the celestial appearances. For in this is victory for the astronomer to be seen, in this, triumph. On the other hand, I seriously doubt that such things can ever be presented by you. However, I am concerned lest this sordid insolence of yours defile the excellent opinion of all good and intelligent men about the late Tycho, and become offensive.[54]

After returning to the technical details, Longomontanus continued his tirade:

For this reason I demand that you undertake this fraud [*imposterum*] with demonstrations, as you ought, unless on account of your badly managed haste in writing to me and the obscurity of your mind you seek both the concealment of truth on such an occasion and perhaps at the same time contempt for me. But however affected toward me you might have been, candor will always be so important to me that, unless strongly provoked, I will never take such papers of yours to court. For this reason apologies and dissimulations [*retorsionibus*] will not be necessary for you in the least, but rather sincere and friendly feelings toward truth and justice directed everywhere, of which virtues Lord Tycho was always most attentive while he lived.

> I hear that you are no longer playing with ovals in order to construct from them hypotheses of the celestial motions for yourself alone. . . . I would like, therefore, under these circumstances finally to recall you from these things and moreover from anything of motor virtue or of divine proportion you might have resolved in each case—which you, by God, would put next to divinity in this profession—to circular motions (since anyone can make use of them), so that the celestial phenomena themselves suffer none of your force nor violence. For unless you desist from previous things and do not touch upon the Platonic music with which the world spirit is endowed,[55] you are going to leave those of us who are doubtful in suspicion, darkness, and error.[56]

The surviving manuscript of Longomontanus's letter shows that it made Kepler angry, for it contains Kepler's marginal notes, which provide a valuable indication of his private thoughts. Some of these merely protested his innocence. For instance, next to the crack about wanting to be recognized as Hercules reborn, Kepler defended himself, writing, "He ascribes these things to me with the clearest injustice."[57] But some of them also registered his contempt. Next to Longomontanus's remonstration that Kepler will defile Tycho's reputation, Kepler wrote, "Rave behind your skillfully constructed mask,"[58] portraying Longomontanus's concern as a studied affectation.

Despite their mutual contempt, Kepler had to be careful in his response, for Longomontanus's letter had presented a clear danger. To begin with, it was obvious that Longomontanus had little tolerance for Kepler and none for his physical astronomy. Longomontanus had not minced words in his warning that Kepler not attempt to convince others of his idiosyncratic research program. Furthermore, Kepler had learned that Longomontanus had been receiving information regarding his research program from Tengnagel. Kepler did not regard Tengnagel as a serious intellectual threat, but the news that they were exchanging information about Kepler's work raised the uncomfortable specter of a conspiracy among the Tychonics to prevent him from publishing his work. This prospect made an attempt to convince Longomontanus of the necessity of his research essential.

Kepler hoped to justify his work in an extraordinary letter written around the beginning of 1605.[59] In some respects, this letter represents the rhetorical structure of the *Astronomia nova* in miniature, in that Kepler appealed to a detailed history of his activities and hard work over the previous five years in order to justify the changes he was proposing in astronomical theory. Furthermore, this letter contained the first reference to the similar account Kepler was preparing in the *Astronomia nova*.

Longomontanus had attacked Kepler for his incursion into lunar theory on the basis of old information that he had received from Tengnagel before a dispute between Kepler and Tengnagel over the lunar theory had been resolved. Accordingly, Kepler's first task was to straighten Longomontanus out as to what had happened. His account of the resolution of their dispute is of particular interest because of the evidence it provides that Tengnagel was actively using his legal rights to censor Kepler's work. Kepler began by informing Longomontanus that the dispute had been resolved, likening Longomontanus's attack to a needless skirmish due to bad communications:

> Without doubt English and Portuguese ships coming upon each other in the Indies today engage each other at a distance with cannon fire in the same way, only to learn shortly thereafter that, with an armistice having been reached at negotiations, there is peace between their kings. And so I will respond to you peacefully, so that you understand there is peace. Nevertheless, I will speak openly, lest I should appear in a fawning spirit to forsake the truth.[60]

Kepler then proceeded to review the situation. Based on what he had heard from Tengnagel, Longomontanus had accused Kepler of an undue refutation of Tycho's lunar theory (in the development of which Longomontanus had played the major role). Not knowing what Tengnagel had written, Kepler was prepared to forgive Longomontanus:

> In fact, you should know that I have undertaken no such refutation. You could have learned this from my sheet, but you ignored this so that your intended indignation would seem just to me. It is one thing to transform, another to refute.[61]

Kepler went on to explain that his transformation of the lunar theory meant only that the same motion could be produced by means of different hypotheses, pointing out that Longomontanus himself had noted six different methods. He affirmed that the transformation was intended merely to simplify the calculation. The innocence of Kepler's explanation was mitigated somewhat when he noted that this had been accomplished at the expense of abandoning Tycho's hypothesis:

> Perhaps someone else will be able to produce the same thing while retaining the Tychonic hypothesis. I was not so lucky, it is true, and I admit to being inclined toward a physical form, which I embrace over others whenever my work makes it equivalent to the other.[62]

Apparently the dispute with Tengnagel stemmed from Kepler's abandoning Tycho's form of the hypothesis. Kepler continued:

> Perhaps something could be selected from the preface to Herwart [von Hohenburg], the Bavarian Chancellor, that would smack of refutation or of any accusation whatsoever, inasmuch as the hypothesis is not clearly discussed. . . . But in the first place, I changed that preface in such a way that it would satisfy Tengnagel. Besides, I said nothing of my own, but took things written to me by Herwart and others, as I can demonstrate at any time from the letters of whichever of the authors you please. But if you doubt that I spoke this way in the Preface, ask Tengnagel for it. There is in his possession such a copy (in Matthias [Seiffart's] hand, but without the slightest changes) of the kind I had written based on my own ideas, not of the kind modified based on Tengnagel's ideas. After you receive it, I will add those things that have been changed. Therefore there is nothing in my transformation that deserves to be called a "refutation."[63]

A brief examination of the prefatory letter to Herwart reveals the probable source of Kepler's conflict with the Tychonics. The letter Kepler referred to is apparently that of 22 April 1603.[64] The judgment of mathematicians regarding Tycho's lunar theory and tables, Kepler wrote, was that they were too complicated and far too tedious to compute. Upon reflection, Kepler noticed that much of this complication was wrought by Tycho's adherence to the methods of strict uniform motion employed by Copernicus. Kepler understood that the same Copernican methods for reproducing the motion of an eccentric with an equant (as he had pointed out in chapter 22 of the *Mysterium*) were represented in Tycho's lunar theory, and that returning to the Ptolemaic scheme would yield some simplification. In addition, Kepler noted that the Ptolemaic form was more amenable to his own physical explanation, as he would soon demonstrate in another work, certainly the *Astronomia nova*:

> And indeed whether a multitude of circles or a half-monthly approach and recession of the center of the orb toward the center of the earth moves something, it is clear that that double Copernican epicycle on an eccentric—which Tycho admired so much that he even introduced this element borrowed from the three superior planets into the theory of the moon—is nothing other than a Ptolemaic eccentric (in the case of the three superior planets), whose motion around some equant point raised above the center of the orb is uniform. I will shortly demonstrate this inequality of motion based on very consistent [*convenientissimam*] physical principles in another work, for which I laid the foundations in my *Prodromus* or *Mysterium cosmographicum*, and I will satisfy Copernicus's objections in full. In the same way, it is demonstrated here that that little circle, which Tycho made cross through the center of the earth, clearly coincides with the second Copernican lunar epicycle, and does not differ except in its position and a certain amount of magnitude, but in

> effect the motion is equal. Tycho himself pointed this out in the explication of his hypothesis. And so, in whole the theory of the moon, in so far as it concerns longitude, will consist of just as many orbs and centers as the number for the three superior planets in Ptolemy, except that in addition the moon's mean motion is also accelerated for a little while in conjunctions and oppositions, which Tycho called the "variation," which, in accordance with the physical cause advanced, does not require a special little circle for the speculation, since it derives from the very same monthly circuit.[65]

This passage described a wholesale recasting of Tycho's lunar theory based on Kepler's physical principles, a process that we have seen reflected also in Kepler's Mars work. Perhaps Kepler had innocently assumed that, since various hypotheses were geometrically equivalent, he could get away with this. It is difficult to make a precise judgment about what Tengnagel found offensive in this letter, since we are not sure whether it was precisely the same or what the amended version contained. That Tengnagel forced Kepler to change anything is significant, for it shows that Tengnagel was concerned not solely with retaining the Tychonic cosmology but also with the Tychonic form of theories. If he thought the reintroduction of the equant-based physical principles would be as damaging to Tycho's intent as abandoning geo-heliocentrism, then Kepler faced a serious obstacle to getting anything in the *Astronomia nova* published.

In his letter to Longomontanus, having argued that the transformation of the lunar tables did not represent a refutation, Kepler became conciliatory. He submitted that the charge of altering Tycho's lunar theory may have been true, and went so far as to thank Longomontanus for the things he brought up to correct him. Still, Kepler urged a unified consideration of planetary theory:

> Nevertheless, all of the planets are so mutually bound together, and the similarity in reasoning from one planet to the others is so strong, that this curiosity wholly merits your indulgence. . . . We are attending to a common labor. Let us not divide all of the labor between us, such that one of us is a maker of axes and the other of wagons.[66]

Kepler had not missed Longomontanus's critical comments regarding his "divine proportions" in the *Mysterium cosmographicum*, and he hastened to deny that he was engaged in such work:

> You suspect that I have revised something to correspond to my divine proportions (yet I do not remember having brought up any such thing in my letter to you). I did no such thing. I acknowledge that to a certain extent I did indeed accommodate the lunar hypothesis to the physical cause of the equations of the eccentric, whose nature I have definitely established in the case of the sun and Mars, but only so much as could be done while preserving the effect of

> the Tychonic hypotheses. But you talk about divine proportions, hinting at my *Mysterium cosmographicum.*
>
> And lo, you praise that work, unless you perhaps secretly brand me as a meddler, who would introduce physics into astronomy. You Tychonic astronomers, having justly stripped physics of the solidity of the orbs, wrongfully left the flight of the planets whirling about in the greatest unbelievability and confusion. Why should not I, on the other hand, assist them by inquiring into the physical forms of their motion through the transparent void? I confess to you, Christian, that during the last five years I have spent at least half of the time that remained for me after my cares at court in the physical contemplation of the motion of Mars. Truly, I think the sciences are intertwined in such a way that neither can be complete without the other. But I see that you are not particularly opposed to this.[67]

Despite this final comment, in which Kepler refers to Longomontanus's expressed willingness to see physical causes in astronomy (post facto), this passage indicates that Kepler knew precisely where the trouble with Longomontanus lay. He was anxious to convince Longomontanus of the necessity of physical contemplation in investigating the motions of the planets. At the same time, he was apparently distancing himself from the *Mysterium* by denying that he was pursuing "divine proportions."

Longomontanus's choice of the *Mysterium* as a target for his attacks may have contributed greatly to the contrast between the *Mysterium* and the *Astronomia nova.* Although the subject matter of the latter was already quite different, the kind of argument that Kepler was willing to use varied greatly between the two works. In the former, Kepler employed metaphysical, numerological, and astrological arguments to support the system of nested polyhedra, but the latter was much more ascetic in the number of extraneous considerations Kepler was willing to bring to bear. This was not because Kepler had abandoned that mode of argumentation. Although there were some aspects of the *Mysterium* that he was willing to disavow in the notes he added to the 1621 edition, much of the work remained valid for him. In addition, the *Harmonice mundi* (1619) was quite similar to the *Mysterium* in character. It is probable that the critical attitude of Longomontanus especially—as Longomontanus clearly did not like the *Mysterium* and could bring pressure to bear through Tengnagel—was responsible for Kepler reining in his speculation in the *Astronomia nova.* Much was at stake in the *Astronomia nova* for establishing the necessity of physical astronomy, and Kepler was not willing to put those arguments in jeopardy for the sake of his other speculations.

We should also note that in the passage above, Kepler made a rhetorical appeal to the sheer amount of time that he had devoted to the physical contemplation of Mars's orbit over the previous five years. Longomontanus had employed a similar technique in his recitation of the laborious steps he and Tycho had gone through in the development of the lunar theory. Kepler continued this line, taking Longomontanus's reference to Rheticus's frustration with the theory of Mars as his cue:

> Now you also mock me with the example of Rheticus. I will laugh with you. Your moon has certainly tormented you miserably, whenever I was a spectator, for as long as I know. Having endured the same thing, you ought to be moved to sympathy if my Mars troubles me. But still, let me give you an account of my time, since you seem to demand it.[68]

Kepler then proceeded with an account of his activities from the beginning of 1600, which remains one of the best biographical resources for this period of his life. He recounted his work with Tycho, Tycho's death, his appointment as imperial mathematician, the conflict with Tengnagel that culminated in Kepler's being commanded to name the works he would undertake, his work on the *Astronomia pars optica*, and finally the state of the Mars work:

> Now see what the result is. The oval path of Mars through the aetherial air is definite, although the cause of the ovalness is uncertain. Yet up to now I had assigned a cause from which it followed that Mars's ingression from a circle was 1300 parts of 152,500, but when I began to elucidate the ingression, I found it was not larger than 800 or 900 parts. All the same, I have included in the 51 chapters everything that I have investigated. If I die, I know all these things will be highly useful for the advancement [of astronomy]. The essence is this: Mars is itself librated along the diameter of an epicycle, and it is driven by a virtue dispersed from the sun into the world. Each motion is nonuniform. The libration is intended and remitted not so much by the regulation of the two circles, as in Copernicus, but those circles themselves are intended and remitted according to the size of the waxing and waning of the solar diameter. The daily motion is larger at perihelion than at aphelion by as much as the appearance of the solar disc is larger at the former than the latter, or, a planet remains in inverse proportion so much longer in one degree of coequated anomaly at aphelion than at perihelion. But the measure of how much it is swept along is the amount of light that has reached from the sun to the planet at whatever distance, namely the size of the disc of the sun's body. *The way in which I, colliding against a thousand walls, was compelled to follow this path alone is described in those 51 chapters*. This is simply certain: a force is propagated from the sun which drives the planets. The other things are uncertain only to the extent that one parallel after another still brings me continually closer to my goal. You have an account of my progress. Apart from you, it may be that no one has seen these *Commentaries* except in my presence. You also have a lengthy account of my time.[69]

Twice in this passage, Kepler referred to the fifty-one chapters he had already composed. The first reference is similar to Kepler's appeal to the University of Tübingen, namely that an account of Kepler's work would be valuable should he die. The second reference is more telling. After describing how the motion of Mars was determined by two nonuniform motions depending on the distance between Mars and the Sun, Kepler asserted that the fifty-one chapters detailed how *that line of research alone* was successful. This was in essence the argument of the *Astronomia nova*. To be sure, if Kepler already had fifty-one chapters of this narrative written, he did not develop the rhetorical form specifically in response to Longomontanus's letter. However, it was in this context that it made its first appearance, and its use demonstrates that Kepler thought that it had utility in mitigating his responsibility for the changes he was proposing to astronomical methodology, and thus to lessen the chances for interference with his publication from the Tychonics.

Kepler was even able to twist Longomontanus's use of the Augean stable around to support his rhetoric of hard work:

> The tone of your reference to my Augean stable sticks in my mind. I entreat you to avoid chicanery, which is wont to be used frequently with regard to unpopular things. So that you might see that I have in mind how the Augean stable provided me with the certain conviction that I have not discredited astronomy—although you can gather it from my present letter—I will use it with the greatest possible justification. But it is to be used as an analogy, not for those things that you or Tycho were responsible for constructing—which either blinded by rage or perverted by malice you quite wrongfully attribute to me—but rather in the comparison of the ancient hypotheses with my oval path. You discredit my oval path. I hold up to you the hundred-times-more-absurd spirals[70] of the ancients (which Tycho imitated by not setting up anything new but letting the old things remain). If you are angry that I cannot eliminate the oval path, how much more ought you to be angry with the spirals, which I abolished. It is as though I have sinned with the oval I left behind, even though to you all the rest of the ancients do not sin with so many spirals. This is like being punished for leaving behind one barrow full of shit although I have cleaned the rest of the Augean stable. Or in your sense, you repudiate my oval as one wagon of manure while you tolerate the spirals which are the whole stable, to the extent that my oval is one wagon. But it is unpleasant to tarry in rebutting this most manifest slander.[71]

A rhetoric of hard work founded on the sheer amount of time that Kepler had spent pursuing his physical astronomy was of course useless without the confirmation of observational accuracy. Longomontanus had said as much when he pointed out to Kepler that no one would prefer his work to Tycho's "unless when all of it has been cleared away, he understands that you have substituted more appropriate things in the heaven and in the celestial appearances." Furthermore, Longomontanus had thrown down the observational gauntlet when he jeered that he doubted Kepler would ever achieve the 3′ accuracy realized in the lunar theory. Kepler noted this. When bringing his letter to a close, he added:

> You warn me off seeking physical causes in this way, lest in the course of this the heaven suffer violence at my hands. My dear Christian, had I wanted to leave eight minutes in doubt, I could have refrained from three times the great labor of the whole of 1604. And so you should know I have done meticulous work in order to agree with the observations to a hair. Had I not done this, I would not already have attempted fully twenty ways of [calculating] the physical equations of the eccentric. So defer your judgment until you have seen my groundwork yourself.[72]

We have seen above that Kepler had already been using a standard of about 2′ accuracy in his work. However, the rhetorical value of Kepler's 8′ stumbling block should not be overlooked. If historians' celebration of this figure can be taken as a measure of its rhetorical success, then it was received exactly as Kepler wished. Note also that Kepler raised his twenty different, insufficient methods of calculating the physical equations of the eccentric. This would also be put to good use in the *Astronomia nova*, in what might be called Kepler's "rhetoric of failure."

Conclusion

That Tengnagel's machinations forced Kepler into publishing his new astronomy in a kind of preliminary form as a commentary on the motion of Mars is clear. The extent to which Kepler's effort to circumvent Tengnagel's editorial control shaped its content is more difficult to establish. The book owes much of the nature of its argument to the resistance of astronomers to Kepler's approach generally, as will be established much more fully in the following chapter with regard to David Fabricius. However, it is clear from Kepler's exchange with Longomontanus that the threat of the Tychonics' interference would have been particularly important. Tengnagel had already used his power to influence at least the presentation, if not the content, of Kepler's work on the lunar theory.

The existence of Tengnagel's editorial power may also shed some light on Kepler's decision to present his findings in the *Astronomia nova* in all three major forms of planetary theory: Ptolemaic, Tychonic, and Copernican. Reiterating all of the proofs in the first twenty-six chapters in all three forms was cumbersome, and certainly did not contribute to Kepler's advocacy of Copernicus, unless we suppose that Kepler might have conceived of it as a means of lulling his readers until the necessity of the Copernican view became apparent. Kepler mentioned in chapter 6 of the *Astronomia nova* that this presentation was Tycho's idea:

> I, in the demonstrations that follow, shall link together all three authors' forms. For Tycho, too, whenever I suggested this, answered that he was about to do this on his own initiative even if I had kept silent (and he would have done it had he survived), and on his death bed asked me, whom he knew to be of the Copernican persuasion, that I demonstrate everything in his hypothesis.[73]

On the basis of this comment, Kepler's decision to use all three forms has been seen as an expression of his obligation to Tycho by carrying out his deathbed plea.[74] It does seem odd, though, that Tycho would have decided on the threefold presentation of his results independently of Kepler, as Kepler described. On the other hand, a letter to Fabricius hints that this agreement between Tycho and Kepler might have been more formal than a last-minute deathbed plea: "The work will be presented, Fabricius, as had already been agreed with Tycho, in such a way that all demonstrations will be expounded in the three forms of the hypothesis."[75] If this agreement were widely known, with Tengnagel's control over Kepler's publications, a purely Copernican exposition would have been out of the question.

In any case, Kepler did recognize that, despite the threefold presentation, he was vulnerable to Tengnagel's interference with respect to his physical ideas, especially because they were in conflict with Tycho's ideas, which Tengnagel was keen to defend. Kepler expressed this fear in a letter to Herwart from around the same time as the letter to Longomontanus:

> There are now 51 chapters. I explain all of my attempts in order that it be all the more certain why I have followed this particular path. I am deeply involved in investigating physical causes. My aim in this is that I describe the celestial machine to be not like a divine being but like a clock (he who believes a clock is alive bestows the artisan's glory upon the work), in that nearly every kind of motion comes from one very simple corporeal magnetic force, just as in a clock all motion comes from a very simple weight. And I show how this physical reason is designated in numbers and geometry. . . . Tycho rejected the orbs. I now show how the planets are moved without orbs, how the eccentricity comes about, etc. Yet I cannot publish without Tengnagel's consent. For perhaps Your Eminence heard from Lord Pistorius about our agreement and mutual obligation. . . .
>
> The astronomical things are set forth in the triple form of the hypotheses [i.e., Ptolemy, Copernicus, Brahe], but the physical things to some extent conflict directly with the Tychonic form. And I would not want to bear Tengnagel

himself as a censor, and thus to say something publicly that I believe differently in private.[76]

The juxtaposition of Kepler's strategy to recount all of his attempts and the threat of Tengnagel's interference suggests that Kepler might also have hoped that presenting his findings in the three major systems might mollify Tengnagel somewhat. But Kepler clearly understood that presenting his results in all three forms would only partially satisfy Tengnagel, for the core of the new astronomy—the consideration of physical causes—made sense only in the heliocentric form. It is perhaps for this reason that, in the same letter, he describes the strategy to circumvent Tengnagel's editorial power by circulating the book in manuscript (described above).

Still, Kepler had reason to be optimistic. Even though Tengnagel had consolidated his power at court by converting to Catholicism (for which Kepler perhaps acidly passed on his congratulations),[77] Kepler saw this as an indication that Tengnagel would be less interested in exploiting Tycho's estate: "All these things give me confidence, that astronomy will be returned to my shoulders, and he will not at all want to interfere with my publications."[78]

When the completion of the *Astronomia nova* was at last in sight, Kepler made one final attempt to wriggle out of his obligation under the contract of 1604. His hope was based on the long passage of time between the signing of the contracts and the publication of the *Astronomia nova*. Kepler presented an unfinished manuscript of the *Astronomia nova* to the emperor as early as Christmas 1604, before the final piece of the puzzle, Mars's elliptical orbit, finally fell into place around Easter of the following year.[79] This early presentation highlights, of course, that that discovery was not essential to Kepler's plan to publish. But it also suggests that Kepler anticipated a delay in the granting of printing costs, for at the same time he requested funds so that the book could be published. He declared the book ready for publication in August 1605. Although the final ten chapters were not completed until 1606,[80] receiving Tengnagel's permission and obtaining printing costs from the emperor were the biggest remaining obstacles.[81] A year and a half after the initial presentation of the manuscript, Kepler had still received no funds from the imperial treasury, and he asked the emperor at least for permission to seek another patron for the work.[82] Six months later, in December 1606, he petitioned again, begging for printing costs.[83] At the end of 1606, the emperor finally ordered that Kepler be given 400 florins for the printing of the *Astronomia nova*—half of what Kepler had been hoping for—and even then, the order does not seem to have been fulfilled until 28 March 1607.[84] At the same time, the emperor reserved for himself complete discretion over the distribution of the copies. Kepler had long received no regular payment of his salary and he was forced to spend much of the money on emergency household expenses. The rest was spent to have the woodblocks for the illustrations cut during the summer of 1607.[85]

The extraordinary delay in publishing the *Astronomia nova* now looked as though it might work in Kepler's favor. In 1604, when he had signed the contract with Tengnagel, Tengnagel had said that he would finish the *Rudolphine Tables* in four years. In the meantime, Tengnagel had become so occupied at court that it seemed as if he would never complete the tables. In the summer of 1607, Kepler looked forward to the fact that in 1608 four years would have elapsed, at which point he hoped to declare the contract void and recapture his philosophical freedom, as he maintained in a letter to their arbiter, Pistorius:

> I do not yet have permission from Tengnagel to publish [the *Commentaries on Mars*]. . . . Tengnagel finally seems to have given up hope of [undertaking] the *Rudolphine Tables*. I will not while I am still alive. If Tengnagel prevents me from divulging anything because of our contracts I will expressly state that I have been deceived, since he himself promised to do the *Rudolphine Tables* within four years. Although ill-disposed toward this promise which was made to the emperor, I promised to wait until those tables appeared, after which my philosophical freedom would remain intact. Now he will never write any tables, therefore our contracts are invalid if they do not include the four-year condition. It will be sufficient that those four years were specified aloud [*nominatos*] to the emperor.[86]

Kepler's hope to escape a contract with a nobleman and imperial councilor on the basis of an oral agreement was probably naïve. Still, Kepler had taken one other precaution to gird himself for the conflict with Tengnagel. When he had petitioned for printing costs in December 1606, he had also asked for an imperial order for the *Astronomia nova* to be published, which he might use to overrule Tengnagel if he tried to interfere with publication.[87] Whether because Kepler used these measures to give himself some legal leverage or not, the parties entered into negotiations in the summer of 1607 over the extent of changes to the *Astronomia nova*, as Kepler reported in a letter to David Fabricius, "Tengnagel seems about to yield, if I permit him to administer corrections, which is hard for me to bear."[88]

In the end, because of Kepler's maneuvering and Tengnagel's courtly preoccupations, the much-feared threat to Kepler's philosophical freedom left only a small overt trace on the *Astronomia nova*. Tengnagel apparently did not exercise his rights beyond attaching a short preface to the *Astronomia nova*, whose harried tone suggests that Kepler gave him as little time to work on it as he could get away with. The date of Tengnagel's "Preface to the Reader" is unknown, but his mention of rebellion and wars suggests the summer of 1608, in the aftermath of Archduke Matthias's march on Bohemia that spring, during which under force of arms he exacted from Emperor Rudolf II the crowns of Hungary, Austria, and Moravia, and the right of succession in Bohemia.[89] Tengnagel also complained in the preface of Kepler's imminent departure for Frankfurt, which was planned for August 1608.[90] Kepler did not actually travel to Frankfurt until April 1609,[91] but he does not seem to have relieved Tengnagel of the pressure to compose the preface as quickly as possible, leaving him a "little scrap of an occasion to write," during which time Tengnagel was also trying to navigate rapidly shifting political seas (shortly thereafter, he left Rudolf to serve Archbishop Leopold of Passau).[92] From internal evidence, there is every indication that Kepler dumped the manuscript on Tengnagel, leaving as little time as he could get away with.

Still, despite his haste, Tengnagel took the opportunity to air his dim view of Kepler's work. In its entirety, his "Preface to Reader" reads:

> I had decided to say many things to you (reader), but the heap of political business, by which I am detained more than usual these days, and the overhasty departure of our Kepler, at this very moment about to go to Frankfurt, has hardly left me even this little scrap of an occasion to write. And so I thought I should give you just three words' warning, lest you be moved by anything of Kepler's, but especially his liberty in disagreeing with Brahe in physical arguments, which has groundlessly complicated the work on the *Rudolphine Tables*. But such liberties are habitual to all philosophers from the creation of the universe to the present. As for the rest, you will come to know from the work itself that it has been constructed upon Brahe's foundation, that is, upon his restitution of the fixed stars and the sun, and that all the heaped up materials (observations, that is) were the works of Brahe. Meanwhile, by this important work of Kepler's, appearing among these tumults of rebellion and wars sprouting again from time to time, while the Republic of letters shows compassion for it all, it is as if we were by this fact blessed with a precursor of the Tables, and after them the observations (seeing the light of day later); and, more zealous for the further progress of the work so fervently desired, also pray with us to God for more felicitous times.
>
> [signed] *Franz Gansneb Tengnagel von Campp.*
> His Imperial Majesty's Councillor[93]

As Donahue points out, the incoherent and jumbled syntax in the original Latin itself makes plain Tengnagel's extreme haste in writing, his incoherence producing "a comical ending to a profoundly aggravating situation."[94] Even so, Tengnagel did make his point, that the reader should not be moved by *anything* Kepler wrote, and especially his impertinence in disagreeing with Tycho in terms of physical arguments, and that Kepler had merely complicated and delayed the important work of the *Rudol-*

phine Tables. At the same time, Tengnagel asserted Tycho's essential role in the undertaking, taking pains to emphasize that this work was based on a Tychonic foundation and that all of Kepler's structural elements were Tycho's as well.

Tengnagel's turn from cashing in on his astronomical inheritance to insinuating himself at the imperial court took the pressure off Kepler as the publication of the *Astronomia nova* finally began. Still, the book retained a rhetorical structure that was influenced by their struggle. During its composition, Kepler had been acutely aware of the danger posed by Tengnagel's right of censorship, and he had sought without success to find ways to escape his legal obligation. Working within the context Tengnagel had engineered, Kepler's response was to seek a mode of exposition that would make his findings public while mitigating his responsibility for the revolutionary changes in astronomy that he was proposing. His solution was a pseudohistorical narrative that depicted him as compelled to follow the path that he had, one that would at the same time reveal the technical challenges he faced and make it appear that he had followed the path into a physical astronomy reluctantly. The Tychonic context thus explains two unusual features of the *Astronomia nova*. The first is Kepler's decision to present his new astronomy based on physical causes in terms of a commentary on the motion of a single planet, Mars. This decision is quite simply explained by the constraints under which he was working, starting with his limited access to the data under Tycho himself and culminating in the snap requirement to present his findings to justify his employment as imperial mathematician. The other feature is the book's wandering narrative. This choice was certainly partly determined by the necessity to shield himself against the Tychonics' accusations that his idiosyncratic approach to astronomical theory was a disrespectful affront to the memory of Tycho Brahe. His departure from Tycho had to be meticulously justified and his own theoretical thrust held in abeyance until he had done so. Although the general features of this narrative correspond to Kepler's positioning of himself vis-à-vis the Tychonics, Kepler's audience shared many of the Tychonics' suspicions; therefore, the narrative was also directed at the astronomical community at large. In addition, it had a further utility: it could be tailored quite specifically to address particular technical issues. We will examine this more specific influence on the composition of the *Astronomia nova* in the next chapter.

⌘ CHAPTER 8 ⌘

DAVID FABRICIUS

Kepler's interaction with the Tychonics determined the subject and influenced the tone of the *Astronomia nova*. But this broader context goes only so far in explaining the *Astronomia nova*'s unusual narrative; it can account for the subject and flavor but not the content of the work. A far more concrete and specific account emerges from analyzing the work within the context of the astronomical community. Focusing on the intellectual and technical objections voiced during the composition of the *Astronomia nova* elucidates Kepler's decisions to include particular material far more clearly. This tailoring of Kepler's presentation required some foreknowledge of the likely reaction of fellow astronomers. While he was composing the *Astronomia nova*, his most significant correspondent in this regard was David Fabricius. During the time Kepler was experiencing difficulties with the Tychonics Tengnagel and Longomontanus, and when even his beloved teacher Maestlin had mysteriously ceased writing in a huff,[1] Kepler maintained a voluminous and wide-ranging correspondence with Fabricius, an East Frisian preacher.[2] Like Kepler, Fabricius had served as an assistant to Tycho for a couple of weeks in Prague, although their paths never crossed; Fabricius worked for Tycho during the summer of 1601 while Kepler was in Graz. If it were not for the short amount of time he spent with Tycho, one might justifiably call Fabricius himself a Tychonic.[3]

Although he made some contribution of his own to astronomy (he discovered the first variable star, Mira Ceti, and was involved in the early telescopic observation of sunspots), Fabricius's importance in the history of science arises from his close relationship with Kepler. Kepler found him a sympathetic and like-minded soul, and they discussed a wide range of astronomical, astrological and natural philosophical issues at great length. Fabricius was, in fact, an even more avid correspondent than Kepler had time for, and Kepler developed the habit of answering five or more letters' worth of Fabricius's queries at one time. Kepler's first statement that the orbit of Mars is an ellipse was embedded in the midst of one such letter, which runs a full forty pages in the modern edition of Kepler's works. The fact that Kepler's replies survive as his drafts make it very likely that he may have gone back through their correspondence to isolate troublesome issues while composing the *Astronomia nova*.

There is evidence that Kepler did just this. There are numerous marginal notes that refer to the composition of the *Astronomia nova*, one of which, for example, reads "N.B. raise [this matter] in the *Commentaries*."[4] At the very least, Kepler recognized the usefulness of the correspondence as a spur to thought; he often worked through problems while writing. He recognized this, as well as acknowledging the accumulating weight of Fabricius's queries:

> I do not write letters to you, dear Fabricius, but the *Commentaries* [themselves], which I put off writing until so much time has elapsed that even I now scarcely grasp the sense of what I have written, unless I reread it carefully.[5] For this reason it is unpleasant, so I do not finish writing. For what you say (in your letter dated 3 April, which I received today 4 June) is true: the work of replying piles up for me, it does not diminish, while one of your letters comes on top of another. Having received one today, I returned to the process of responding.[6]

Fabricius's Influence on the *Astronomia nova*

Because Kepler was more open with Fabricius than with any of his other correspondents, his letters to the Frisian astronomer are a great resource for studying the development of his thought. But we will be concerned here as much, and perhaps more, with the other, less-studied side of the correspondence. Fabricius was as able an astronomer as almost any of Kepler's correspondents—Kepler called him the greatest observer after Tycho[7]—and because of his irrepressible curiosity and candor, it is from him that Kepler heard critiques of his work that were representative of those of the astronomical community. Thus for the purposes of this study, we find in Fabricius a rare resource for studying the reception of Kepler's work during its developmental stages. In Fabricius, Kepler found a test case for the acceptability of his new astronomy. Because there is a distinct correlation between the objections Fabricius raised and the type of demonstrations Kepler offered in the *Astronomia nova*, we must also consider their correspondence an extremely important influence on the rhetorical structure of that book.

Fabricius's good-natured resistance to Kepler's work was evident from the very beginning of their correspondence. Fabricius had occasion to write to Kepler after a visit he received around March 1602 from the Tychonics Tengnagel and Eriksen. Fabricius heard from them about Tycho's death and Kepler's appointment as imperial mathematician, and hastened to write lauding Kepler and asking to be kept informed of his progress:

> Most distinguished sir, M[aster] Kepler, more than a Uranian friend,
>
> I want you to know that in these days the singular alumni of the Tychonian Urania, the noble sir, Franz Tengnagel, G[ent]. and Johannes Eriksen of Hamburg, have come to us in East Frisia. With their most happy arrival, they made me faint again at the untimely death of the late great Tycho, in such a way that, carried off in a kind of ecstasy, I believed I was not here but rather conversing with you in Prague. They most honorably made mention of you and your sincere, heartfelt emotion at the interment of Lord Tycho, and they proclaimed before me in more than superlative words your tireless zeal in pushing forward and preparing the triumph of Urania. From this declaration I drew as much hope and joy in turn as I had felt sorrow and melancholy before over Lord Tycho's death. And so, my most agreeable Kepler, see to it that in addition to Lord Tycho's expectation, you strive to live up to the commendation of these Uranian visitors, and finally to the sincerest hope we newly formed of you. Strive also to carry on to your utmost the course that has been begun, which will undoubtedly render immortal glory. I pray, support noble sir Franz's Herculean effort, put the common good first, and [together] under Tychonian arms lead the exiled Urania back into her ancestral kingdom. I pray also for the sake of our Uranian fraternity that you do not disdain to share your thoughts on Uranian matters with me quite often through letters.[8]

This letter interestingly foreshadows Kepler's conflict with the Tychonics. Tengnagel and Eriksen had left Prague in the summer of 1601 and were probably then returning for the first time since Tycho's death. They knew of Kepler's appointment as imperial mathematician, and yet evidently—since Fabricius speaks of "Sir Franz's noble effort" and exhorts Kepler to "put the common good first"—Tengnagel had informed Fabricius of his intention to carry on Tycho's work.

Despite the laudatory beginning, in this introductory letter Fabricius already questioned Kepler's work. Fabricius undoubtedly discussed Kepler's research with Tengnagel and Eriksen during the course of their visit, although he might have learned some of what he knew when he visited Tycho in the summer of 1601. Fabricius's interest was certainly stimulated by Kepler's curious findings, for he mentioned that he had just begun his own investigation of Mars's motion using his own observations and he mentioned a series of problems—"Whether one and the same eccentricity comes forth by means of calculation from different acronychals, whether Mars directs its course with respect to the mean or true sun, and finally what is the cause that the acronychal latitudes of Mars are not in the same circle"[9]—all of which are apparently related to Kepler's research. Fabricius's parallel researches on the motion of Mars certainly made him a most suitable sounding board for Kepler's ideas.

Fabricius's initial objections to Kepler's work had to do mostly with Kepler's innovations with regard to the solar theory. Fabricius knew that Kepler's use of the true rather than the mean sun in the theory of Mars, as well as his bisection of the earth's eccentricity, would cause a serious upheaval in Tycho's solar theory. He mentioned that he learned from Eriksen about Kepler's method of determining the elements of the earth's orbit from the parallax to the same position of Mars from three locations of the earth (as would be described in chapter 24 of the *Astronomia nova*). He questioned Kepler's method, and asked him to send an example of it. Furthermore, Fabricius questioned whether the earth-sun distances, which were Kepler's reason for bisecting the earth's eccentricity, were really responsible for the annual variation in Mars's orbit, and he mentioned that his own observations had shown that the sun could by no means be the center of Mars's eccentric. Fabricius's volley of questions, along with his disclaiming request—"But perhaps we have not sufficiently followed your reasoning. For that reason I desire a more extensive explanation of the causes etc."[10]—set the tenor for his subsequent correspondence with Kepler. Fabricius was ever good-natured and polite but tenacious in questioning every aspect of Kepler's innovations.

From the beginning, Kepler took pains to justify his work to Fabricius. His first letter to Fabricius of 18 July 1602 has been lost,[11] although we can infer from Fabricius's response that Kepler either attempted to quiet Fabricius's objections or at least welcomed the discussion, for he opened the floodgates to a further elaboration of Fabricius's objections in subsequent letters. We can also infer from one of Fabricius's later comments that, even with his initial letter, Kepler was perhaps sounding out where Fabricius stood with regard to the Tychonics, for Fabricius wrote back, "You write that I was perhaps put up to praising you by Lord Tengnagel. This would be unlike me, sir Kepler, and I would not want you to believe any such thing about me."[12] Regardless of what Kepler might have said in his initial letter, he sought to seal his confidence with Fabricius when he signed off his next letter, "Swear by your beloved Urania to keep these things to yourself, and do not expose me to the Tychonics and so beget trouble for me."[13] Fabricius hastened to assure him that their correspondence would remain confidential, and mutual secrecy remained their understanding from then on.[14]

We mentioned above that the kind of objections that Fabricius raised against Kepler's work clearly influenced the content of the *Astronomia nova*. Let us consider an example from early in their correspondence. Fabricius wrote in a letter of 28 April 1602 (o.s.) that he was himself assiduously investigating the question of Mars's eccentricity and was troubled by the wide variety of values of the parameters different methods produced. He finally exclaimed:

> And I am especially amazed that the eccentricity of Mars taken from Tycho could provide acronychal prosthaphaereses so close to the truth even though the eccentricity is not true. I ask you to explain this question to me skillfully, and lead me out of this labyrinth. I have nothing here that I can consult. You are as Theseus to me, otherwise I shall not easily extricate myself.[15]

Kepler conflated this question with an inquiry from a later letter about how his method of determining the eccentricity from four acronychals always produced the same eccentricity,[16] and addressed the question of "how four acronychals always produce the same commensuration and yet that it is false."[17] Kepler's demonstration in his letter to Fabricius, although garbled by a bad copyist, is clearly carried out on the same basis as chapter 21 of the *Astronomia nova*, titled "Why, and to what extent, may a false hypothesis yield the truth?"[18]

Another example of the influence of Fabricius's objections on the form of the *Astronomia nova* is the set of worked examples that Kepler included in the book, notably in chapters 16 and 26 to 28. Kepler's motive for giving these in such detail was almost certainly Fabricius's specific requests for such worked examples. For instance, Fabricius was baffled by Kepler's method for solving the vicarious hypothesis by means of the "quadrate *regula falsi*":[19]

> I ask most of all that you briefly demonstrate your method of inquiry for four acronychals (e.g., in an example) and disclose [it] in a numerical transcript through a sequence of the solution of the triangle, so that in this way I too, following your tracks, can also see and examine the same truth. For I am extremely interested in the investigation of these matters. You indeed showed the angles by means of letters in the diagram you laid out [in the lost letter of 18 July 1602], [but] you did not add the true method of calculating by means of the quadrate *regula falsi*. Nor are there any mathematical books now at hand from which I can learn the method for the sake of my exiled Urania (alas!). Consequently, *if you want faith bestowed on your thoughts*, see to it that I come to know your method.[20]

Kepler provided a detailed explanation of his method of solving the vicarious hypothesis in a letter to Fabricius,[21] but the extensive worked example in chapter 16 of the *Astronomia nova* also corresponds precisely to this request.

Similarly, after attacking Kepler's bisection of the earth's eccentricity, Fabricius allowed that he may not have understood Kepler's method of triangulating from Mars's position to the orbit of the earth and requested a worked example:

> However it can be (in view of which I will nevertheless bring this up as well) that my procedure for finding the eccentricity of the sun from three annual

> parallaxes to the same location of Mars on the eccentric is different from your method, and also that the true distances of Mars are not given in this way. For this reason I desire that you now send a single such example computed by you or at least a diagram of the demonstration and [an outline] of the whole declaration, so I can compare it with my method. You will also do me a favor in this. For a comparison of the investigation brings truth more completely to light, or perhaps I do not understand your mind sufficiently.[22]

Again, Kepler provided a detailed explanation in a letter,[23] but the example corresponding to this request found its way into the *Astronomia nova* in chapters 26 and 27 as well.

In both cases, Fabricius made it clear to Kepler that his consent to Kepler's innovations could be obtained only if Kepler gave him a worked example, and in both cases these worked examples also found their way into the *Astronomia nova*. Given the specificity of the correlation between Fabricius's requests and the content of the *Astronomia nova*, Fabricius's conditional acceptance of Kepler's findings must be seen as a major influence on Kepler's rhetorical argument in the *Astronomia nova*.[24] In general, Kepler took account of where in their correspondence Fabricius expressed his confusion or requested clarification and tailored the presentation of his results in the *Astronomia nova* to avoid the same confusion in his readers. In this way, Fabricius acted as a representative member of the astronomical community, expressing what demonstrations would be necessary to convince him of the truth of Kepler's findings. One might call this an early instance of peer review.

Fabricius's objections thus shed light on Kepler's use of a variety of different demonstrations, in chapters 22 through 29 of the *Astronomia nova*, that the eccentricity of the earth needs to bisected. This redundancy has been noted by historians of science. Koyré attributed it to Kepler's appreciation of the importance of this finding.[25] Wilson attributed it to the great uncertainties in the values of the distances derived, and suggested that Kepler was merely establishing a range of values around the value of half of Tycho's eccentricity, which he finally assumed.[26] Both these interpretations are certainly true and arguable, but neither takes into account the immediate context of the composition of the *Astronomia nova*, namely, that Fabricius's first major objection to Kepler's work was his abandonment of Tycho's solar theory.

Fabricius objected to the use of the true rather than the mean sun and the bisection of the earth's eccentricity in his initial letter to Kepler, and subsequently elaborated his objections in a more detailed letter.[27] Fabricius's objections to Kepler's innovations with regard to the earth's orbit were probably ultimately based in his faith in Tycho's solar theory. As he wrote, "We ought not heedlessly to depart from Tycho's well-developed solar hypothesis."[28] Fabricius was also distrustful of using observations of Mars to investigate the solar theory. As well as involving observations of Mars in the consideration of the sun's theory (observations "surely brought in not from its own but from a foreign domain"[29]), Kepler's work involved changing a well-investigated theory based on an unknown one ("We should judge the uncertain based on the certain, and not the contrary"[30]). Furthermore, Fabricius insisted that bisecting the eccentricity would have observable consequences, which Tycho himself certainly would have found or which would be apparent in the size of the sun and eclipses. (It is significant in this regard that Kepler dispatches the latter objection in the *Astronomia pars optica* and cites this finding in chapter 29 of the *Astronomia nova*.[31]) The variety of methods Kepler adduced to justify bisecting the eccentricity were most probably meant to allay Fabricius's (and the Tychonics') criticism for abandoning Tycho's solar theory. Because of Fabricius's attachment to this—Tycho's only well-developed planetary theory—and his systematic series of objections in its defense,[32] Kepler needed to provide an unusually secure justification for abandoning it.

The Continuation of Kepler's Physical Research

Regardless of his knowledge of Fabricius's connection with the Tychonics and his concern to keep their correspondence confidential, Kepler was evidently forthright from the start about the nature of his research program, and the Kepler-Fabricius correspondence therefore contains a valuable record of the progress of his research. He informed Fabricius in his first letter, now lost, that Mars's orbit was an oval.[33] In response, Fabricius warned him, "My dear, most learned Kepler, see that you do not make too much of your speculations and do violence to Mars with your new hypotheses."[34] This statement, another discussion of the advisability of abandoning "Tycho's little circles,"[35] and a mention of "an astronomy without hypothesis"[36] suggest that, in his initial letter to Fabricius, Kepler had laid out his opinion of the absurdity of Tycho's circles and the necessity of proceeding with an astronomy based on natural principles in much the same way as in his earlier letter to Archduke Ferdinand.[37]

Kepler continued his explanation of his research program with direct reference to physical astronomy in his first extant letter to Fabricius. Thus when countering Fabricius's main objection to bisecting the earth's eccentricity and addressing the question of the near observational equivalence of the simple and bisected eccentric theories (at least for the position of the sun), Kepler conceded, "The main reason for my daring this was my speculation, about which see Maestlin's table in [chapter 15 of] my

Mysterium."[38] And although Fabricius denied that the earth-sun distances in the simple eccentric solar theory caused the annual variation in Mars's orbit,[39] Kepler still proclaimed the bisected solar eccentricity to be "the key to astronomy."[40] As we saw in the previous chapter, "The Key to a Universal Astronomy" was one of the preliminary titles of the *Astronomia nova*.

After he had addressed Fabricius's specific objections and questions, Kepler informed him more broadly about the physical basis of his inquiry and about his most recent thoughts. Drawing a small epicycle on the end of a radius from the sun, he bade him, "Take this diagram of the theory of Mars and the sun and speculate with me about how whatever equation you demand might be computed individually. I do not yet know how to do this calculation except by using the vicarious hypothesis."[41] This was one of the earliest manifestations of Kepler's epicyclic distance model, which we will discuss more completely below. It was a simple and conventional mathematical way to model Mars's changing distance from the sun. If Mars were not swept around by the sun, Kepler wrote, it would uniformly traverse the small epicycle, whose size was 0.9165 in comparison to the radius of 10. That is, the position of Mars on the epicycle would be determined from the mean anomaly. Imagine now, he added, that the time it takes the planet to traverse one degree on a large circle around the sun is proportional to the distance from the sun, "because the strength in a narrower circle is more compressed and denser and for that reason it is also stronger in its effect than in more spacious, higher circles."[42] This idea is a manifest continuation of Kepler's thought from the time he was writing the *Mysterium cosmographicum*.[43] At that time, Kepler had examined the idea more qualitatively. Now he was attempting to calculate the motion precisely, although he admitted that he did not yet know how to do the calculation. He could say definitively only that the orbit of Mars was some kind of oval, due to the fact that it would spend so much time near aphelion that by the time it reached quadrature, midway between aphelion and perihelion, it would have come closer on its epicycle than would be required for the equation of a circle.[44] He was thus referring to an egg-shaped oval, with the fat part around aphelion.

As soon as Kepler set out to derive the motion of a planet by seriously attempting to calculate the effect of the planet-moving force, he ran into an intractable problem. If one wants to know how quickly Mars is moving, this is easily found in the epicyclic distance model that Kepler was discussing by using the mean anomaly (which is simply its position on the epicycle) to find the distance from the sun, from which the speed can be derived reasonably simply. If, on the other hand, one wants to know Mars's location, one needs to know the cumulative effect of the changing speeds. This might be found by adding all these distances, but first, there is—in principle—an infinite number of them, and second, there would be more of them accumulated where the planet was moving more slowly and fewer where it was moving more quickly. "Here you see the difficulty of calculating the place of the eccentric from the true and physical hypothesis," Kepler wrote after describing the problem.[45] As a way around it, Kepler described his approximate method of dividing the orbit up into 360 (or more) equal divisions, over each of which the planet was assumed to travel at a constant speed. By setting the sum of all of the distances equal to the planet's period, the time it would take to arrive at any particular division (its simple or mean anomaly) could be determined in proportion to the entire period.

Kepler's problems with this initial distance model were compounded by a lingering ambiguity about the nature of the epicycle he was using to model the change in distance. In chapter 39 of the *Astronomia nova*, when discussing how a planet might traverse an eccentric orbit, Kepler would discuss at length the absurdities inherent in the idea of the planet actually traveling on the epicycle. However, despite Kepler's care in the *Astronomia nova*, it is clear in his letter to Fabricius that he had not yet come to that understanding of the epicycle's role. It is clear that part of his problem was that he himself had not broken with the traditional conception of an epicycle. On the one hand, he clearly recognized that the epicycle was merely a representation of some faculty of the planet:

> When I say that the planet heeds the sun's virtue in its striving, I express the physical cause of the planet's motion. When I say it heeds the center [of the epicycle], I express our way of understanding this. For I do not believe that the planet heeds some imaginary center, as we, for whom a chart and a paper is at hand, might. I concede that the planet heeds the sun in maintaining the magnitude of its little circle, because the sun, the source of virtue, is perceived from one or another angle, and is comprehended by the planet in its imagination. For I believe these intelligences to be disposed materially. In the same way we see things with our eyes, they comprehend through some other perception.[46]

But on the other hand, it is clear from the way he described the problem that he was dissatisfied with the notion of the planet's being moved by the force coming forth from the sun and at the same time being swept forward by the epicycle at some points and backward at others.

While writing the letter he began to untangle the problem:

> Let me now say what occurs to me while writing, about how we might picture this physical reason more neatly, while maintaining everything that stands out to us as needing to be done. For it seems hard to believe that the planet is borne along by the sun's force, which is a natural faculty, and at the same

> time strives to diminish or intensify [its motion] in proportion to the requirements of a circle describing equal spaces in equal times, which is an animal faculty. It would be neater for us to transfer every natural faculty to the sun, to which we already assign the greatest share [of natural faculties], and which is the heart of the world. But to the planet we would attribute merely some intellectual faculty.[47]

Kepler was still not able, however, to divorce the epicycle from the representation of the orbit completely, for his solution was to imagine the epicycle as being at right angles to the plane of deferent. In this way, the planet could approach or recede from the sun without adding to or subtracting from its lateral motion around the sun. Nevertheless, he hastened to specify that the motion that the epicycle would now cause in latitude is imaginary.

This provisional model found Kepler in the midst of grappling with the problem. He could not quite dispense with the epicyclic representation completely—hence the right-angled epicycle—and yet he was anxious to develop a cause for the planet's libration toward and away from the sun that was purely natural. He wanted to assign the entire motive faculty for moving the planet laterally to the sun, and some faculty for librating radially to the planet. Because the planet's libration was complex, Kepler endowed it with an intellectual faculty to regulate its motion. In order to provide the planet's intellectual faculty with some means of doing this, he introduced the idea of the solar *species*. "Let [the planet]," he wrote, "traverse equal degrees in equal times [i.e., around its epicycle], let it measure its descent and ascent according to the appearance of the solar *species*."[48] Rather than regulate the uniformity of its motion based on some fictional epicycle, Kepler wanted the planet to be able to regulate its approach and recession from the sun based on this measure that it could conceive.

The solar *species* occupied a prominent position in Kepler's account of the physics of planetary motion in the *Astronomia nova*, where the idea was introduced in chapter 33. The term itself is so ambiguous that it is generally left untranslated.[49] The word has connotations of "form" and "emanation," but, in agreement with Stephenson, I believe it is best understood as "image" or "appearance."[50] Kepler meant, at least initially in this letter to Fabricius, that the planet can perceive the angular size of the sun, and thus know its distance from the sun and regulate its motion in such a way as we might understand with the epicyclic distance model. But Kepler quickly came upon another way of making use of the concept.

Although he had found a way for the planet's intellectual faculty to make its contribution to the planet's motion, attributing all the lateral motive power to the sun introduced a problem of its own. If all of the planet's lateral motion were to be attributed to the sun, half the equation of the planet's motion previously attributed to the epicycle would have to be made up. For in a Ptolemaic eccentric, there were two halves to a planet's "equation," or two sources of nonuniformity of motion. The optical equation was the *apparent* nonuniformity of motion caused by the displacement of the observer from the center. Thus, for example, the earth, which had an eccentric but not an equant prior to Kepler, had only an optical equation. The physical equation was the real nonuniformity of motion around the eccentric caused by the displacement of the center of uniform motion from the center of the eccentric to the equant point. Kepler had previously imagined the motor virtue to explain only the action of the equant—the physical equation. The other half of the equation—the optical equation—had still depended on the orbit being eccentric. But the epicycle he was now doing away with was simply an equivalent form of that simple eccentric. In order to compensate for the loss of the optical equation, he had to double the proportion (that is, square) the effect of the motor virtue. By making it that much weaker at apogee and that much stronger at perigee, he sought to compensate for the loss of the lateral motion of the epicycle. This line of reasoning is clearly akin to his reconsideration of the use of the motor virtue hypothesis in finding the ratio of the periods of the planets.[51]

In addition to providing some tangible measure by which the planet could regulate its motion, using the solar *species* resolved the problem of the loss of the lateral motion of the epicycle. In order to compensate for the half of the equation previously provided by the epicycle, the effect of the solar virtue had to be doubled, or squared. By using the solar *species*—or appearance—which is a two-dimensional quantity, rather than simply considering the sun as a point source, Kepler believed he accomplished this:

> Previously, I pictured a circular line and the fount of virtue flowing out to the line as a mathematical point. I said that the measure of the degrees of motor virtue emanating from the sun is determined by these things. But the virtue is more neatly conceived of to emanate not from the center of the sun alone but from the whole body. With this proposed, a double proportion [i.e., square] of virtues to the proportion of the distances is brought about. For the ratio of light and virtue emanating from the sun would now most plainly be the same (which before I had seized upon only in part). If someone came twice as close to the sun, how much larger would the diameter of the sun appear? Roughly doubled. . . . But because the sun is not a diameter but a circle, then since the proportion of figures to the proportion of diameters is double, therefore the visible area of the sun will be four times its previous size.

> This is the measure of the increasing of light with proximity; let it be the same with virtue. The sun, therefore, will bring about twice the equation that could be [brought about] by the 9165 circle [i.e., the epicycle, in comparison to a deferent of 100,000], and the compounded path of Mars around the sun will clearly remain an oval as before.[52]

This solution was extremely pleasing to Kepler. It separated the roles of the sun and the planet into circular and radial elements. It obviated the need to explain how the planet would navigate an eccentric orbit. And by making the proportion of motor virtue identical to that of light (which was on his mind due to his optical investigations), this idea made motor virtue all the more analogous to light, as he had always assumed it to be.

Kepler expressed his enthusiasm to Fabricius:

> I hope this speculation pleases you. I am exceptionally pleased with it myself, and I thank you for urging me to write. . . . Consider whether I have made a step toward establishing a physical astronomy without hypotheses, namely fictitious ones. That the virtue is in the sun is certain. The ascent and descent of the planets is likewise certain from the larger and smaller apparent *species* of the sun. For there is a large difference between an opposition of the sun and Mars in Leo [i.e., where Mars is at apogee] and an opposition of the same in Aquarius [where it is at perigee]. These things are therefore not hypotheses (or as Ramus calls them "figments") but the very truth, like the stars themselves. I truly suppose nothing beyond this.[53]

Although he had the greatest hopes for his new line of research, he did not yet, of course, have a means of calculating the compound motion. In the meantime, he suggested that for his astrological work Fabricius should calculate Mars's heliocentric longitude by means of the vicarious hypothesis and use the sun-Mars distance from the epicyclic distance model to determine the annual parallax.[54] This would turn out to be a contentious suggestion. He finally closed by asking Fabricius to keep these things to himself.

Fabricius did not comment on these speculations immediately, but we can perhaps gauge his initial response from a brief letter he wrote acknowledging the receipt of Kepler's letter of 1 October 1602. In his letter of 8 November 1602 (o.s.), Fabricius wrote about the method of four acronychals, "The things you attested to about [deriving] the relative proportions in the hypothesis of Mars by means of observing four acronychals, I will investigate and work through by means of a calculation at the earliest possible opportunity."[55] Regarding the bisection of the solar eccentricity, he exclaimed:

> The things you bring up to prove the bisected solar eccentricity wash away my objections for the most part sufficiently strongly and completely. Nor do

> I find anything further to oppose. I rejoice that I have been freed from the groundless fear arising from the alteration of the solar hypothesis.[56]

And yet regarding Kepler's physical speculations, he wrote only, "I cannot go into my judgment about your thoughts concerning a motor virtue for Mars now on account of a lack of time."[57]

It was Kepler's suggestion that he use the vicarious hypothesis for the longitude and the "true" hypothesis for the distance that gave Fabricius pause. On reflection, he decided that he was not convinced that the earth's eccentricity needed to be bisected. On the contrary, he concluded that the bisection was the cause of the difficulties Kepler was encountering. After having some time to think about it, he wrote:

> My dear, most learned Kepler, you are working to investigate the motion of Mars using two hypotheses, one vicarious, the other "true." This is evidence that this hypothesis is not true in all respects, or at least that it is defective. You seek the prosthaphaeresis of Mars from the vicarious, and the distances from the "true" hypothesis. If the vicarious were true, then both would be given. If, on the other hand, your "true" hypothesis were true in every respect, the vicarious would not be needed. Therefore it is especially necessary to work it out so that some such hypothesis is devised that corresponds as closely as possible to everything in the heaven itself, which certainly does not seem so very difficult to me, if you give up your fancy about the sun's double eccentricity. Believe me, if I judge rightly, this is the sole and truest cause that it is necessary for you to construct a double hypothesis of Mars as well. . . . I would want you diligently to consider these things and give up your fancy regarding the sun. I pray, cleave this knotty objection with the sword of truth for me. Do not rush to judgment, lest a late repentance befall you.
>
> I write these things in a good spirit in defense of the truth, because I considered everything more diligently since my last letter. You bind Mars to the sun with too tight a leash.[58]

Fabricius's objection to bisecting the earth's eccentricity was based on the fact that this would cause the (perigeal, presumably) distances to be too large. On the other hand, in his next query, "What in the world might the truest cause of the double eccentricity in Mars and also in the other superior and inferior planets be?"[59] he proceeded to note, "I am utterly convinced that the sun's eccentricity is itself mixed in with Mars's eccentricity as with the others as well, which intricacy renders the eccentricity intricate as well."[60]

This must have been exasperating to Kepler. The bisection of the earth's eccentricity (made possible by the introduction of an equant) was the foundation of his new astronomy, in part precisely because it enabled him to remove the annual variation in Mars's eccentricity, which was an artifact of the simple eccentric solar theory. Fabricius understood that the eccentricities were intermingled, but refused to accept the solution. Kepler wrote in the margin here, "He himself presses the point and involves himself in what I was about to reply to his objections about Mars."[61]

Despite his suggestive understanding, Fabricius was in fact badly mistaken. When he asked about the twofold eccentricity of the planets, he was referring to the alternate double-epicycle form of Copernicus's theory, apparently thinking that the larger of the two epicycles corresponds to the planet's eccentricity and the smaller to the amount of the sun's eccentricity mixed into the planet's theory.[62] If this were the case, he asked, why should these two motions both begin from the planet's apogee, when the smaller circle ought to be referred to the sun's apogee? If he does not concede that the epicycles ought to be independent, Fabricius wrote, Kepler cannot maintain that halving the earth's eccentricity will remove the one corresponding to the sun. Kepler could only respond by writing in the margin, "This is, of course, not the case."[63]

Even before receiving this revealing critique, Kepler had begun to suspect that Fabricius did not fully understand what was going on with respect to the sun's eccentricity and the theory of Mars. In a letter of 2 December 1602, which crossed with Fabricius's letter just discussed, in addition to providing detailed explanations of the method of deriving the vicarious hypothesis and of triangulating to the earth's orbit from a stationary position of Mars, Kepler tried to untangle the problem for Fabricius. "There follows a question of yours," he wrote, "in which you mix up the heaven with the earth, confusing yourself and me, and veiling the real state of affairs so that I discern just a little light in these shadows, and even that poorly."[64] Part of the confusion, again, was that Fabricius was considering the double-epicycle version of Copernicus's theory favored by Tycho. He had thus asked whether the center of Mars's eccentric was really fixed at the sun—perhaps a slip of the pen (he should have written "deferent" or "concentric"), but still a reasonable query for this form of the theory. Kepler began pedantically by pointing out that this would be a concentric, not an eccentric. He then proceeded to the matter at hand, whether motions should be referred to the mean or true sun. He went to some length to explain that the mean sun is a truly imaginary point and expressed his doubt that this was a suitable point of reference for the theories of the planets:

> If the angles of the coequated anomaly depended on the point of the mean motion of the sun, I am afraid that if that point were not the sun itself it would not be suitable for the great mass of the sphere carrying Mars in Tycho's hypothesis. Or also, in Copernicus, roughly speaking, I am afraid that Mars would lose that point from sight if it were not the sun itself.[65]

That is to say, in Copernican and Tychonic theory, the center of the planetary orbits is the center of the earth's (or sun's) orbit, not the sun itself, since the earth's (or sun's) orbit is eccentric. Kepler justifiably wondered how Mars in its orbit would keep track of the center of the earth's orbit if there were nothing there.

Ironically, Fabricius responded to this letter by proposing the first of his own theories of Mars, which almost completely ignored Kepler's patient explanations. Still firmly convinced that Kepler's use of the true sun and bisection of the earth's orbit were responsible for the difficulties he was experiencing, Fabricius proposed a theory that flew in the face of Kepler's admonitions about the imaginary nature of the mean sun and its unsuitability as a center for planetary spheres. In essence, Fabricius's theory consisted of a Copernican double epicycle theory for Mars whose center was carried around on a smaller deferent 5400 parts in radius (i.e., 3/2 the solar eccentricity) such that the position on the smaller deferent was always parallel to the position of the mean sun. The center of this small deferent was eccentric from the earth by 1800 parts, located toward perigee.[66]

As Fabricius described it, his theory is confused. This is entirely understandable; he was living under siege at Esen at that time and lamented having none of his books with him. However, we can surmise what Fabricius hoped to accomplish. The change in the distance from the earth to the larger Copernican apparatus caused by Fabricius's small deferent as the sun moved from perigee to apogee is 3600 units, but it is *closer* at apogee. Apparently this deferent was set up to mitigate the excessive change in earth-sun distances that had convinced Kepler to bisect the earth's eccentricity. With a simple eccentric, the change in earth-sun distance from apogee to perigee is 7200 units. In Kepler's bisected form, it is 3600 units. Fabricius's deferent is thus apparently designed to counteract half the change of distance in the simple eccentric form of the solar theory to bring it into line with Kepler's distances. Still, Fabricius's theory is confused. Although it is apparently Tychonic (geo-heliocentric), which form Fabricius favored, the small deferent does not correspond to the motion of the sun (i.e., earth), and it lacks such a deferent to account for the second anomaly.

Despite its tentative nature, Fabricius was having fun developing his theory. "I await your thunderbolts," he wrote, "with dauntless spirit. I am as pleased with my hypothesis as you are with yours. . . . I pray, reveal your judgment and censure to me freely and candidly."[67] Kepler was not impressed. In the margin of Fabricius's letter, he wrote, "It is easy to transform the path through the multiplication [of circles]."[68] Furthermore, Fabricius's methodology annoyed Kepler. Fabricius had written that although he was without his books, he had put the theory together over several days and nights of thinking and was confident that it would, in principle, obviate the need for the vicarious hypothesis.[69] He had checked it only roughly, but thought it worthy of further checking. Kepler paraphrased Fabricius in his refusal to assess the hypothesis:

> You also describe your hypothesis to me, and ask that I check it with a calculation. But because you add that you "hope that it will satisfy observations" and that it was "fashioned from your thoughts," I will not do it; it is a useless labor. You must indeed be annoyed. But I do not, like you, preconceive an entire hypothesis, which I check later with good hope. For this hope will deceive us forever. I posit a few things by way of principles, from there I follow the observations.[70]

Kepler's response expresses well his annoyance with Fabricius's method and the gulf that separated Kepler from his contemporary theorists. It is also highly significant because it expresses precisely the argument of the *Astronomia nova*: For Kepler, it is the *process* that establishes the validity of the results.

Kepler's helpful suggestion from his letter of 1 October 1602 that Fabricius provisionally use his two deficient hypotheses in combination had engendered far too much misunderstanding. Fabricius had harped on this point ever since, pointing out that Kepler's failure to reduce the motion of Mars to a single hypothesis indicated that neither of his hypotheses was true.[71] Furthermore, this failure had spurred Fabricius on to producing the first of his own Mars theories to compete with Kepler's. Kepler had never meant Fabricius to take his suggestion so seriously. He had meant all along that the vicarious hypothesis was only a helper for the "true and physical hypothesis." In the margin of one of Fabricius's letters, next to Fabricius's complaint about the twofold nature of his hypothesis, Kepler had written, "It is already set up as one."[72]

Kepler tried to dispel this confusion once and for all in a letter of 4 July 1603, by pointing out that his "true hypothesis" was not untrue but—because of the infinitude of different distances of Mars from the sun—was simply so difficult to calculate that the vicarious hypothesis was brought in, but merely as a helper:[73]

> The bisection of the sun's eccentricity had pleased you, but now you accuse the hypothesis of Mars of being misguided. The hypothesis does not strike you as being sufficient in itself, but brings in as though for assistance another substitute hypothesis, which is false. Listen, Fabricius, the sole true hypothesis is sufficient in itself, but you and I could not follow its too laborious dictates. The essence of the matter is this: Mars approaches the sun according

> to a circular regulation [*lege*] traversing equal arcs of its epicycle in equal times. But at each stage of its approach to the sun, it changes the modulus of its speed. For at apogee it is turned around slowly (around its particular center of the eccentric). And yet at each stage it exhibits a precise modulus of its speed, according to its distance from the sun. Therefore the sum of the virtue sent out over a definite time, and therefore also of the path measured out around the center of the eccentric (and so also around the sun as well), is contained in the collection of all the distances—which are infinite. Now you go and compute either the infinite points of its path or arcs divided into the smallest parts, of ten minutes, say. You will discover the same thing I discovered from the vicarious hypothesis, that it differs by scarcely half a minute. But I will be ready more quickly. For you, it will have to be begun for apogee and taken through the smallest parts, and for that reason you will have taken on an enormous task. In fact, for just a quick calculation of the genuine hypothesis and of whatever location you like of the eccentric independently—without division through the smallest parts all the way up to apogee—I lack something: knowledge of the geometrical generation of the oval path or of the facial shape [*facialis* μετωποειδοῦς], and of the section of its plane into a given ratio. If the figure were a perfect ellipse, Archimedes and Apollonius would already have sufficed.[74]

Kepler's problem was that although he had known that the orbit was some kind of oval for some time, there was still no easy way to calculate the cumulative effect of the solar virtue. Even with the area law, without the knowledge of "the section of [the oval's] plane into a given ratio," he could not find Mars's position easily for any given time. This was the key to Kepler's early use of an ellipse (an incorrect one) as an approximation for the oval path. It was not the ellipse's shape that attracted him; it was its areal properties. If he could approximate the oval generated by Mars's uniform motion around an epicycle with an ellipse, then he could use the ellipse's geometrical properties straightforwardly to calculate the mean anomaly (the sum of the distances, or area) for any position. For this use it did not matter, looking ahead, whether or not the sun was at one focus. The significance of this fact did not become clear until later.

In spite of Fabricius's tenacious objections to Kepler's use of two hypotheses, his critiques were not wholly misplaced or unhelpful. He played a great enough role in disproving Kepler's first epicyclic distance model that Kepler gave him credit in the *Astronomia nova*, overgenerously suggesting even that Fabricius might have arrived at the proper ellipse before him.[75] Fabricius accomplished this in the course of doing what Kepler had suggested, using the vicarious hypothesis for the heliocentric longitude and the "physical hypothesis" for the distance. In a letter of 27

October 1604 (o.s.), Fabricius reported that this method in one instance yielded an error of 13′ in geocentric longitude.[76] He concluded that this was the result of two compounded errors. The first was Kepler's exact bisection of the earth's eccentricity (Fabricius had finally come around to abandoning a simple eccentric for the earth, but the exact bisection was apparently too convenient), and the second, major error was Kepler's modeling of the distances:

> The distances themselves do not increase or decrease as you want, but in a far different way. For all of the distances ought to be longer, and around mean distances the difference to be added to your hypothesis will be at its greatest around 9′. . . . For this reason, see to it that your distances, or in your way, the collected distances, are exaggerated little by little from aphelion on (in relation to a point on the semicircle). . . .
>
> Therefore I give you this advice: you should consult more examples near the mean distances. . . .
>
> I brought these things up not for the sake of carping on your most ingenious hypothesis, but so that with my petty labor, such as it is, I might in some measure assist you, who would throw himself alone into such Herculean labors.[77]

Kepler had independently begun to doubt his distance model earlier. He reported in a letter written around February 1604 that he had resorted to calculating as many distances as possible in the hope that from these he might learn the correct shape of the ellipsoid:

> But I press on with the distances of the eccentric. As soon as I have traced out many distances throughout the whole circuit of the eccentric, it will be readily apparent where the apogee is, how large the eccentricity is, and whether the path is an oval. Then accordingly the hypothesis is to be found, which represents all those distances.[78]

Modern scholarship has discounted the role of such a collection of distances in the derivation of the ellipse on the grounds that the uncertainty of the distances was too large to determine unambiguously the orbit's shape.[79] This is indeed true, but that Kepler even expressed this hope is an indication of his despair at this point with the intractability of the problems with which he was dealing. He explained to Fabricius that he did not think it was necessary (or possible) to be able to calculate the longitudes by means of this hypothesis immediately, for "[w]e perceive many things from afar which we cannot reach due to the absence of means," but nevertheless, he strove to have one day a single hypothesis by means of which he could also calculate the equations.[80]

The Elliptical Orbit and Its Challenges

The ellipse, when it came to Kepler, had two essential features that released him from the problems facing him. First, it had—or Kepler was able to convince himself that it had—a natural cause. And second, he learned to manipulate it so that he could directly calculate the equations. Let us examine these features in this order.

Kepler finally discovered that the orbit of Mars was an ellipse in the spring of 1605. Exactly when Fabricius received this news is unclear. The long letter in which Kepler related the elliptical orbit is dated 11 October 1605, but it was made up of several sections written over a long period of time. The first of these was written sometime after Easter 1605. After a pause of some weeks, he resumed writing, finishing the second section on 4 June, when another letter from Fabricius arrived. Kepler sent some of this material to Fabricius before traveling to Graz that summer.[81] The final section was not completed until his return at the beginning of October.[82] In addition, it is not clear exactly how much of this letter Fabricius received,[83] but regardless of any confusion in the presentation, it is obvious from Fabricius's reply that he had received enough information to understand that Kepler had decided that the orbit was an ellipse and, more important, roughly how to use it.

Of the two important final stages in discovering and working with the ellipse, the first was Kepler's realization that the distance of Mars from the sun was a function of the sine of the eccentric anomaly. As we have seen, Kepler began his physical analysis of the motion of Mars long before with the assumption that Mars moved on a perfectly circular eccentric, with its mean anomaly determined by the distance or area law. This initial assumed orbit produced positions of Mars that were acceptable at apsides and quadrants, but unacceptable in octants. In the superior quadrant, as Mars left apsides, it surged ahead of where Kepler knew (by using the vicarious hypothesis) it ought to be. In the inferior quadrant, it lagged behind. In his first thoughts that led him to an oval orbit, he concluded that in order to slow the planet as it left apsides, more distances (or a greater area) had to be heaped up there, and similarly to speed it through quadratures so that it would not fall behind in the inferior quadrant, fewer distances (or less area) were needed there. Although Kepler stated the problem in these terms, he proceeded effectively by subtracting distances in quadratures with respect to a circumscribing circle rather than extending them beyond the circle at apsides. These approaches are essentially equivalent; the conclusion was that the orbit was some kind of oval. He had then tried to model this decreasing distance in quadratures with an epicyclic distance model in which the epicycle rotated uniformly with time. Because the mean anomaly increased more rapidly than the eccentric anomaly (which would produce an eccentric circle in this kind of distance model), this would have the effect of pulling the position of Mars inside an eccentric circle, making an oval orbit of the kind he desired.

Unfortunately there were several problems with this model, not the least of which was the fact that he had been thinking of time as a dependent variable. If the sum of distances or area determined the time the planet spent on any part of the orbit, then it was nearly meaningless to use a distance model in which the distance was dependent on time. Kepler became mired in the difficulty of trying to apply this model before finally making use of the fact that he had the vicarious hypothesis, which could provide accurate heliocentric longitudes for any given time. In conjunction with the epicyclic distance model, which gave distances with respect to time, he could now at least see where points on this orbit fell. It had the oval shape he desired, coming within an eccentric circle at quadrants by 858 parts of 100,000. At this point, he substituted an ellipse for the troublesome oval, certainly because an ellipse has an area that can be calculated manageably. He could now return to the same physical analysis he had done on the eccentric circular orbit.

The results of this were suggestively opposite: again adequate at apsides and quadrants but off in octants. The planet now lagged behind rather than surging ahead of the position predicted by the vicarious hypothesis in the superior quadrant, and, similarly, it flew through quadratures too quickly and ended up ahead of the position it ought to occupy in the inferior quadrant. Apparently he had erred in precisely the opposite sense: too much area was being accumulated near apsides and too little near quadrants. In both cases, the error had been about 8′. The true orbit therefore ought to lie midway between this approximating ellipse and the eccentric circle, or about 429 parts of 100,000 in from the eccentric circle at quadrants.

Kepler had relied greatly on the use of his epicyclic distance model in his analysis of what was happening. An epicycle that would produce an eccentric circle had dissatisfied him, when he considered it more carefully, because he noticed that the planet approached the sun nonuniformly, more quickly when it was in the inferior half of its orbit and more slowly above (see figure 8.1). If the planet moved around the center of the epicycle β such that the angles γβδ, γβε, γβζ and so on increased as the eccentric anomaly (as would be the case for the epicyclic representation of a simple eccentric), its change in distance would be as γα, δα, εα, ζα, or equivalently, γα, ια, λα, ζα. Thus its change in distance over equal amounts of eccentric anomaly near apogee (γι) and perigee (λζ) would not be equal.

The oval model seemed more natural because in this case the planet at least moved uniformly with time around the center of the epicycle β, even

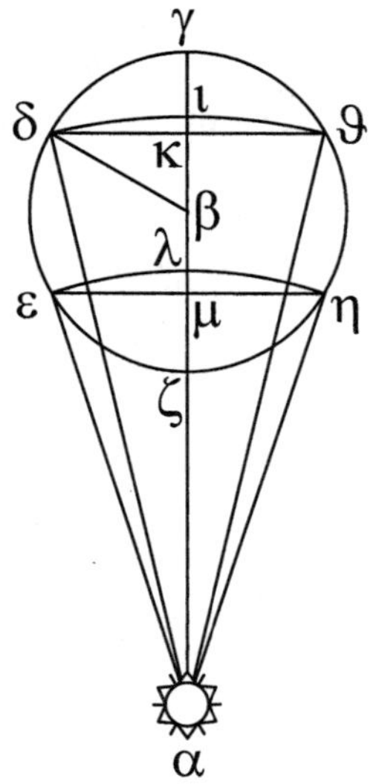

Figure 8.1 The epicyclic distance model
Kepler used this epicyclic distance model as an aid for his analysis of how Mars's distance from the sun changes. The sun is at α. The distance $\alpha\beta$ is Mars's mean distance. The radius of the epicycle, e.g., $\beta\delta$, represents Mars's eccentricity. Mars's distance at aphelion is $\alpha\gamma$, at perihelion $\xi\alpha$.

if it too suffered the same kind of nonuniformity. In this case, Kepler believed, since its change in distance around aphelion ($\gamma\iota$) would still be smaller, it would spend more time there, tracing out a broad, circular path. Likewise, the distances changed most quickly around perihelion, where the planet was spending the least time. The resultant oval orbit would be egg-shaped with the pointy end toward perihelion.[84]

This is how he summed up these considerations to Fabricius when introducing the ellipse (note that he was using an historical presentation of his discoveries with Fabricius as well):

> When I saw that the distances accumulated using a perfect eccentric circle erred nearly as much in excess—both so far as the distances themselves and their effect in the prosthaphaereses of the annual orb, and the equations of the eccentric—as my ellipse that I described to you earlier in numbers (which differs from the oval by an insignificant amount) erred in deficiency, I could quite correctly have reasoned as follows. The circle and the ellipse are from the same genus of figures and they err equally in opposite senses, therefore the truth lies in between, and between elliptical figures lies only an ellipse. And thus, Mars's path is absolutely an ellipse, with the lunula trimmed to half the width of the lunula in the original ellipse. Now, the lunula was 858

> parts of 100,000 wide. Therefore, it ought to be 429 wide, which is the exact diminution of distances accumulated in the mean distances using a perfect circle. Here, I say, is the very truth. But behold how in the meantime I was in turn deluded and thrown into new labors. Or rather, behold how miserably I hesitate in the presence of the discovered truth, in accordance with the saying that he who never doubts is never certain about anything. The original ellipse with the shortening of 858 parts had this natural cause, namely, that it was said that the center of the epicycle proceeded slowly when the planet is turned in the apogee of the epicycle, and more quickly when underneath. Moreover, the epicycle itself approached equally in equal times. This was moderately in accordance with nature. But now, if it were an ellipse with a shortening of 429, I would lack a natural cause. For it was absurd that the center of the epicycle proceed non-uniformly, while the circumference of the epicycle proceeded neither uniformly nor with the non-uniformity of the center itself, but with a peculiar non-uniformity, which would be at least half of the inequality of the center.[85]

Kepler's first essential realization was that there was a "natural cause" that would get him half the reduction in distance of his earlier distance model. "[I]t was," he wrote in the *Astronomia nova*, "as if I were awakened from sleep to see a new light."[86] What Kepler found was that precisely at quadrature, the secant of the optical equation 5°18′ was 100,429.[87] He realized that to get a reduction of 429 parts, he needed only to substitute the adjacent side for the hypotenuse in this triangle. Or, in terms of the epicyclic model he was using to analyze how the distances changed, the distances varied as the projection of the epicycle onto the line connecting the center of the epicycle to the sun (as γα, κα, μα, ζα in figure 8.1). Specifically, in terms of its greatest distance, Mars's distance from the sun could then be expressed as

$$1 + e - e\,(1 - \cos\beta) = 1 + e - e \cdot \operatorname{vers}\beta,$$

where $\operatorname{vers}\beta$ is the versed sine (one minus the cosine) of the eccentric anomaly. Or, as Kepler would say, the distances varied as the versed sine of the eccentric anomaly.

Let us defer for the moment the question of how this can be thought of as a naturally caused variation, to consider the other important feature of the ellipse. Kepler had been troubled by the inexactness of the area law for some time. In his letter of February 1604, he had described to Fabricius a geometrical analysis of the error introduced when calculating the area of a circle as a collection of distances from an eccentric point. (This analysis of the conchoid also found its way into the *Astronomia nova*, chapter 40.) Kepler had based the area law on Archimedes's calculation of the area of a circle as a sum of an infinite number of lines drawn

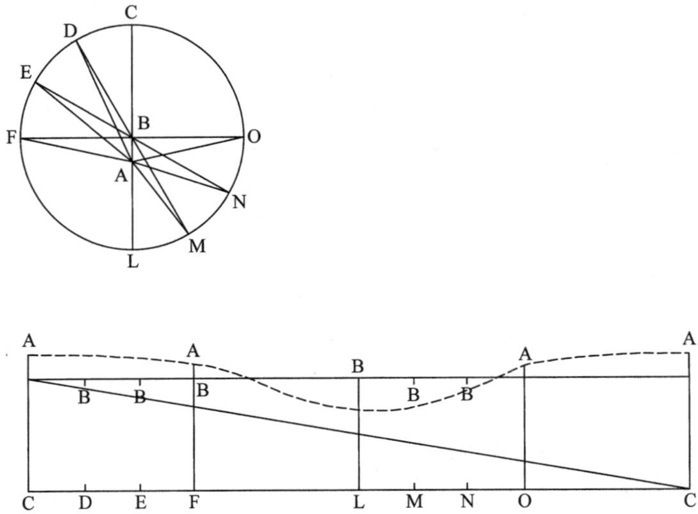

Figure 8.2 The conchoid
Kepler used this diagram to analyze the shortcomings of measuring the area of a circle from an eccentric position. The sun is at A. The circle is "unrolled" below. The rectangular area bounded by B and CDEFLMNOC is the area as measured from the center B. The area bounded by CDEFLMNOC and the "conchoid" curve A is the area as measured from the sun A.

from the center to the circumference. When he attempted a brute force calculation of the distance/area law by calculating the distance from an eccentric point to the circumference for 360°, he found that the sum was greater than 36,000,000, the value that would have been obtained from the center (based on a radius of 100,000). See figure 8.2.[88] In his geometrical analysis, Kepler "unrolled" a circle to get a rectangle, whose bottom CDEFLMNOC represented the circumference of the circle and whose upper surface BBBBBB represented the center. The sum of the distances from an eccentric location was represented by the conchoid curve AAAA, which started above the center (i.e., at aphelion C, where the distance CA is longer), passed below BBBBBB (in the inferior quadrants, where the distances LA, MA, NA, are shorter than LB, MB, NB), and then returned above. Because the distance from the eccentric is longer at quadrants (i.e., FA and OA are longer than FB and OB), the parts of the conchoid curve above and below the center do not entirely cancel. The area of a circle measured as the sum of the distances from an eccentric point is therefore larger than the sum of the distances from the center.[89]

Put another way, Kepler considered the sum of pairs of points opposite the center of the circle. At apsides, the sums of the distances from the circumference to the center (CB + BL) and from the circumference to the eccentric (CA + AL) were the same, but as one moved along the circumference toward quadrants, the sum from the points on the circumference to the eccentric point (e.g., DA + AM) became increasingly larger than the sum to the center (DB + BM). This was particularly disturbing because of the increasing importance of the distances at mean distances around quadrature. As Kepler explained to Fabricius, "Therefore I have erred up to now in supposing that the distances are counted if they are collected in a sum from the plane of the ellipsoid which the planet describes. For it lingers least where it accumulates [too] large distances."[90] That is, Kepler had been concerned to speed the planet through mean distances F and O, precisely where the error in the area law would be largest, accumulating distances, and therefore times, that were too large as AF and AO were to BF and BO.

The Via Buccosa and the Didactic Function of Error

Kepler made one more major error after coming to the ellipse, from which his analysis with the conchoid helped to extricate him. This error resulted in the *via buccosa*, or "the puff-cheeked oval." The *via buccosa* is also an interesting error because its inclusion in the *Astronomia nova* illustrates the didactic function of Kepler's recitation of his errors, which brings us back to the influence of the Kepler-Fabricius correspondence on the argument of the *Astronomia nova* discussed earlier in this chapter.

The *via buccosa* derived from the error Kepler made, once he knew how Mars regulates its distance from the sun, of where to place the planet. When he realized that the secant of the optical equation provided the distance he needed, he reasonably assumed that this distance ought to be marked out on the radius from the center to the orbit to the planet.[91] See figure 8.3.[92] The epicyclic distance model is shown in dashed lines to remind us that it should not be thought of as a conventional epicycle. For any given value of eccentric anomaly, β_1, β_2, Kepler knew that Mars's distance from the sun would be as the projection of the epicycle on the diameter, that is, its distance would be SD_1, SD_2. The question now was: Where is Mars at that distance? Since the epicyclic distance model yielded a distance for a given value of eccentric anomaly, and the eccentric anomaly was the angle measured from the center, C, Kepler quite reasonably supposed that Mars's position ought to be constructed on the radius from the center for that value of eccentric anomaly. That is to say, Mars's position should be constructed by swinging the distance SD around the sun to the radius CR_1, CR_2, so that Mars's position would be given by B_1, B_2. As is clear from the diagram, the distance R_2B_2 is greater than the distance R_1B_1, so Mars will come in from a circle more in the inferior quadrant

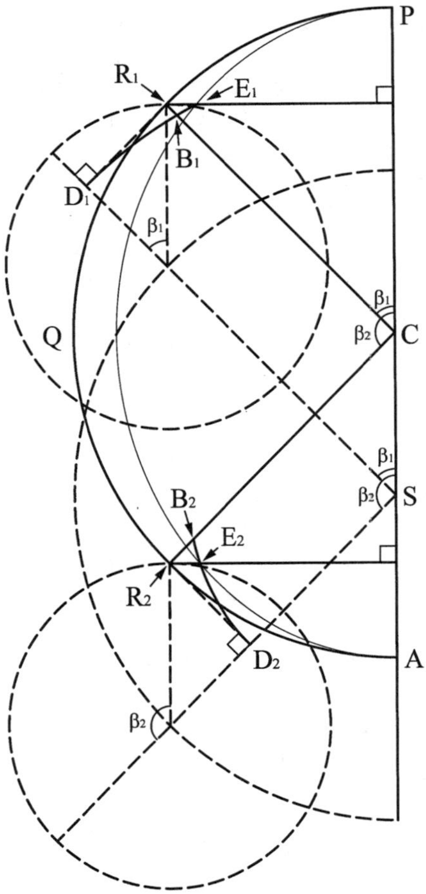

Figure 8.3 The *via buccosa*

than in the superior, and the result will be another asymmetrical oval that is wider toward aphelion. Why Kepler called this shape "puff-cheeked" when it might more aptly have been called "fatheaded" is not clear, but I have reproduced the diagram in the same orientation as it is given in the *Astronomia nova*.

When Kepler came to check the Martian longitude produced by this construction against the vicarous hypothesis he found the result was an error in octants by about 4′ or 5′. It was consideration of the area law that showed him how correctly to place the planet. He most probably recalled that the conchoid erred because it did not intersect BBBBBB, which represented the circle's center, at quadrants F and O; for the parts of the conchoid above and below the line to cancel and the areas thus to be equal, it should have equaled the radius there (i.e., AF would have to be equal to BF). But his new distance *was* the radius at quadrants, that is, according to the epicyclic distance law, at 90° of eccentric anomaly, the distance along AF would be equal to BF. He might then have realized that the sum of the distances given by the epicyclic distance law do correctly measure the area of a circle from an eccentric point. But they measure the area of the circumscribing circle, not the ellipse. To get from the area of the circle to the proportional area of an inscribed ellipse, it was necessary to drop a perpendicular to the line of apsides, as shown in figure 8.3. The planet would find itself at the intersection of the perpendicular and the ellipse, E_1, E_2. This placement would ensure that the area, as measured by the sum of the distances from the sun, would be correct. Alternately, to construct the location of Mars for a given eccentric anomaly, (1) get the distance from the epicyclic distance law, SD; (2) drop a perpendicular to the line of apsides from R, the position on the circumscribing circle corresponding to that eccentric anomaly; and (3) swing an arc around S with length SD. Where the arc intersects the perpendicular will be the position on Mars on the ellipse PE_1E_2A. Indeed, it may not have been until far into this line of reasoning that Kepler realized that the ellipsoid was and had to be an ellipse.

Unfortunately, although the answer carried the weight for Kepler of having suddenly crystallized out of a number of troublesome issues he had been turning over in his mind for some time, it was suddenly and confusingly thrust on Fabricius. After Kepler had worked through the error of the *via buccosa*, its significance faded. The flood of realizations regarding the elliptical form of the orbit, tying up as they did numerous threads of thought, served to convince Kepler that he had indeed hit on the correct answer, and that the *via buccosa* was a short-lived and irrelevant folly. Therefore, although he had initially thought to recount the error to Fabricius, he thought better of it. He struck through the section of the page in his draft detailing the *via buccosa*, wrote across the diagram, "This was a total waste of time. See below," and made a note lower in the margin indicating where the copyist should resume.[93] Thereafter, he launched into a quick and somewhat incoherent recipe for calculating Mars's position using the ellipse, and then into a curt redefinition of fundamental astronomical terminology.

With the ellipse and the area law, Kepler introduced an intractable problem into astronomical theory, one that he could accept but that Fabricius and other contemporary astronomers would find terribly confusing. In the letter in which Kepler announced the ellipse, he provided Fabricius with the following instructions for calculating Mars's position, considerably simplified in this paraphrase: For a given mean anomaly, i.e., for any given time, the eccentric anomaly must first be determined, which can be done either with a table or indirectly. In the latter case, one must first "guess" a value and then refine it iteratively. The reason for this confusion was that Kepler had run into a problem known still as "Kepler's problem." In fact, it is impossible, in this case, to calculate directly the position on a circle corresponding to a given area measured from an eccentric point. That is to say, one cannot begin with mean anomaly, or time, expressed as an area according to the area law, and proceed directly to the eccentric anomaly. One must instead guess a point on the circle, calculate the area, correct the guess, and repeat until any desired level of accuracy is achieved. Once this point on the circle circumscribing the ellipse has been found, one can proceed to Mars's position on the ellipse using the construction above.

With the ellipse and the area law Kepler had also wrought havoc with the traditional definitions in astronomical theory. Let us begin with what Kepler understood as the eccentric anomaly (see figure 8.4). It was the angle β (or angle ACN) at the center of the orbit (midway between the foci) measured from aphelion to a point N on a circle circumscribing the ellipse. It was analogous in some ways to the old eccentric anomaly, the angle measured at the center of the orbit from aphelion to the planet, AC♂. But recall that Kepler's analysis with the conchoid had shown that the area measured from the sun was to a circle, not an ellipse. To find the proportional area on the ellipse, it was necessary to drop a perpendicular from N to the line of apsides. The point where this line intersected the ellipse was the location of the planet, from which the coequated or true heliocentric longitude, γ, could be calculated. Although this method was undeniably accurate and had a compelling internal consistency to Kepler, it had the effect of throwing into confusion a number of time-honored astronomical quantities. The mean anomaly, previously an angle, α, measured around the equant point, Q, had become an area, AN⊙, in the circumscribing circle measured from the sun, ⊙. The eccentric anomaly suffered most of all. Previously it had been the angle β from aphelion, A, to the planet measured from the center, C, or equivalently the arc along the circle from aphelion. It retained these meanings for the circumscribing circle, but it no longer had a clear meaning for the ellipse, for the angles ACN and AC♂ were no longer the same. And thinking of it as an arc along the orbit no longer made sense, because at various points along the ellipse one degree of eccentric anomaly no longer measured a constant arc.

We will take up in a moment Kepler's analysis of how his model of the variation of Mars's distance could be thought of as natural. But first we will consider the function of the *via buccosa* in the account of Kepler's discovery in the *Astronomia nova*. The *via buccosa*, is presented in chapter 58, "In what manner the reciprocation discovered and demonstrated

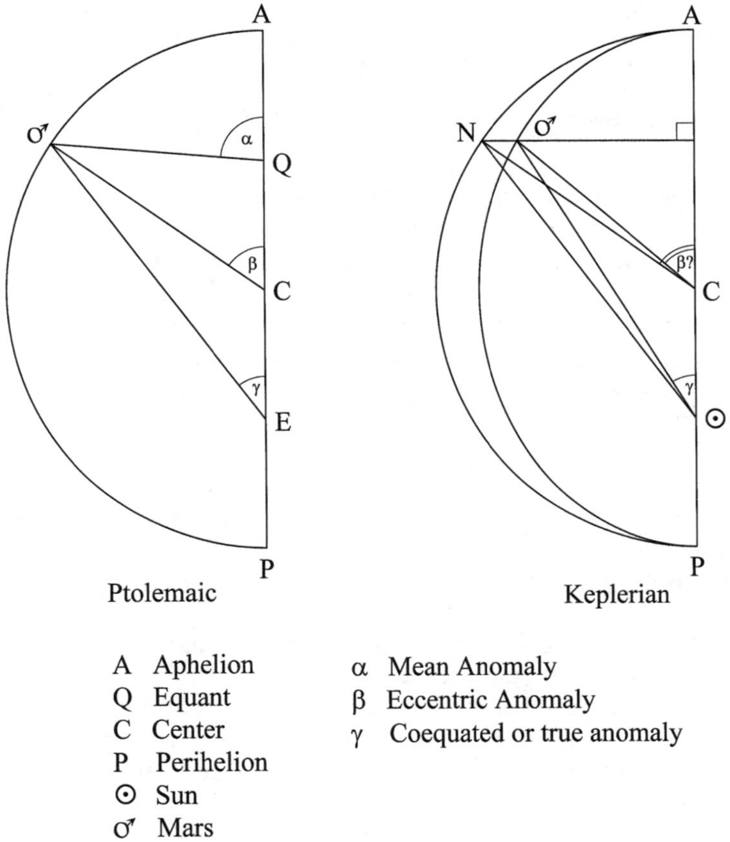

Figure 8.4 Comparison of anomalies

in chapter 56 may be accepted, and nevertheless an error may be introduced in a wrongheaded application of the reciprocation, whereby the path of the planet is made puff-cheeked."[94] The *via buccosa* is the final error presented in part 4 of the *Astronomia nova*. Since part 5 concerns latitude, the *via buccosa* is effectively the last error in Kepler's long "battle with Mars."

What purpose did it serve to introduce this final, admittedly "wrongheaded" error into Kepler's account of the discovery of Mars's orbit? His account was gathering steam in the final chapters of part 4. In chapter 56, he introduced the correct epicyclic distance relation: the planet varies its distance as though it librates along the diameter of a small epicycle, such that $d = 1 + e - e\,\mathrm{vers}\beta$. In chapter 57, he described at length how such a variation in distance can be thought of as "natural," a topic that exercised Fabricius more and more once Kepler had introduced the final distance law and his bewildering construction of Mars's position using the ellipse and the area law. In chapter 58, he treated the *via buccosa*. Chapters 59 and 60 end part 4 with clear, formal numbered "protheoremata" (in a prominent, larger typeface, no less) proving the relationship of the area of the ellipse to a circumscribing circle and giving explicit instructions on the construction of Mars's position using the new theory. So, what purpose does chapter 58 fulfill? It serves as a justification for Kepler's very odd method of calculating Mars's position and for the fundamental redefinitions of astronomical terminology it entailed by demonstrating that the construction of Mars's position is *not* straightforward. Since the distance relation gave the distance in terms of the eccentric anomaly, it was only natural to think that its position could be constructed on the radius from the center (as, indeed, Kepler himself had done). But such a construction, proceeding as it does directly from the eccentric to the true anomaly, overlooks the mean anomaly altogether, which was governed by the area law. It is the bizarre construction of Mars's position at the intersection of the ellipse and the perpendicular dropped from a circumscribing circle to the line of apsides that mediates the relationship between the mean and eccentric anomalies. Kepler had shattered the standard definitions of astronomical terms and their relationships to one another, most especially the eccentric anomaly; it now referred to a circle that was not the planet's path in a direction the planet was not on. The *via buccosa* showed that trying to hold on to the old definitions, especially of the eccentric anomaly, would only lead to error. The necessity of recounting the error of the *via buccosa* became clear from Fabricius's total confusion about how to construct Mars's position and his bewilderment at Kepler's redefinition of astronomical terminology (to be discussed shortly) that arose when Kepler *omitted* this error in their private correspondence.

But the construction of Mars's position was not Kepler's only challenge in 1605 after he had discovered that its orbit was an ellipse. Having found the right distance model and the means to calculate the equations, Kepler lacked only one thing: a natural cause that he could substitute for the intellectual faculty. He shared this frustration with Fabricius:

> But there is also another thing that I desire in this hypothesis; namely that, though driving myself right to insanity, I cannot fashion a natural cause why Mars, to which a libration along the diameter should with so much probability be given (for the matter was being beautifully reduced totally to magnetic virtues), should rather want to go on an ellipse or a path very close to it. Nevertheless, I believe magnetic virtues probably do not entirely regard the sine but something or other.[95]

We find then another wonderful instance of Kepler trying to come to grips with a problem while writing:

> The eccentricity smacks of magnetic force in every respect, as it is in my *Commentaries*, such that if Mars had a magnetic axis striving toward the sun with one pole and fleeing with the other, and it were laid out along its axis in the middle distances, then as long as it is turned about in the descending semicircle it presents the attracting pole toward the sun, to the greatest extent in the middle distances, and so always approaches the sun, but to the greatest extent in the middle distances, and not at all in apsides. And then in the ascending semicircle it flees from the sun equally. And so perhaps (for it should be permitted me, most congenial Fabricius, while I talk with you to be kept working, and while working, to make progress) there is some law or another by which a magnet flees and follows something as the sine.[96]

Kepler's first analysis of how this might come about was unsuccessful. Treating Mars's globe as a circle, he considered how equal attracting and repulsing areas of this circle might cancel one another, leaving a segment whose area—and therefore strength—varied somehow as the sine of the eccentric anomaly. But then he broke off, saying that he did not see how it might come about.

After some weeks, though in the same letter, he returned to the Mars work, and immediately took up the same problem again with more success. Crediting Gilbert with the finding that one could treat a magnetic sphere not only as a collection of infinitesimal plane circles but also each circle as a collection of infinitesimal straight lines, Kepler proceeded with part of the familiar analysis of chapter 57 of the *Astronomia nova*. Viewing Mars's magnetic axis now as a line that retains its orientation in space, Kepler was able to show straightforwardly that the component along the magnetic axis, that is, the measure of the attractive or repulsive force, would vary as the sine of the angle from aphelion (not as the eccentric but as the coequated anomaly, but perhaps this was close enough). We should note in passing that although Kepler addresses and resolves several objections to this idea in the letter to Fabricius (for instance, that the earth's magnetic axis is not aligned perpendicular to the line of apsides), he does not employ the circular river analogy of chapter 57 of the *Astronomia nova*.

The major features of the *Astronomia nova* were now all in place, as Kepler triumphantly announced to Fabricius:

> Furthermore, at the same time you see both that that most earnestly desired union is now finally complete, and that that adulterated vicarious hypothesis has been eliminated. Everything I sought has been accomplished; the causes of each eccentricity are given. You have an astronomy without hypotheses.

> Of course it seems that up to now it had been an hypothesis when I said that Mars's eccentric is a perfect ellipse. But this was previously concluded from physical causes; it is not therefore a hypothesis in my *Commentaries*. It is indeed in the calculation, but it is also a true supposition of the true path of the planets, giving the distances and the equations.[97]

But Fabricius would not prove so easy to convince. Indeed, it is with the discovery of the ellipse that the seeds were sown for the ultimate breaking off of their correspondence.

The Collapse of the Kepler-Fabricius Correspondence

Fabricius lodged his first criticisms of Kepler's breakthrough in a letter of 11 January 1606 (o.s.).[98] When he had no answer from Kepler for more than a year, he repeated and greatly elaborated them in a letter of 20 January 1607 (o.s.). After lamenting not having heard from Kepler for so long, Fabricius got right to the point, first criticizing Kepler's hypothesis for violating the principle of uniform circular motion altogether:

> I see that the motion of Mars in the sky agrees with your new hypotheses in every respect. But the procedure for calculating is intricate and difficult. Moreover I would bring up something against your hypotheses generally. First, by means of your oval or ellipse you do away with uniform circular motion, which more than anything else seems to me unworthy of further consideration. As the sky is round, so it has circular motions, both regular and uniform around their centers to the greatest degree. The celestial bodies are perfectly round, as is manifest in the sun and moon. Therefore there is no doubt that their every motion also comes about by means of a perfect circle, not an ellipse or a departure [from a circle]. And they are likewise moved uniformly about their centers. And since in your ellipse the center is not everywhere equally distant from the circumference, a uniform motion will certainly be non-uniform to the greatest degree about its proper center. Therefore if while retaining a perfect circle, you can justify the ellipse by means of another small circle, it would be more appropriate. It is not enough to be able to save the motion, but one must also put together the kind of hypotheses that differ in the least extent from natural principles.[99]

Fabricius, like his contemporaries, thus showed himself deaf to Kepler's long-argued stand that his use of physical principles was natural. Once Kepler's labors had resulted in an orbit with no reference to uniform circular motion, Fabricius withdrew to the astronomers' orthodox position that only uniform and circular motions were appropriate.

Even beyond this objection to Kepler's failure to adhere to uniform circular motion, however, Fabricius had specific objections to Kepler's means of arriving at the ellipse, to his assertion that the variation in distance according to the sine was somehow natural:

> I am utterly convinced that the simpler something is the closer it will be to the truth, and the truth per se is itself simple. Do not take this liberty of mine badly. You know that I am tenacious and most passionate in seeking the truth. Nor can I be satisfied before the reasons for every unknown thing in the hypotheses thus far have truly been given. For I am of the opinion that we will never arrive at the discovery of the true hypothesis unless the causes of the motions are discerned more deeply. . . . And do it in such a way that circularity and equality of motions are retained astronomically and geometrically. I have worked four years seeking these things, and I work even now, and seek the astronomers' stone (as I would call it).
>
> You apply the secant in [the orbit of] Mars in accordance with the ellipse, but I want to know from you what cause might be furnished for constructing it with the secant, not larger or smaller than the sine or secant, for it is necessary for there to be a definite underlying cause, if you want to fashion the ellipse by means of the secant of the halved eccentricity and want it to correspond to the truth. . . . You ought not to interpret this to mean that I do not concede that the true motions can be given from your hypotheses. Not at all, for I see and experience this. But I scarcely believe those hypotheses to be natural, since your ellipse takes away circularity and equality of motion. It is absurd that the planet is really moved non-uniformly per se.[100]

He restates the objection similarly in a following letter:

> I want to know why the secant of half of Mars's eccentricity is taken for the lunula, or for the diminution of Mars's path, and not more or less. If you give me a truest cause, you will resolve many doubts for me. . . . Perhaps a semidiameter of the circle from the proper center, and from the sun (for I accommodate everything to the Tychonic hypothesis), drawn parallel to the circle makes that difference of motion, but nevertheless I do not see how it would be done.[101]

Fabricius's objection that this motion could not be "natural," I would maintain, was the origin of Kepler's circular river analogy in the *Astronomia nova*, chapter 57: "By what Natural Principles the Planet May Be Made to Reciprocate as if on the Diameter of an Epicycle."[102] The burden of this curious chapter was to show that the variation in distance Kepler was proposing had a "natural" analog in the case a boat traveling on a circular river. To my knowledge, Kepler had not developed this analogy to any significant extent anytime prior to Fabricius's objections.[103]

In addition to these general objections, Fabricius had a number of specific technical objections to Kepler's use of the ellipse. The largest part of these stemmed from Fabricius's utter confusion as to how to calculate the position of the planet with the ellipse and the area law. He had misunderstood Kepler's instructions and in fact overly complicated the procedure. He evidently believed that after determining the mean anomaly on the circumscribing circle, he should drop a perpendicular to the ellipse, extend a line through that point to the circle to get the eccentric anomaly, and drop another perpendicular to the ellipse from that point to find the true anomaly. Kepler's instructions admittedly had been brief, but how they gave rise to this level of misunderstanding is unclear. Nor is it clear why Fabricius wholeheartedly agreed that Kepler's theory was entirely accurate when he obviously had no idea how to use it. However, Fabricius's extreme confusion over what the true position and true distance were clearly contributed greatly to his refusal to accept the theory.

In addition, Fabricius demanded that, if possible, a method should be provided by which the parameters of the ellipse could be derived from three acronychal observations, analogously to classical astronomical theory:

> If your ellipse is conformable to a true hypothesis, you will also give from it the way in which the eccentricity and the apogee are to be found using three acronychals, or you will show the reason why those things cannot be found for your ellipse with three acronychals. If the motions correspond to the ellipse from every position, then you ought to show reciprocally, as though a priori, how the motions can be constructed with three acronychals, so that I persuade myself in all respects that it certainly can and should be done. As long as you cannot construct those things from three acronychals, the reason and hypothesis of the true motions will remain unknown. Neither the ellipse nor another fictitious form will satisfy the spirit, regardless of how it provides consonant motions in the sky.
>
> For this reason, my dear Kepler, exert yourself in showing that you can construct the eccentricity and apogee from three acronychals immediately and just as though a priori, and you will without hesitation cast aside your ellipse, and you will find the hidden truth rather in the departure from a circle.[104]

We should note Fabricius's continuing conditional acceptance of Kepler's findings (which we saw also in his query regarding the secant): if the hypothesis were indeed true, then Kepler would have to address these things to convince him. But there is another important undercurrent to Fabricius's queries. His insistence that Kepler should be able to provide a method to derive the hypothesis "just as though a priori"—that is, to derive the parameters starting with the proposition that the orbit is an ellipse—was part of an effort to get Kepler to detach the result of the ellipse from the methodology by which he derived it. This illustrates again the tension between traditional, mathematical, a priori astronomy and Kepler's new astronomy based on causes.

This query of Fabricius's was more than incidental, for it derived from Fabricius's rival effort to produce his own theory of Mars, one that would preserve uniform circular motion and give the mathematical causes he sought. His efforts evidently stemmed from his extreme dissatisfaction with Kepler's account of the ellipse, for Fabricius's renewed speculations began in the first letter after Kepler's triumphant announcement in which Fabricius gave his attention to the Mars theory. In this letter of 11 January 1606 (o.s.), Fabricius expressed his continuing frustration that different combinations of three acronychals did not produce the same aphelion and eccentricity, and he suggested that this might be accounted for by considering that the position of the aphelion and eccentricity vary with Mars's position on the eccentric. He proposed that the center of Mars's eccentric librates along the diameter perpendicular to the true line of apsides, in such a way that the variable line of apsides has its maximum deviation around quadrature. At first sight, the theory is indeed confused, but it did have the effect of speeding Mars through quadrature, for the deviation of the line of apsides to the side contributes to the motion of the planet most there. The theory also had the beneficial attribute, for Fabricius, that it increased the distance of the planet in the middle distances, as he had earlier pointed out to Kepler. However, the effect is absurd, in that it creates an orbit wider than it is long. For this reason, it is clear why Kepler did not give serious consideration to Fabricius's theory for some time.

Fabricius initially offered these thoughts to Kepler in the hope that Kepler might make something of them; he was himself too overcome with melancholy to pursue them.[105] Not having heard anything from Kepler, over a year later he developed the idea much further in letters of 5 April and 1 June 1607 (o.s.).[106] Fabricius's revived effort to develop his own theory may have spurred Kepler finally to answer his objections to the ellipse. Kepler's first letter since the announcement of the ellipse in October 1605 dates from 1 August 1607, shortly after receiving these two letters from Fabricius. And significantly, unlike most of Kepler's letters, this one is devoted exclusively to winning Fabricius over; it addresses no other issues from their correspondence.

Kepler begins "fighting with Fabricius," as he playfully puts it, with an extended defense of his principles against Fabricius's insistence on uniform circular motion:

> You think uniform motion is eliminated in an oval figure. By all means. But spiral figures eliminate the same thing for you, and the Ptolemaic equant eliminates it. In fact, even if Copernicus strives to restore uniform motion,

> nevertheless he does not reproduce that which is seen in the orderly movement of a planet. For in it, a planet proceeds non-uniformly, and furthermore it deviates from the path of a circle, which Copernicus himself acknowledges. But the principles, you say, by which that motion is brought about, namely circles, separately have uniform motions?
>
> I grant this, but not the motions, which bring about something in agreement with the phenomena. Besides, the principles by which the motion of a planet is brought about remain constant for me as well. The only difference is in that for you they are circles and for me they are corporeal virtues. Among other things, the rotation of the body of the sun is constant to me, and it is most uniform; the circulation of the immaterial and magnetic solar *species* is constant; the perception of that *species*, or motive virtue, in a remote planet at a fixed distance is constant; constant and most circular, albeit very slow, is the rotation of planet's body's axis, whence comes the progression of the apogees; constant is the magnetic virtue bringing together or separating the bodies of the sun and the planets, according to the angle of inclination of the axis of the planet to a line from the sun. Moreover, the way in which the planet passes from one level of virtue to another happens with an extraordinary agreement with the principles just now put forward.[107]

Having defended his physical principles, Kepler then attacked uniform circular motion. If Fabricius means the resultant path, Kepler wrote, then Copernicus's path was admittedly not circular, and Ptolemy's and Tycho's actual paths were complicated spirals. And if he meant the component circles, then he spoke of a fiction, for there were really no epicycles or orbs, as Fabricius ought to know as "an initiate of Brahean astronomy," that is, as one who knew the implications of the comet of 1577. Almost tauntingly, Kepler then sketched out a Copernican construction of his ellipse, but dismissed it, saying, "How does it help to feign the causes of the planet's oval motion?"[108] Finally, he appealed to their shared belief in natural principles, closing ranks against the instrumentalists:

> Finally, you say that it is not enough to be able to save the motions, but hypotheses must also be constructed that are in agreement with natural principles. I embrace this dogma of yours with complete unanimity, and it was the cause of much labor for me in the *Commentaries on Mars*. As regards the admonition, I want you to have done with Osiander, who wrote the unsigned preface to Copernicus's work, and also with Christian Severin [Longomontanus], both of whom think that it is enough that hypotheses satisfy the observations, notwithstanding that they are false.[109]

The bulk of Kepler's letter was devoted to rehearsing over and over again the technical details that baffled Fabricius, gently at first but with obvious and growing frustration as he systematically addressed four of

Fabricius's letters in which the same confusion was repeated. Most elementary was clarifying his use of the ellipse. Noting at first, "The ellipse is the natural hypothesis; the circle surrounding the ellipse is merely for the sake of calculation," Kepler proceeded to explain how to calculate the location of Mars on the elliptical orbit, and how and why it is necessary to use the circumscribing circle.[110] His explanation, as he went through it again and again, was more lengthy and much clearer than that which accompanied his previous announcement of the ellipse. He described repeatedly the redefinition of terms the ellipse required, until he finally exclaimed, "My dear Fabricius, if I were to hand over astronomy de novo, in such a way that it was not necessary for me to speak with the ancient words, I would use other ones. I would speak of delay, arc, angle, circle, the designation [*nomen*] of the elliptical arc, the measure of delay, the area of the circle."[111] With Kepler's redefinition, we witness a particularly vibrant example of Kuhn's incommensurability of paradigms, in which Kepler had to make clear the new ways in which traditional astronomical quantities had to be understood.[112] The necessary change in vocabulary became only one aspect of the difficulty Kepler knew he would face with the reception of his work.

Beyond Fabricius's failure to comprehend what Kepler was doing, there was at least one fundamental theme behind some of his criticisms. It had to do with the indeterminacy of Kepler's techniques. In the derivation of the vicarious hypothesis, and more importantly in the determination of the eccentric anomaly from the mean anomaly, Kepler had been dealing with mathematical operations for which there were no direct procedures for solving. His only recourse had been to approach the solutions of these problems iteratively, gradually refining values until the desired level of accuracy had been reached. Particularly in the case of solving for the position of Mars for a given time—perhaps the most standard problem in astronomical theory—the indeterminacy introduced by "Kepler's problem" was a most fundamental barrier to the acceptance of Keplerian astronomy throughout the seventeenth century. Indeed, much of the history of the reception of Kepler's work involved various geometrical approximations of the area law, which were addressed to overcoming this difficulty.[113]

Kepler acknowledged the difficulty, writing with respect to the derivation of aphelion from acronychals and the vicarious hypothesis, "It is artless, but the outcome is wholly determined and unique."[114] Still, this problem had concerned him. Back in 1600 when he had first run into the problem, he had appealed through Herwart to François Viète for a direct solution that would free him from this "gambler's artlessness."[115] Later, when he had run into the similar problem trying to calculate areas on ellipsoids and Fabricius had objected, Kepler challenged him:

> Now, what more is there that offends you? Is it because it cannot be calculated by a short cut? We lack nothing, my dear Fabricius, but geometry. Teach me geometrically to construct, square [i.e., find the area], and to cut ellipsoids in a given ratio, and straightaway I will teach you to calculate with the genuine hypothesis. So now, call upon those Netherlandish geniuses of yours to help me with this.[116]

By 1607 Kepler had apparently given up on a solution, but defended himself against Fabricius's charge that he could not simultaneously solve each part of the equation by invoking the example of Ptolemy, who also required two steps.[117] Still, he acknowledged, "The way [from the coequated anomaly to the mean anomaly] is geometrical albeit long, although on the other hand the way from the mean anomaly or time to the coequated anomaly is ungeometrical, [proceeding] by means of a fictional rule."[118]

Despite the acknowledged "artlessness" of some of his geometrical techniques, Kepler returned to the issue of the "naturalness" of his theory after having dispatched various of Fabricius's technical objections. Having addressed Fabricius's critiques may have reminded Kepler that they were one face of Fabricius's attempts to create his own theory in accordance with the "natural principles" of uniform circular motion. What is significant is that Kepler tried to turn Fabricius's criticism around on him. By emphasizing the simplicity of his own approach, Kepler sought to induce Fabricius to abandon work on his rival theory:

> You consider my subtlety worthy of praise, if it were not contrary to Nature. My dear Fabricius, I condemn all subtlety whether contrary to Nature or unnecessary. . . . Truly, you philosophize too broadly about the simplicity of truth. Nature is simple; it is also complex. Its simplicity is not to be judged according to our opinion but to itself. And, in fact, it would be astonishing if any simpler hypothesis had been brought forward than I put together, in which first, the planet first stays directed to the magnetic axis by an animal faculty, and with the advance of years does not turn at all; second, at the same time due to the magnetic corporeal virtue the planet approaches the sun according to the size of the angle of inclination of the axis to the sun; and third, the sun drives the planet in the orbit according to the *modulus* of its approach. This is genuine simplicity, manifest in the principles themselves. If from these very few things many things now follow—the physical part of the equation, the optical, the distance, the elliptical path—then in consideration of those many effects, will you deny that the principles are simple?
>
> It is truly astonishing, what kind of laws you describe to me for establishing the hypothesis taken from your mind and not from the heavens.[119] In plain words you demand from me as many things as I have refuted in chapters of the *Commentaries*. And so I pity your labor, in which now so many years

you attempt the impossible, and in a way pursue something already accomplished. For I claim most resolutely that the true cause of the motions has been found, to the extent that they can be comprehended by man.[120]

He restated this point more emphatically later in the letter:

Since your speculation errs on so many accounts, having been picked up from the gutter, do not wish to seek a revision from me. On the contrary, I apply the death blow, I will cut it down at the roots and destroy it with fire. I pity you your journey, who with such labor, as you say, finally attained this point where you are, that is, as I understand it, although in the meantime you might have slept. But you also freely supply suggestions as to what needs to be done so that those things of yours might be brought to their conclusion. You advise foolishly; Alexander taught me to untie the Gordian knot. And so I advise you not to attempt what is already accomplished. I tell you, if a hundred planets remained, provided that we had such observations of them as we have for Mars, it would be possible to arrive at their hypotheses by means of my invention, if only they mimic the nature of the seven known planets.[121]

Fabricius did not prove so easy to deter. The sheer incomprehensibility of Kepler's theory kept him from heeding Kepler's explanations. In spite of Kepler's assertion that the true theory was found, and perhaps reinforced by sixteenth-century astronomers' skepticism and knowledge that there were a number of ways to obtain equivalent results, Fabricius pushed on with the development of his own theory. The most remarkable aspect of Fabricius's theory of Mars was that he apparently arrived at an independent construction of an ellipse based on a wholly different derivation. Kepler had supplied him with a sketch of the construction of an ellipse based on uniform circular motion,[122] but it does not appear that this had any decisive influence. The principles on which Fabricius's derivation was predicated predated this construction, and it is not clear that he understood that they were the same in principle.

I will not here offer a detailed examination of Fabricius's Mars theory.[123] Its development was tentative and confused, and it is not clear whether Fabricius himself understood well what he was doing, but there are some general features that should be discussed. We saw above that the theory that Fabricius began to work out early in 1606 was founded in his frustration that three acronychals do not always provide the same aphelion and eccentricity. His response had been to assume that in addition to the real, fixed apsidal line, there was a mobile apsidal line, whose direction and eccentricity varied with the planets motion away from apsides, in such a way that the center of the planet's eccentric librated along a perpendicular to the line of apsides. The eccentricity and deviation of the apsidal line were at their greatest when the planet was at quadrants.

As we mentioned, the development and presentation of his ideas was confused, and Fabricius appears to have flipflopped on whether the deviation of the apsidal line was toward or away from the side the planet was on. He ultimately settled on the latter. This made a great deal of difference: If the deviation were toward the planet, the planet would be located farther than a simple eccentric at quadrants, and consequently Kepler would not have taken this idea seriously for an instant; but if it were away from the planet, Fabricius's construction could turn out to be an ellipse.[124] Moreover, because Fabricius's construction was expressed in terms of the action of a mobile apsidal line, it might not have been clear to either of them that Fabricius's theory was, in principle, a different construction of the same theory Kepler had given him.[125]

It is clear, however, that Fabricius was delighted with the results. It is of particular significance to this study that Fabricius found his theory superior to Kepler's on the grounds that it was easier and more *acceptable*:

> My dear, most learned Kepler, look now at each of our hypotheses, compare them, and decide which one would be easier and more convenient to prove and persuade. I think this: no easier hypothesis for the superior planets has ever been offered or can be offered. . . . What is more I also add this, that I show by means of this same construction of the hypothesis the truest reason why the true aphelion and eccentricity could not be given up to now using three acronychal observations. . . .
>
> I do not want, my dear Kepler, to call these inventions of mine more than dreams. I pray do not bother to test them, if you do not find the truth with equal ease and pleasantness. Only then sharpen the pen, and glance over them more imperiously. I did not want nor could I be quiet before I discovered this hypothesis and explored wholly and utterly the true causes of many mysteries hidden up to now, which I finally found now thanks to God after a thousand cares, innumerable calculations, and laborious wakefulness over the space of six years.
>
> I do not want my libration to call you to account, my dear Kepler, for which reason, is this not as agreeable to nature as your extraordinary speculations regarding your hypothesis? Even if your hypothesis saved the motions to the highest degree, still the method of operating by means of your hypothesis is so perplexing and laborious that it can frighten someone off even at first glance.[126]

Although Fabricius was convinced that his theory was better than Kepler's in principle, he still wanted Kepler to test it against observations and comment on it. It is a measure of Fabricius's confidence and delight in the novelty of the theory that he asked Kepler not to reveal it to anyone. In fact, in order to avoid word of the theory getting out, Fabricius mentioned that he had not even shown it to his own son, Johannes.[127] Nevertheless, he offered it for inclusion in the *Astronomia nova*, saying that if Kepler found it to his satisfaction (as he did not doubt) and wanted to append it to his *Commentaries*, then he would write out everything in more finished form, with demonstrations and so forth.[128]

Kepler did not reply to Fabricius's theory for several months. If we suppose that he did not get around to assessing the theory for some time because he did not believe it would work, we must imagine that he was stunned when he found it was sensibly equivalent to his own. (We must note that Kepler erred in assessing Fabricius's theory. Although it was elliptical, Fabricius's theory did not reproduce the area law as Kepler mistakenly thought it did.[129] What is important is that Kepler *thought* Fabricius's theory was equivalent.) The implications were immediately and devastatingly clear to Kepler:

> You have triumphed over this course, Fabricius, and you can brag that with your libration hypothesis you made things sensibly equal to my physical hypothesis. And so now I cease to ridicule your triumphal songs, I cease to give up hope for hypotheses from which the physical causes have been expunged. I cease to deny to the Emperor's mechanician [Joost] Bürgi, that it is impossible that the motion of Mars, which is brought about by physical causes, expresses itself in circles. For this libration of yours is most opportune to mechanicians, since it kills two birds with one stone. Now what? Should I now abandon my limitless schemes concerning physical causes? Should I transfer the nature of the heavens from the area of my triangle to your librations, confirming that your hypothesis is natural and rational because it is simple, and mine are fictitious and alien to nature because the area of the triangle measures times but an angle measures the optical equation? I will not do it, for that would not be to philosophize. Rather, I am glad for myself that this libration of yours remained concealed up to now, until the business of the physical causes was investigated. For I confess frankly, that if you had preceded my physical with your libratory form, and I had seen its agreement with observations, I would never have cut [my way] into the causes of the motions. For who, I ask, turning the librations over in his mind and seeing them to agree with observations, who, I say, would suspect that anything else needed to be investigated than the cause of this libration, inasmuch as it appears most true?. . . Who would think that all of those things can be brought about in an entirely different way by magnetic virtues?[130]

Who indeed? Perhaps even while writing this, Kepler saw the danger that Fabricius's theory held for his program of physical astronomy. He knew that the preponderance of astronomers would not follow him if a simpler mathematical alternative were available. He recognized that he could not allow this, and turned to attack Fabricius's theory:

> You say a daughter was born to you of mother geometry? I saw her; she is beautiful. But she will be a most mischievous whore, who will seduce the husbands away from my many daughters born of mother physics. Your hypothesis will attract to her lecturers and philosophers, she will give means of escape to the enemies of celestial physics, the defenders of ignorance, architects of solid orbs, dense mechanicians, those who would release themselves from the bonds of my physical demonstrations.[131]

Fabricius's libration hypothesis represented a clear danger to Kepler. He refused to append it to the *Astronomia nova*, despite Fabricius's subsequent sending of a neatly composed manuscript and his repeated suggestion that the theory be included.[132] Indeed, with his attack on Fabricius's theory, Kepler effectively broke off the correspondence.[133]

Kepler believed that Fabricius had created an adequate theory in the classical form based on uniform circular motion from which consideration of physical causes had been "expunged."[134] Such a theory flew in the face of what Kepler had been pursuing for so long, and threatened to make meaningless Kepler's astronomical work, in which he strove to establish the truth of heliocentrism for the glory of God. Such a theory was thus simply incompatible with Kepler's rhetorical appeal in the *Astronomia nova*. The argument of the book is predicated on the fact that only by consideration of the physical causes of the motion was Kepler able to find an adequate theory. This may have been true enough, but it did not preclude the possibility of a classical theory existing. Appending Fabricius's theory would have eviscerated the argument.

Kepler had found in Fabricius a kind of test case for the acceptability of the new astronomy, one who played a decisive role in determining the content and rhetorical character of the *Astronomia nova*. The discovery of how difficult it was to make Fabricius understand his work, compounded by Fabricius's success at recasting Kepler's work into compounds of circular motion, reinforced Kepler's resolution so to intertwine physics and astronomy that they could not be separated. As he explained to Fabricius:

> Indeed I will rather do this, I will deeply interweave and entwine Copernicus into the amended astronomy, and so also into physics, such that either each will perish at the same time or both will survive.[135]

⌘ CHAPTER 9 ⌘

THE RHETORICAL CHARACTER OF THE *ASTRONOMIA NOVA*

The Context of the *Astronomia nova*

Kepler's idea of a physical astronomy, of a defense of Copernicus based on physical principles, long preceded the publication of the *Astronomia nova*. We have seen that the foundations of Kepler's thought existed as early as his student disputations in defense of Copernicus from Tübingen, that he worked out the principles of his planet-moving force in the *Mysterium cosmographicum*, and that he had an eye on these principles from the start of his research with Tycho Brahe through the discovery of the elliptical orbit of Mars. We have seen also that Kepler encountered resistance to his ideas all along the way. This resistance took the form of both theological objections to his disputation and the material on the reconciliation of heliocentrism and Scripture he had wished to include in the *Mysterium*, and astronomers' objections to his introduction of physical considerations into planetary theory.

His succession to the post of imperial mathematician had made the situation more acute. On the one hand, Tycho's death had given him the opportunity to pursue his research free from Tycho's hostile oversight, and his success in accounting for the most comprehensive collection of observations in existence would undoubtedly give great weight to the truth of Kepler's theories and his program. On the other hand, the interference of the Tychonics posed a serious threat to Kepler's being allowed to publish anything at all based on Tycho's observations. Furthermore, in the jockeying for imperial favor, Kepler had lost responsibility for the high-profile *Rudolphine Tables* and had been put in the position of having to get something published to justify his employment. He had thus chosen to publish his "key to a deeper astronomy" in the form of a commentary on Mars.

In addition, throughout the period during which Kepler completed his research for the *Astronomia nova*, he had been receiving feedback from David Fabricius. In the course of their correspondence, it became increasingly clear to Kepler that even a sympathetic astronomer with experience in the problem of Mars's orbit, like Fabricius, ultimately wanted nothing to do with his physical astronomy. When Kepler finally discovered how to employ the area law properly, the complexity of his physical astronomy greatly increased and his differences with Fabricius became irreconcilable.

In the light of the constraints under which the *Astronomia nova* was produced, we must examine the work with two important questions in mind: What was Kepler trying to establish, and how did he hope to accomplish this?

At its most fundamental level, Kepler's astronomical work was devoted to the glory of God. This is clear from the time he resolved the conflict of his religious calling and his astronomical vocation,[1] through the *Harmonice mundi*, which closes with the prayer:

> O Thou, who by the light of nature increases in us the desire for the light of Thy mercy in order to be led by this to Thy glory, to Thee I offer thanks, Creator, God, because Thou has given me pleasure in what Thou hast created and I rejoice in Thy handiwork. See, I have now completed the work to which I was summoned. In doing so I have utilized all those powers of my mind which Thou hast loaned me. I have shown man the glory of Thy works, as much of their unending wealth as my feeble intellect was able to grasp.[2]

For Kepler, the Copernican system was the ultimate manifestation of God's design of, and presence in, the world.[3] All his astronomical work was thus to a large extent concerned with establishing the truth of heliocentrism.[4]

That the *Astronomia nova* was devoted to establishing the physical truth of the Copernican system becomes clear if we review Kepler's decision to publish the work. Kepler undertook to write the *Astronomia nova* in October or November 1602, long before the discovery of the ellipse and elucidation of the area law. Furthermore, those "laws" were sufficiently buried in the book that we must dismiss the idea that the book was devoted to their announcement. At the time he resolved to write the book, he had written to Herwart von Hohenburg that his work on Mars had allowed him to solve the problem of the orbit of the sun (or earth). And in many ways, the *Astronomia nova* was as much about the earth's orbit as it was about Mars's orbit.

To understand why the problem of the earth's orbit occupied such a central position in Kepler's thought and his defense of Copernicus, we must bear in mind that Kepler had been concerned with physical astronomy ever since he wrote the *Mysterium cosmographicum*. In the *Mysterium*, he proposed that consideration of a planet-moving force could explain the planets' orbital motion. He showed the close equivalence of the Ptolemaic and Copernican planetary models, and recognized the promise of his own model of the equant for accounting for planetary motion in the absence of material planetary spheres. Still, there were inconsistencies with his model, the most important of which was that the earth had never been given a bisected eccentric, which his physical model naturally required. Since the time of Hipparchus, the orbit of the sun had always been treated as a simple eccentric. The discovery Kepler made while working under Tycho, that the earth's orbit could appropriately be represented with a bisected eccentric like the other planets, was a major confirmation of Kepler's physical ideas. Attendant discoveries regarding the orbits of Mars, Venus, and Mercury—such as that the variable eccentricity Tycho had proposed for Mars was an artifact of the simple eccentric he had used for the sun's orbit—offered further confirmation. At the time he pledged to undertake the *Astronomia nova*, Kepler was prepared to publish a highly significant simplification of planetary theory. All planets, he would have shown, had the same orbit: a simple bisected eccentric. Changing the theory of the earth was the essential step in his simplification. Furthermore, the changes Kepler now proposed were directly in line with his physical principle of a planet-moving force coming forth from the sun and pushing the planets around their orbits.

During the course of reworking the orbit of Mars to fit precisely into this scheme, Kepler was forced to refine his physical principles. It had become clear that Mars's orbit was not precisely circular and that the calculating of the cumulative effect of the planet-moving force was a complicated procedure. By the time he resolved to publish the *Astronomia nova*, he knew that Mars's orbit was some kind of oval, but the nature of that oval was then of little consequence. The fundamental argument of the book remained focused on the reformation of the theory of the earth and the establishment of the truth of heliocentrism through physical principles.

Establishing the physical truth of heliocentrism, however, required more than the suggestive finding that the earth had the same type of orbit as the other planets. Kepler was fully aware of the facility with which astronomers could transform hypotheses from heliocentric to geo-heliocentric to geocentric. Furthermore, as is clear from his *Apologia pro Tychone*, he was aware of the general ambivalence of astronomers toward the truth of astronomical hypotheses, especially at the level of mathematical planetary theory rather than at the cosmological level. Thus the *Astronomia nova* was as much about astronomical methodology as planetary theory. Kepler's proof of heliocentrism depended on the validity of physical astronomy. However, establishing the validity of physical astronomy itself entailed a redefinition of astronomy. Although cosmology was subject to physical considerations, astronomical theory was still widely regarded as a purely mathematical pursuit.

The Rhetorical Nature of the *Astronomia nova*

The constellation of social and intellectual pressures we have just reviewed provided the context for the book that Kepler was to write. Early in its composition, he concluded that a historical presentation of his ideas would be necessary, a technique that he himself acknowledged to be rhetorical. He expressed this quite plainly in the *Astronomia nova*:

> The scope of this work is not chiefly to explain the celestial motions, for this is done in the books on Spherics and on the theories of the planets. Nor yet is it to teach the reader, to lead him from self-evident beginnings to conclusions, as Ptolemy did as much as he could. There is a third way, which I hold in common with the orators, which, since I present many new things, I am constrained to make plain in order to deserve and obtain the reader's assent, and to dispel any suspicion of cultivating novelty.
>
> No wonder, therefore, if along with the former methods I mingle the third, familiar to the orators; that is, an historical presentation of my discoveries. Here it is a question not only of leading the reader to an understanding of the subject matter in the easiest way, but also, chiefly, of the arguments, meanderings, or even chance occurrences by which I the author first came upon that understanding.[5]

Kepler himself invoked the contrast between his work and standard astronomical texts. The *Astronomia nova* is indeed markedly different from classical astronomical works, such as Ptolemy's *Almagest* and Copernicus's *De revolutionibus*, in which the exposition of planetary theories proceeded systematically with few clues as to their source or development. Furthermore, he apologetically justified borrowing the style of the "orators," a term that in Latin refers to rhetoric,[6] for whom narrative was a conventional technique.[7] It must be stressed that Kepler never claimed that what he was offering was a true historical account of his research. He stated explicitly that a full account of his work "would be boring and pointless to recount," and that "I shall describe only so much of that labour of four years [1601–1605] as will pertain to our methodical enquiry."[8]

Given that rhetoric in its broadest sense encompasses all human communication,[9] including all levels from the structure of a work to word choice and meter, the mere statement that the *Astronomia nova* is rhetorical might seem a commonplace. However, given Kepler's avowal of the rhetorical nature of the book, at least some consideration must be given to what this meant to him. At the same time, let me acknowledge that my examination of the rhetoric of the *Astronomia nova* will be neither comprehensive nor grounded in a thorough study of the art.

It is important to note that the *Astronomia nova* was rhetorical not only in the sense of being persuasive but also in a classical sense. Kepler, like anyone with a sixteenth-century university education, was no stranger to classical rhetoric. He had taught rhetoric in Graz,[10] and the *Apologia* was tightly organized along the lines of classical judicial rhetoric.[11] Nor should Kepler's open acknowledgment of the rhetorical character of the *Astronomia nova* be overlooked.

According to Aristotle, dialectic—the art of probable reasoning through syllogisms—was the proper form of philosophical discussion. Rhetoric differed from dialectic in its play on the emotions of the audience. A rhetorical philosophical book, especially a mathematical one, thus required some apology.[12] Although Kepler felt it necessary to legitimate the rhetorical nature of the *Astronomia nova*, he might have taken comfort in Aristotle's opinion that "rhetoric is useful [first] because the true and the just are by nature stronger than their opposites, so that if judgments are not made in the right way [the true and the just] are necessarily defeated [by their opposites]. And this is worthy of censure."[13]

Aristotle defined three persuasive elements in rhetoric derived from *ethos*, the character of the speaker, *pathos*, emotion awakened in the audience, and *logos*, the argument.[14] All these elements must be borne in mind while considering the *Astronomia nova*.

The ethos Kepler cultivated was that of an honest seeker after truth. In considering the rhetorical character of some of Kepler's letters, we have alluded to his appeal to his long labor when defending his work. This was a major element in the *Astronomia nova* as well. But he was also anxious not to be seen as a "cultivator of novelty," one who would change astronomy for change's sake. It was thus his intention to portray the changes he was defending as necessary. For example, although we have seen that physical considerations guided his work from the outset, the argument of the *Astronomia nova* presents the illusion that he resorted to physical causes only after the utter failure of classical theory.

Kepler's emotional appeal to his audience has long been seen as noteworthy, for it makes the *Astronomia nova* engaging. However, it has not always been recognized for what it is: the deliberate use of pathos. Examples abound. One of the most often cited had to do with the iterative derivation of the vicarious hypothesis:

> If this wearisome method has filled you with loathing, it should more properly fill you with compassion for me, as I have gone through it at least seventy times at the expense of a great deal of time, and you will cease to wonder that the fifth year has now gone by since I took up Mars.[15]

There was much that Kepler had to include in his argument to show why he had proceeded as he had. By involving his audience in the frustration of this effort he hoped, on the one hand, to establish the necessity of what he had done and, on the other, allow them to share in the joy of his triumph:

> Thus, in telling of Christopher Columbus, Magellan, and of the Portuguese, we do not simply ignore the errors by which the first opened up America, the second, the China Sea, and the last, the coast of Africa; rather, we would not wish them omitted, which would indeed be to deprive ourselves of an enormous pleasure in reading. So likewise, I would not have it ascribed to me as a fault that with the same concern for the reader I have followed this same course in the present work. For although we by no means become Argonauts by reading of their exploits, the difficulties and thorns of my discoveries infest the very reading—a fate common to all mathematical books. Nevertheless, since we are human beings who take delight in various things, there will appear some who, having overcome the difficulties of perception, and having placed before their eyes all at once this entire sequence of discoveries, will be inundated with a very great sense of pleasure.[16]

While designed to arouse the emotions of the audience, and—in a sense—entertain, Kepler's narrative also played the role of unfolding the logos. Although it has beguiled many of its readers, the argument of the *Astronomia nova* is actually very purposefully constructed. Kepler provided a number of avenues at various levels of detail to approach the book's argument, or to recover it if the reader had become lost. He wrote that he would include only material that contributed to his argument; we must therefore approach all of it with an eye toward how it serves its purpose.

It will not be our intention to review in detail the astronomical aspects of Kepler's exposition.[17] We will rather seek to explain what Kepler hoped to establish and to comment on the means he employed to do so. Consequently, we will focus first on the book's front matter, i.e., the title page and dedications, to illustrate how Kepler turned this publishing apparatus to his advantage. Then we will turn to the book's introduction and how Kepler hoped to appeal to his audience of "physicists." Finally, we will consider the book's line of argument in detail and how it was meant to be read by Kepler's audience of mathematicians.

Front Matter

In full, the title of the *Astronomia nova* read: *New Astronomy Based Upon Causes [ΑΙΤΟΛΟΓΗΤΟΣ], or Celestial Physics, Treated by Means of Commentaries on The Motions of the Star Mars, from the Observations of Tycho Brahe, Gent., by Order and Munificence of Rudolph II,*

Emperor of the Romans, &C., Worked Out at Prague in a Tenacious Study Lasting Many Years by His Holy Imperial Majesty's Mathematician Johannes Kepler.[18] Lest a lengthy examination of the title page seem picayune, we should recall Kepler's discussion with Maestlin regarding the title of the *Mysterium*, when he expressed his concern that it not "deceive the community of mathematicians."[19]

When he first conceived of the *Astronomia nova*, Kepler had mentioned as a possible title "The Key to a Universal Astronomy," a title that reflects the unifying character of Kepler's discoveries regarding the earth's orbit. (This title appears to have survived in part as the subtitle of part 3—the crux of the book—"The Key to a Deeper Astronomy.") Two years after conceiving the book, Kepler reported to the University of Tübingen that he had discovered such things "that astronomy can simply be regarded as new." Kepler's "New etiological astronomy, or celestial physics" succinctly proclaimed the ultimate goal of the book: that astronomy could and must include consideration of physical causes. Here he was squarely attacking the objection Maestlin and Tycho had raised against his work, that he not extend his physical ideas into planetary theory itself. Astronomy would no longer be insulated from physics.

Astronomy had occupied an uncomfortable position in the classification of the sciences since Aristotle, who had classified it as mathematics in the *Metaphysics* but more like physics in *Physics* and *De caelo*. This classification was further complicated by the changeless nature of the heavens, which made them more a metaphysical than a physical object.[20] The arrival of Ptolemaic theory in the second century A.D., which was fundamentally incompatible with Aristotle's account of the heavens, perhaps decisively rent astronomy into separate mathematical and physical domains. There were, however, serious attempts to reconcile Ptolemaic theory with the idea of celestial spheres by Ptolemy himself and notably by medieval Arabic scholars, whose influence on Copernicus seems clear.[21] The flowering of scholasticism in Europe, on the other hand, entrenched Aristotelian philosophy and gave strength to the dichotomy forged by ancient Aristotelian commentators between the physical and purely mathematical aspects of astronomy. By the end of the sixteenth century, the distinction was eroding as astronomers adopted a variety of views toward the status of astronomy.[22] However, even if astronomers held differing views over the reality of heliocentrism or the existence of epicycles, what Kepler was proposing was more radically disturbing, for his approach subordinated astronomical theory to physics. In this sense, his etiological astronomy was truly new.

The primacy of Kepler's new astronomical methodology is further emphasized by the statement that it is "treated by means of commentaries" on the motions of Mars. We mentioned above that the *Astronomia nova* should not be seen as a treatise devoted primarily to the elucidation of Mars's orbit. The commentaries on Mars in one sense were merely the vehicle to support his findings. This is not to say that the relation of the book to the theory of Mars was incidental. The word *Martis* was equaled in size on the title page only by Rudolf II's name. And Kepler did state in the introduction that he would cheerfully pursue the reformation of astronomical theory. However, the emphasis on Mars in the title does obscure the fact that some of the book's most important findings had to do with the theory of the earth, although observations of Mars supported them (which, as we saw, caused Fabricius some misgivings).

Kepler was obliged, of course, to mention that he was using Tycho's observations and that his work was supported by the beneficence of the emperor. But mentioning Tycho also had the desirable effect of drawing attention to his own work. Tycho had publicized his labors and the promise of his observations in his *Astronomiæ instauratæ mechanica* (1598; reprinted 1602). Even before the publication of the *Astronomia nova*, the publication of the *Rudolphine Tables* was avidly awaited.[23] As Tycho's successor and the only one working from his observations, Kepler's work would command attention. Finally, let us note that even on the title page, Kepler began establishing his *ethos* with the statement that his work was "worked out at Prague in a tenacious study lasting many years."

Immediately on the verso of the title page, Kepler set out to defend the realism and truth of heliocentrism. He began by quoting Ramus's condemnation of the manifest absurdity of Copernicus and his offer of his professorship at Paris to anyone who could construct "an astronomy without hypotheses." In his response to Ramus, Kepler claimed to have succeeded in the *Astronomia nova*, if one did not "exclude [from astronomy] the support of Physics, which it can by no means forgo."[24] And Kepler was already preparing to defend himself against the charge that admitting physics into astronomy was absurd and Copernicus against the implication that he had not considered his hypothesis real:

> Thus with the same facility I, too, admit something commonly considered most absurd philosophically, defending it, not with a gigantic effort, but with the best arguments. For when it functions, it effects nothing new, nothing unaccustomed, but only fulfills the function for which it was invented.
>
> It is a most absurd business, I admit, to demonstrate natural phenomena through false causes, but this is not what is happening in Copernicus. For he too considered his hypotheses true, no less than those whom you mention considered their old ones true, but he did not just consider them true, but demonstrates it; as evidence of which I offer this work.[25]

Kepler then proceeded to identify Andreas Osiander for the first time in print as the author of the anonymous preface to *De revolutionibus*,[26] and asserted, "Thus Copernicus does not mythologize, but seriously presents paradoxes; that is, he philosophizes. Which is what you wish of the astronomer."[27]

We have seen in his comments to Maestlin and Fabricius that Kepler had long framed his work with reference to Ramus,[28] but why did he confront these issues on the back of the title page? It is possible that Kepler conceived of his comments as some form of advertisement to the reader, or that he remembered to insert them only late in publication, but we can also consider a more intriguing interpretation. Tengnagel was exercising his right to attach a preface to Kepler's work, and Kepler knew of the confusion engendered by Osiander's anonymous preface to *De revolutionibus*, which had misrepresented Copernicus's views about the reality of heliocentrism for more than sixty years. From these facts, we can only suggest that Kepler's comments on the verso of the title page were meant in some way to inoculate his work against such a fictionalist preface and to establish his own realist stance beyond any doubt. If this interpretation is correct, the act was warranted for we saw above that Tengnagel's preface urged the reader not to "be moved by anything of Kepler's, but especially his liberty in disagreeing with Brahe in physical arguments."

Rhetorical flourishes were to be expected especially in the *Astronomia nova*'s letter of dedication to Rudolf II. But one should not be so entranced by Kepler's detailed humanistic allegory of his warfare with Mars to overlook its purpose. For in the dedication, Kepler set about establishing his ethos. He began, of course, extolling the glory that Mars's bondage brought to the House of Austria; this was simply good politics. Mars's restraint by the "bonds of Calculation" would bring great joy, Kepler wrote, in "that part of my profession in which strife arises between me and my fellow soldiers," that is, astrology.[29] Kepler would leave the astrologers to their rejoicing, and instead "occupy myself with Astronomy, and, riding in the triumphal car, will display the remaining glories of our captive that are particularly known to me, as well as all the aspects of the war, both in its waging and in its conclusion."[30] Even in his allegory, Kepler would present his triumph in the form of an historical account.

Kepler began his allegory with his own theological motivation for devoting himself to astronomy, thinly disguising God as Jupiter,

> the eternal Architect of this world, and the Father of Heavens and Humans in common, Jupiter, located [Mars] in the front lines of the visible bodies . . . so that he might raise human minds, lulled to sleep by a deep somnolence, from the slanderous reproach of idleness and ignorance, arouse them to venture forth, and provoke them forcefully to carry out investigations in the heavens for the praise of their Creator.[31]

Kepler then described the worthiness of his foe. While taking a stab at the artifices and fictitious hypotheses of previous astronomers, he also made the point that previous theories were incapable of describing

Mars's motion, which was to be a central feature of his argument: "It is he who is the most potent conqueror of human inventions, who, ridiculing all the sallies of the Astronomers, escaping their devices, and striking down the hostile throngs, kept safe the secret of his empire."[32] The key to Kepler's triumph, in contrast, was precisely his refusal to accept conventional theory:

> Nevertheless, here too, as in other kingdoms, the ruling influence of our enemy has been sustained and supported, more than any other thing, by the persuasion and confusion of the multitude of people, the defiance of which I have always considered the path to victory. Indeed, when I was but indifferently well versed in this theater of Nature, I formed the opinion, with practice [*usus*] as my teacher, that, just as one human being does not greatly differ from another, neither does one star differ much from another, nor one opponent from another, and hence, no account is to be received easily that says something unusual about a single individual of the same kind.[33]

Kepler alluded here to the fact that the secret of his triumph lay in his conviction that the planetary theories are the same, which we have seen to be an essential feature of his work stretching back to the *Mysterium*.

Kepler did not fail to emphasize Tycho's role as "commander-in-chief in this war," who "explored the habits of this enemy of ours nearly every night for twenty years, observed every aspect of the campaign, detected every stratagem, and left them fully described in books as he was dying."[34] From these observation books, Kepler ceased to fear his enemy as he came to know him. It seems that Kepler even alluded to the utility of the completeness of Tycho's observations in a charming description of what must be his technique of triangulating to a single position of Mars from different positions of the earth:

> [H]aving diligently noted the moments of time at which he was accustomed to arrive at his former positions, as if going to bed, I directed the Brahean machines thither, equipped with precise sights, as if aiming at a particular target, and besieged each position with my enquiry as the chariots of the great Mother Earth were driven around in their circuit.[35]

Having earlier raised the importance of his own unorthodox approach in bringing Mars to submit, Kepler now emphasized his tenacity in continuing his pursuit in the face of innumerable defeats, which he would relate at length in the chapters that followed:

> The campaign did not, however, succeed without sweat, since it frequently happened that machines were lacking where they were most needed, or that they were transported over muddy roads by inexperienced charioteers at great expense of time and material, or that the launching of some of them,

where I had not yet investigated the matter, occurred in other directions than I had had in mind. Often the brightness of the sun or of the moon, and often an overcast sky, cheated the commander's eyes; and more often the interposition of vapourous air deflected the globe, forcing it from the straight path. Also not infrequently, the walls, where they were presented most obliquely, received ineffectual blows, however numerous they might be. Add to this the enemy's enterprise in making sallies, and his vigilance for ambuscades, while we were frequently asleep. Also, his constancy in defence: whenever he was driven or fled from one castle, he repaired to another, all of which required different means to be conquered, and none of which was connected to the rest by an easy path—either rivers lay in the way, or brambles impeded the attack, but most of the time the route was unknown. Each of these things is thoroughly described in its own place in this commentary.

Meanwhile, in my camp, is there any sort of defeat, any kind of disaster that has not occurred? The overthrow of the Most Distinguished Leader, rebellion, plague, pestilences, domestic matters both good and bad, destined in either case to take time, . . . desertion and poverty of the soldiers; the inexperience of novices; and, at the head of all, the extreme deficiency of provisions.[36]

The next passage is obscure, but alluded to Kepler's discovery of the correct distance law (ch. 56) and his subsequent justification of how this might be considered the result of natural principles (ch. 57):

At last, when he saw that I held fast to my goal, while there was no place in the circuit of his kingdom where he was safe or secure, the enemy turned his attention to plans for peace: sending off his parent Nature, he offered to allow me the victory; and, having bargained for liberty within limits subject to negotiation, he shortly thereafter moved over most agreeably into my camp with Arithmetic and Geometry pressing closely at his sides.[37]

Finally, Kepler referred to the error of the *via buccosa*, the incorrect application of the distance law (ch. 58), and the resolution of this problem (chs. 59–60):

However, from the time when, after surrendering, he abode by our house's fair laws of friendship, he, through hidden illusions (being unaccustomed to rest), did not cease to incite among us I know not what further fears of war, and if we happened to become terrified, we would give him much to laugh at. But, seeing us strong in spirit, he agreed to live with us in earnest, and, dropping the appearance of hostility, confirmed his faith with us.[38]

Not missing an opportunity to solicit further patronage, Kepler closed by passing on a plea from his captive to the emperor, that he might supply

Kepler with the funds to enlist an army to bring Mars's relatives to share his captivity.

Kepler continued pressing his point even in the epigrams following the dedication, another element of humanist publishing apparatus.[39] In his response to Tycho's "Hortatory Ode" (a typical Tychonic exhortation to youths to lift their minds from earthly pursuits to assist in the reformation of astronomy), Kepler honored Tycho's accomplishments—of course—but pointed out his own as well. Taking as his cue Tycho's call to "Firm up the coffered ceiling of heaven with sturdy new crossbeams," Kepler recalled that he had filled those "Great gaping spaces that never were braced by the placement of columns" with "Famous timbers from Samos, the five regular solids."[40] Even in praising Tycho, Kepler subtly alluded to Tycho's conservatism, the sense only Kepler himself could make of the observations, and the adequacy of Kepler's theories alone:

> Wonderful is your daring, O Brahe, and sweet your labour,
> Even though you preferred not to stray from received opinion,
> Doubting much above earth, and many things in the heavens.
> Nevertheless I was pleased to be numbered among your disciples,
> Spreading your nights out before me, and secrets you've found in your searching,
> Over long years, and to shine a clear light on your great undertakings.
> Would that you had lived on, that the Fates had never snatched from you
> Prizes matching such deeds, and such well-merited triumphs.
> No other orbits spread themselves out for your vision or subtle
> Instruments, other than those that are buttressing my new crossbeams:
> You would have been the expert on "great heaven's innermost secrets."[41]

Kepler's "new crossbeams" referred to his motive forces. He had found no other orbit than that which supported them.

Similarly, in the elegy that followed, Kepler described in poetry his disagreement with Tycho, his physical astronomy, and drove home the uniqueness of his solution.

> Though for me the earth swims starlike in airy gyrations,
> You see the same earth stand fixed in the place that is central:
> I am inclined to ascribe these beliefs to the Ancient Masters.
> They were not mine; they displeased you while you were living. . .
> Without you repeatedly looking up carefully using your diopter,
> I could not look down from there on the earth's course,

Nor measure its swift paces and reining in around Capricorn [i.e., aphelion],
And how many parts give you, Phoebus,[42] the center of your way.[43]
That by a similar step it might seek and flee from the sun, while
Nonetheless not being spun with a spin like the Lord's, but
Gathering forces as it approaches its source, and in turn,
Languishing as it retires to more distant places;
Whence the seven Globes are borne by the sevenfold Minds in
Sequence, and by the eighth soul, from Father Sun.
Nature thus is exempted from numberless twistings and turnings:
Nine times five [Aristotelian orbs] thus depart from the family of God.
Cheat against reason in tens, O Tycho; cheat it by minutes,
Which none but you would number: the whole would collapse.
O human cares—the amount of vanity in our affairs! To
Think that the stars can't be reached by a different road.[44]

The *Astronomia nova*'s Introduction

After the dedication and poems, Kepler presented his long introduction, which began with the curious lament, "It is extremely hard these days to write mathematical books, especially astronomical ones."[45] The difficulty Kepler confronted from the very beginning of the book, and indeed in his own work generally, was that it did not have a clear audience. On the one hand, there were astronomers, who—whatever their views toward Copernicanism—were not prepared to accept physical reasoning into astronomical theory. Both Maestlin and Tycho had made their opinions clear in this regard. On the other hand, there were natural philosophers, whom Kepler wanted to convince of the truth of heliocentrism but who he knew were unlikely to be mathematically competent to follow the details of his argument. The *Astronomia nova* was thus presented in such a way as to be accessible to both audiences.

The introductory matter to the *Astronomia nova* was meticulously crafted to lead the reader through Kepler's complicated argument. In addition to the introduction itself, it included a fold-out synoptic table of contents and lengthy summaries of the individual chapters. That the introduction itself was meant primarily for "physicists" and the synoptic table and chapter summaries for astronomers was made clear in the introduction, especially at the end, where Kepler wrote, "All this has been said for the sake of the physicists. The astronomers and geometers will find the rest in the following summaries of the individual chapters, each in its proper place."[46] To accommodate his audience of "physicists," Kepler therefore began the introduction by playing down the mathematical aspect of the treatise, acknowledging that rigorously mathematical books were tedious and that few were suitably prepared to read them. He even offered the ingratiatingly candid admission that "I myself, who am known as a mathematician, find my mental forces wearying when, upon rereading my own work, I recall from the diagrams the sense of the proofs, which I myself had originally introduced from my own mind into the diagrams and the text."[47]

Since he could not count on nonmathematicians to follow the burden of his proofs, Kepler recounted his argument in the introduction in general terms. And "for the sake of those professors of the physical sciences who are irate with me, as well as with Copernicus," he declared that he would openly lay out the physical principles of his proofs. As "a professional mathematician," he vouched for the mathematical validity of the proofs. (He further simplified the reading of the book for those not interested in geometrical demonstrations by having the demonstrations set in italic type, leaving those parts where physics was mixed in with geometry in plain type,[48] presumably to facilitate skimming over the purely geometrical passages.) He left the physicists to the examination of his principles with the knowledge "that unless they are refuted the proof erected upon them will not topple." He admitted that because he was mingling celestial physics with astronomy there would be "a certain amount of conjecture" and the drawing of probable conclusions, but such dialectical reasoning was entirely in accordance with physical science. One of the notable features of the *Astronomia nova*, in contrast to Kepler's youthful *Mysterium*, was that he became much more careful about the distinction between astronomical and physical arguments. Kepler was more circumspect in the *Astronomia nova*, always acknowledging when he would apply a physical conjecture to an astronomical finding.

After a brief description of the primary world systems, in which he acknowledged that, with a few adjustments, they "are for practical purposes equivalent to a hair's breadth," he proceeded to the twofold goal of the work and, indeed, a summary of the fundamental argument of the work:

> My aim in the present work is chiefly to reform astronomical theory (especially of the motion of Mars) in all three forms of hypotheses, so that our computations from the tables correspond to the celestial phenomena. Hitherto, it has not been possible to do this with sufficient certainty. In fact, in August of 1608, Mars was a little less than four degrees beyond the position given by calculation from the Prutenic tables. In August and September of 1593 this error was a little less than five degrees, while in my new calculation the error is entirely suppressed.
>
> Meanwhile, although I place this goal first and pursue it cheerfully, I also make an excursion into Aristotle's *Metaphysics*, or rather, I inquire into celes-

> tial physics and the natural causes of the motions. The eventual result of this consideration is the formulation of very clear arguments showing that only Copernicus's opinion concerning the world (with a few small changes) is true, that the other two are false, and so on.
>
> Indeed, all things are so interconnected, involved, and intertwined with one another that after trying many different approaches to the reform of astronomical calculations, some well trodden by the ancients and others constructed in emulation of them and by their example, none other could succeed than the one founded upon the motions' physical causes themselves, which I establish in this work.[49]

Despite Kepler's announcement that his calculations eliminated previous errors and his statement that his chief goal was the reformation of astronomical theory, he had little more to say about observational adequacy in the introduction. Instead he focused entirely on how his work confirmed the physical causes of the planets' motions and therefore the Copernican system.

The essential feature of his argument, which we have seen expressed already in his correspondence, was that "[no] other [approach] could succeed than the one founded upon the motions' physical causes themselves." The argument that Kepler was unable to produce an adequate theory of Mars without resort to physical causes did not have demonstrative certainty. (Nor was it necessarily true, especially if we consider that Kepler believed Fabricius's theory was adequate.[50]) However, it was a good rhetorical argument, provided that Kepler established that he had made a good-faith effort to find a theory in the classical form. Of course, in a sense, Kepler's only recourse was to rhetorical argument, for the pursuit of astronomical theory through consideration of physical causes had no place in contemporary divisions of knowledge.

The most marked difference between Kepler's research as it was actually carried out and as it was presented in the *Astronomia nova* was in its sequence. This was undoubtedly partially in the interest of clarity—jumping around as he had from the theory of Mars to the theory of the earth, from longitudes to latitudes, would only have confused the reader. But it also served the valuable purpose of establishing Kepler's good faith in pursuing traditional methodology. Thus the question of physical causes was held in suspension until the failure of the vicarious theory.

In the introduction, Kepler presented four steps toward a "physical hypothesis of the motions," which correspond only loosely to the five parts into which the *Astronomia nova* was divided. The first step was the shift of reference from the mean to the true sun. "Here the Braheans could have raised the objection against me that I am a rash innovator,"[51] he wrote, because their theories based on the mean sun were adequate.

Again, Kepler was anxious to avoid the implication that he was cultivating novelty: "Therefore I have to look again and again at what I am doing, so as to avoid setting up a new method which would not do what was already done by the old method."[52] In part 1, he concluded at most that his method was as good. When he moved on to the motion of Mars in part 2 (where he constructed the vicarious hypothesis), he found that by using true opposition rather than mean opposition, his theory was superior. Nonetheless, he left hanging the question of which was preferable:

> Meanwhile, throughout the entire second part (as far as concerns geometrical demonstrations from the observations) I leave in suspense the question of whose procedure is better, theirs or mine, seeing that we both match a number of observations (this is, indeed, a basic requirement for our theorizing). However, my method is in agreement with physical causes, and their old one is in disagreement, as I have partly shown in the first part, especially chapter 6.[53]

Because the resolution of this question required an accurate theory for the earth, it was only decisively resolved in parts 4 (from longitude in chapter 52) and 5 (from latitude in chapter 67). In his correspondence, by contrast, Kepler was sure about both the use of the true sun and the bisection of the earth's eccentricity even before he had an adequate form of the vicarious hypothesis.[54]

Kepler's second, decisive step toward a physical hypothesis of the motions came with the establishment in part 3 that the earth needed an equant and that its eccentricity was to be bisected. He now had a strong argument for which of the cosmological systems would be more likely, for he was able to show that in the Ptolemaic system, each of the epicycles in the models for the superior planets would now need a line of apsides that retained its orientation in space and a point of uniform motion on the line of apsides opposite the center of the epicycle from the point of attachment, and that eccentricities of the epicycle's equant and point of attachment for all the superior planets would be in exactly the same ratio as the theory of the sun. Still, he knew that the absurdity of this arrangement in no way undermined its mathematical validity, so he labeled this in the introduction with a marginal note as first of his "physico-astronomical arguments":

> With these things thus demonstrated by a reliable method, the previous step towards the physical causes [the use of the true sun] is now confirmed, and a new step is taken towards them, most clearly in the theories of Copernicus and Brahe, and more obscurely but at least plausibly in the Ptolemaic theory.
>
> For whether it is the earth or the sun that is moved, it has certainly been demonstrated that the body that is moved is moved in a nonuniform manner,

> that is, slowly when it is farther from the body at rest, and more swiftly when it has approached this body.
>
> Thus the physical difference is now immediately apparent, by way of conjecture, it is true, but yielding nothing in certainty to conjectures of doctors on physiology or to any other natural science.
>
> First, Ptolemy is certainly exploded. For who would believe that there are as many theories of the sun (so closely resembling one another that they are in fact equal) as there are planets, when he sees that for Brahe a single solar theory suffices for the same task, and it is the most widely accepted axiom in the natural sciences that Nature makes use of the fewest possible means?[55]

Once Kepler committed himself to this foray into physics, he immediately found himself in a philosophical arena where dialectical arguments would result at most in probable conclusions. Still, he "yield[ed] nothing in certainty to . . . any other natural science."

Arguing against the Tychonic system raised the specter of Tengnagel's interference. We have seen above that Kepler was distressed with the prospect of having Tengnagel as a censor, especially with regard to his physical arguments;[56] Tengnagel had in the past exercised his rights to protect Tycho's legacy. In the introduction to the *Astronomia nova*, however, there are few signs that Kepler felt constrained in his arguments against the Tychonic system, except for an apologetic marginal note,[57] and a left-handed commendation of the Tychonic system in the section devoted to Scriptural objections to Copernicus. Tengnagel's preface shows that Kepler gave him little time to prepare it, and also that he was distracted by his activities at court, so we can conclude that Tengnagel had perhaps lost interest in the matter.

Arguing against the Tychonic system was more difficult than arguing against Ptolemy, for in the Tychonic system, there was only one theory of the sun that would have to be accommodated to Kepler's new discoveries. Ptolemy could be dismissed out of hand on the principle of simplicity. Tycho required a larger array of considerations. The first of these further "physico-astronomical arguments" was that in the absence of celestial orbs, the motion of a planet orbiting the sun while the sun orbited the earth became too complex for planetary intelligences and moving spirits responsible for the planet's motion (Kepler required such intelligences to direct the motion in the absence of real spheres). The necessity of intelligences itself became an argument against Tycho, when Kepler declared that in his system, "most of this can be done with physical rather than animate faculties, namely, magnetic ones."[58]

To this argument, Kepler added more specific conclusions that were drawn from chapter 33. These arguments were more powerful but also required more latitude from his readers with regard to the "physical conjecture"—which had not been clearly stated at this stage of the introduction—that because the motion varied with distance, the source of the motion was at one of the endpoints. From his incontrovertible "very reliable [geometrical] demonstration" that the earth, like the other planets, moves more swiftly along its orbit when closer to the sun (a straightforward conclusion from the presence of an equant), using his "physical conjecture," he concluded the source of its motion was in the sun, as it was for the other planets. As the source of motion for the planets (and the earth), then, the sun was more probably at rest; the mover was more appropriately unmoved.[59] The Tychonic alternative would be the absurdity that the sun, while still moving the planets, was moved by the earth, which was at rest:

> Upon this most valid conclusion, making use of the physical conjecture introduced above, might be based the following theorem of natural philosophy: the sun, and with it the whole huge burden (to speak coarsely) of the five eccentrics, is moved by the earth; or the source of the motion of the sun and the five eccentrics attached to the sun is in the earth.
>
> Now let us consider the bodies of the sun and the earth, and decide which is better suited to being the source of motion for the other body. Does the sun, which moves the rest of the planets, move the earth, or does the earth move the sun, which moves the rest, and which is so many times greater? Unless we are to be forced to admit the absurd conclusion that the sun is moved by the earth, we must allow the sun to be fixed and the earth to move.[60]

To this Kepler added the further consideration, from Copernicus, that the period of the earth falls between those of Mars and Venus:

> Does not the nature of things cry out with a great voice that the circuit in which these 365 days are used up also occupies a place intermediate between those of Mars and Venus about the sun, and thus itself also encircles the sun, and hence, that this circuit is a circuit of the earth about the sun, and not of the sun about the earth?[61]

Kepler was quickly being drawn into the more metaphysical type of arguments he presented in the *Mysterium*. We have seen above that Kepler was warned off the approach of the *Mysterium* by Longomontanus. At this point in the introduction to the *Astronomia nova*, he quickly backed off, writing, "These points are, however, more appropriate to my *Mysterium cosmographicum*, and arguments that are not going to be repeated in this work should not be introduced here."[62] However, he could not resist quickly referring his readers to further metaphysical arguments from the sun's dignity and illumination in the *Mysterium*, *De revolutionibus*, *De caelo* (in reference to the Pythagoreans) and the *Astronomiae pars optica*.

Having proposed that the sun was more suitably placed in the center of the system, Kepler had now to establish the suitability of the earth for motion. He thus introduced a number of excursions addressing various objections to the earth's motion, most notably a "true theory of gravity" to counteract physical objections to a moving earth and an exegetical framework for resolving Scriptural references to a stable earth. Such topics were entirely appropriate to the introduction, where prejudicial objections were to be preemptively addressed.[63] But these two excursions were so long that they completely interrupted the flow of his argument.

Having completed these long excursions, Kepler returned to the argument he had been outlining earlier. His first, tentative step toward his physical hypothesis of the motions had been the proposed use of the true sun. The second, decisive step was the introduction of an equant into the theory of the earth. When he came, toward the end of the introduction, to the third and fourth steps, Kepler's exposition became more abbreviated. Rather than the earlier tone of the introduction, in which he presented his physical propositions more rigorously and pointed to the body of the work for demonstrations, here he began vaguely pulling conclusions from the body of the work without such a clear foundation. And at the same time, he began making use of the rhetorical appeal to the amount of work invested in the research.

There are at least two plausible explanations for this change in tone. On the one hand, as we have seen, the *Astronomia nova* was conceived with the reformation of the earth's orbit as its primary finding and much of the book was composed before the discovery of the elliptical orbit, so Kepler may have added his final two steps toward a physical hypothesis to the introduction only late in the book's composition, after they had revealed themselves. On the other hand, it can also be argued that the introduction was composed more purposefully, and that the change in tone was due to Kepler's reaching the limit of what he could expect the natural philosophers (for whom the introduction was composed) to appreciate. Furthermore, Kepler did not have as secure a physical foundation as he had earlier. The argument of simplicity, say, and the argument that a mathematical point could not be a mover had a commonality with conventional natural philosophical doctrines. His distribution of a motive power through a rotating solar *species* could not be so straightforwardly argued in the introduction.

The third step was analogous to the second. The theory of the earth had required bisection of the eccentricity, and Kepler's work on Mars showed that—contrary to Copernicus and Tycho—its eccentricity was precisely bisected as well. With this finding, Kepler proposed a common physical analysis of the motions for all of the planets:

> [S]ince there are (of course) no solid orbs, as Brahe demonstrated from the paths of comets, the body of the sun is the source of the power that drives all the planets around. Moreover, I have specified the manner [in which this occurs] as follows: that the sun, although it stays in one place, rotates as if on a lathe, and out of itself sends into the space of the world an immaterial *species* of its body, analogous to the immaterial species of its light. This *species* itself, as a consequence of the rotation of the solar body, also rotates like a very rapid whirlpool throughout the whole breadth of the world, and carries the bodies of the planets along with itself in a gyre, its grasp stronger or weaker according to the greater density or rarity it acquires through the law governing its diffusion.[64]

To make the planets alternately approach and recede from the sun and so to achieve their nonuniform motion as they were carried by the denser or rarer *species*, Kepler had to assign each planet an individual mover. For establishing the existence of these movers, he appealed vaguely to the amount of work they required. By this time, Kepler had reached the point where he simply could not explain the exigencies of his various unsuccessful distance models to nonmathematicians. Indeed, he could scarcely offer any rational explanation at all for the doggedness with which he had pursued his physical explanation. As he hastened to absolve the movers of fault, he made a pathetic appeal to the exhaustion his misguided efforts had cost him:

> The amount of work [the movers] occasioned me in part 4 is incredible, when, in producing the planet-sun distances and the eccentric equations that were required, the results came out full of flaws and in disagreement with the observations. This is not because they should not have been introduced, but because I had bound them to the millstones (as it were) of circularity, under the spell of common opinion. Restrained by such fetters, the movers could not do their work.
>
> But my exhausting task was not complete: I had a fourth step yet to make towards the physical hypotheses.[65]

This final step was, of course, the discovery of the elliptical orbit. But now Kepler was completely beyond the competence of his target audience. As to how he came to this conclusion, he could offer only, "By most laborious proofs and by computations on a very large number of observations, I discovered that the course of a planet in the heavens is not a circle, but an oval path, perfectly elliptical."[66]

We have seen how the ellipse simultaneously resolved a number of questions that Kepler had been pursuing. The mechanism by which it was produced and the way in which it resolved the problems with the application of the area law were two important confirmatory features. But it was impossible to address these issues meaningfully in the introduction. Kepler could only state that it worked:

> Geometry gave assent to this, and taught that such a path will result if we assign to the planet's own movers the task of making the planet's body reciprocate along a straight line extended towards the sun. Not only this, but also the correct eccentric equations, agreeing with the observations, resulted from such a reciprocation.[67]

"Finally," Kepler wrote, "the pediment was added to the structure, and proven geometrically: that it is in the order of things for such a reciprocation to be the result of a magnetic corporeal faculty."[68] Reducing all of a planet's motion to a corporeal faculty was for Kepler an essential step in his physical astronomy. By showing the planetary movers "with great probability to be nothing but properties of the planetary bodies themselves, like the magnet's property of seeking the pole and catching up iron,"[69] he succeeded in making astronomy a truly physical science. It was the final step in showing the mathematical regularity of the planets' motions to be subject to physical (bodily) explanation:

> As a result, every detail of the celestial motions is caused and regulated by faculties of a purely corporeal nature, that is, magnetic, with the sole exception of the whirling of the solar body as it remains fixed in its space. For this, a vital faculty seems required.[70]

Although he did not specify it in the introduction, the reduction of every celestial motion to the result of corporeal natures represented the dissolution of the boundary between natural philosophical cosmology and mathematical astronomical theory. Astronomers should no longer be able to embrace Copernicus's theories while refusing to commit themselves to the physical truth of heliocentrism. (He did equivocate somewhat for those "who do not wish to put such a great trust in the nature of bodies,"[71] allowing that the planets could be endowed with rational and animate faculties that could accomplish the same thing as his natural corporeal reciprocation.)

The Second "Introduction" and the Line of Argument

Thus far, we have considered the outline of the argument of the *Astronomia nova* as it was presented in the introduction "for the sake of physicists." The argument of the book as it was presented to astronomers in the summaries of the individual chapters was different in some respects.

It was not a coincidence, I believe, that the open acknowledgment of the rhetorical character of the book quoted above came at the beginning of the chapter summaries. A mathematical work should rightly have been demonstrative rather than narrative. Because Kepler knew the resistance his reformation of astronomical theory would face, he worked hard to demonstrate the necessity of each step and to maintain as long as possible the impression that his research was not prejudicial in any way.[72]

We have seen that Kepler was careful to establish the suspension of his commitment to physical causes. In the introduction, it was couched in terms of the use of the true versus the mean sun. He also established such a suspension of belief in the body of the work but in more detail. On a large scale, it corresponded roughly to the book's division into parts. The first part of the *Astronomia nova* established the near equivalence of using the true or mean sun, at least so far as acronychal observations were concerned. The second part was devoted to the investigation of Mars's orbit "in imitation of the ancients," which failed. In the third part, Kepler turned his attention to the earth's orbit. Having reformed the earth's orbit, he returned in part 4 to Mars's orbit and solved it in terms of longitude. Finally, part 5 was devoted to Mars's motion in latitude.

Part 2, "On the Star Mars's First Inequality, in Imitation of the Ancients," was particularly important in establishing Kepler's suspension of belief in physical causes. (He did not stress the importance of imitating the ancients in this regard in the introduction, presumably because he did not think that physicists would understand or appreciate the significance of the vicarious hypothesis.) The construction of the vicarious hypothesis was "in imitation of the ancients" in the sense that using an eccentric and an equant for Mars and a simple eccentric for the earth was entirely orthodox. Furthermore, the empirical investigation of how the total eccentricity was to be divided contributed to the sense that Kepler had allowed complete flexibility in the derivation of the model.

Kepler's refusal to work from physical causes continued through part 2. I have argued above that Longomontanus's warning not to proceed along the line of the *Mysterium* contributed to Kepler's caution with regard to his physical ideas. In the *Astronomia nova*, he did not ignore or disavow the connection between the work he was doing and the *Mysterium*, but he refused to draw conclusions or work from considerations of the motive force until he had exhausted all other possibilities and established through a variety of different demonstrations that the earth's orbit needed an equant and a nearly bisected eccentricity. Thus, although Kepler discussed the preferability of heliocentrism and the use of the true sun in part 1, "[f]or the present, I am postponing a more careful consideration of whether this sort of [heliocentric motive force] hypothesis is in accord with physical principles."[73] Similarly, when preparing the vicarious hypothesis, Kepler remarked with regard to Ptolemy's introduction of the perfectly bisected eccentric:

> Now I myself could also have taken the bisection of the eccentricity as certainly established, and with better reason than Ptolemy, because in ch. 22 of my *Mysterium* I had brought forward a physical cause for the bisection. Indeed, it was for this very reason that I had come to Tycho, that I might use his observations to inquire further into my opinions expressed in that book. I of course did this without prejudice, and continue to do so.[74]

Perhaps as evidence for his lack of prejudice, however, Kepler did not follow the bisection:

> And so, since I thought then that this was altogether too much to assume, and also saw it pointedly called into question by Copernicus . . . , I envisaged a method which would lead me to a knowledge of the ratio of the two eccentricities (which, as I said, is not indubitably 2:1).[75]

As his discoveries are presented in the book, the subsequent failure of the vicarious hypothesis becomes the decisive justification for abandoning strictly mathematical astronomical theory. The generality of his solution supported the good faith of his effort.

The complete generality of Kepler's argument was further supported by his consideration of all possibilities. This generality was to some extent embodied for roughly the first half of the book in his repeating most demonstrations in all three world systems. But as we have seen, he was legally obligated to do so, and Kepler must have understood that these demonstrations were redundant. After a point, readers competent enough to follow his proofs could transform the hypotheses themselves. Indeed, when he tired of the threefold demonstrations in chapter 26, he announced:

> Let it be enough for now, to present this form of demonstration in the three hypotheses. In what follows, whenever there is need of the same demonstration, I shall use Copernicus's form alone, it being the simplest, so as not to be too long-winded. Here, in contrast, the industrious reader has seen how any of these diagrams can be transformed into either the Ptolemaic or the Copernican form using parallel lines.[76]

More functional in terms of exhausting possibilities was Kepler's continuing use of the mean sun. Although he argued in part 1 for the use of the true rather than the mean sun, he nonetheless repeated many of his demonstrations in part 2 for both, and part 3 is rigorously divided into demonstrations involving the mean sun and again the true sun. We must remember that Fabricius objected both to the use of the true sun and the bisection of the earth's eccentricity, and implied that the difficulties Kepler encountered with Mars were due to these innovations. Kepler's continual resort to repeating demonstrations with regard to the true sun was done specifically to counter such an objection. For example, after disproving the vicarious hypothesis in longitude in chapter 20, he repeated the demonstration with respect to the mean sun:

> Now, as in the previous chapter, I shall accommodate the proof that the eccentric's eccentricity has been falsely determined to the Brahean rendition as well, which depends upon the sun's mean motion. This is done so that no one will think the discrepancy is a result of my having wrong-headedly transposed the observations from the sun's mean motion to its apparent motion.[77]

The argument of the *Astronomia nova* was thus set up very carefully not to neglect any possibility. Viewed in this light, the synoptic table of contents, which Kepler presented "as a thread leading through the labyrinth of the work,"[78] was an exhaustive map of a maze whose every outlet was a dead end.

Although Kepler did not commit himself fully to deriving the motions from his planet-moving force until midway through it (in chapter 32), part 3 represented the beginning of the construction of a uniquely Keplerian astronomy. Part 2 had established the inability of classical theory to cope with the motions of Mars at the Tychonic level of observational accuracy. In bringing part 2 to a close, he reiterated the generality of this finding:

> Up to the present, the hypothesis accounting for the first inequality (in which Brahe and Copernicus are in agreement, both differing somewhat in form from Ptolemy) has been presented using the sun's mean motion, which all three authors had substituted for the sun's apparent motion. Thereafter, it was shown that whether we follow the sun's apparent motion and the hypothesis found in chapter 16 [the vicarious hypothesis], or the sun's mean motion and the hypothesis proposed in chapter 8 according to Brahe's rendition [Longmontanus's Mars theory], in both instances there result false distances of the planet from the centre, whether of the sun (for Copernicus or Brahe) or of the world (for Ptolemy). Consequently, what we had previously constructed from the Brahean observations we have later in turn destroyed using other observations of his. This was the necessary consequence of our having observed (in imitation of previous theorists) several things that were plausible but really false.
>
> And this much of the work is dedicated to this imitation of previous theorists, with which I am concluding this second part of the *Commentaries*.[79]

In part 3, "Investigation of the Second Inequality, that Is, of the Motions of the Sun or Earth, or the Key to a Deeper Astronomy, wherein There Is Much on the Physical Causes of the Motions,"[80] Kepler established that the earth's eccentricity needed to be bisected and an equant introduced. This finding then became the point of departure for the derivation of the earth's and Mars's orbits from physical principles:

> In this third part I first approach the second inequality. Here I shall use unquestionable observations to demonstrate, with either a confirmation or a refutation, all that I have hitherto supposed as principles but had doubts about. Once this is found it will be like a key: the rest will be opened up.[81]

Note that he continued to stress the certitude of the observations and the openness with which he was approaching the result. Still, he was candid about what he expected to find. He recounted the model of the equant in chapter 22 of the *Mysterium* and the qualification he had been forced to make for the absence of an equant in the theory of the earth. Although "I decided to leave that speculation open, until the matter were clearer to astronomers," he wrote, "I nevertheless entertained a suspicion that this theory might perchance also have its equant."[82] He went on to discuss how, when he heard Tycho's suggestion in his letter of 1 April 1598 that the annual orb varies in size,[83] "a sprite whispered to me" that this was an artifact of the simple eccentricity in the theory of the earth. In a clever rhetorical stroke, he accused Tycho's solution of "novelty":

> For what physical cause could make the circuit of the centre of the planetary system (Tychonic) or of the circuit of the earth (for Copernicus) or of the epicycle bearing the star (for Ptolemy) grow and shrink? What, I ask, is this novelty unprecedented in astronomy, this unlikely absurdity? Wouldn't it seem more worthy of belief that the sun (for Copernicus) or the centre of the planetary system (Tychonic) or the body of the planet (for Ptolemy) would in certain places be farther from, and in others nearer to, the selected point of uniform motion (at rest for Copernicus and Tycho, and moving around on the circumference of the eccentric for Ptolemy)—and this especially on the line of apsides? And for this, that suspicion of mine arising from my *Mysterium cosmographicum*—that an equant might be introduced into the theory of the sun (or, as I call it, the theory of the Ptolemaic epicycle)—provides a convenient occasion.[84]

Regardless of his narrative justification for investigating whether the earth's eccentricity needed to be bisected, Kepler was very careful to make his investigation incontrovertible. He approached the question first using the mean sun, then the true sun, and for each case provided a number of independent demonstrations. As I argued above, this unprecedented innovation, clashing as it did with Tycho's solar theory and meeting Fabricius's deep skepticism, needed an unusually sure foundation, especially because it was the ultimate justification for his physical astronomy.

Kepler was thus careful to suggest that he had been driven to the conclusion. Reviewing his findings in chapter 32, he recalled that Ptolemy had bisected the eccentricities of the superior planets, that the demonstrations that refuted the vicarious hypothesis had shown that Mars's orbit indicated bisection, and that he had now shown the earth's eccentricity was also bisected. Venus and Mercury, he suggested, could also be treated the same way, and thus all the planets had a bisected eccentricity. Only then did Kepler commit himself to the motive force hypothesis from the *Mysterium*:[85]

> Now in my *Mysterium cosmographicum*, published eight years ago,[86] I postponed arguing this case of the cause of the Ptolemaic equant for the sole reason that it could not be said on the basis of ordinary astronomy whether the sun or earth uses an equalizing point and has its eccentricity bisected. However, now that we have the confirmation of a sounder astronomy, it should be transparently clear that there is indeed an equant in the theory of the sun or earth. And, I say, now that this is demonstrated, it is proper to accept as true and legitimate the cause to which I assigned the Ptolemaic equant in the *Mysterium cosmographicum*, since it is universal and common to all the planets. So in this part of the work I shall make a further declaration of that cause.[87]

But even though taking up the motive forces from the *Mysterium*, Kepler was still careful to contrast the necessity of the conclusion to which he had been driven in the *Astronomia nova* from the metaphysical arguments of the *Mysterium*, although he did not dismiss them. In the following chapter, after concluding that the sun is the true center of the system and the source of motive power, he remarked:

> But indeed, if this very thing which I have just demonstrated a posteriori (from the observations) by a rather long deduction, if, I say, I had taken this as something to be demonstrated a priori (from the worthiness and eminence of the sun), so that the source of the world's life (which is visible in the motion of the heavens) is the same as the source of the light which forms the adornment of the entire machine, and which is also the source of the heat by which everything grows, I think I would deserve an equal hearing.[88]

Kepler finally laid down the principles of his planet-moving force in chapters 32 through 39 of part 3. As noted above, he proceeded much more carefully and systematically in the *Astronomia nova* than in the *Mysterium*. Thus, for instance, chapter 32 is devoted solely to a demonstration that an equant causes a planet to move more slowly at aphelion than at perihelion in proportion to the distance from the sun. (This relation holds strictly only at apsides; Kepler skirted around the problem of the strict relation in intermediate points.) Chapter 33 then proposed that because the motion was proportional to distance, that the latter was the cause of the former, that the source of motion must be in one of the endpoints, and only the sun was appropriate to be the cause of the motion. Here Kepler finally committed himself to the use of the true sun; continuing to consider the mean sun no longer made sense in the physical analysis of the motions. (Nevertheless, he did refer the reader to chapter 52 for the observational proof that Mars's line of apsides passed through the sun, where "nothing is assumed but the bare observations."[89]) Chapter 33 was the kernel of Kepler's proof of heliocentrism. The essential argument was that the sun was the source of the planet's motion. What Kepler had to say about the nature of the planet-moving force and how it could be used to derive the motions of the planets only supported this point.

Kepler began to consider the nature of the planet-moving force from the second half of chapter 33 through chapter 35. I will not attempt to give a précis of these arguments.[90] Let us note only that it was important for him to establish that this force was amenable to physical analysis, which was an uphill battle. Because the force was celestial, he encountered the problem of the division between sublunar and supralunar realms. His extensive use of analogy with light and magnetism should be seen as an attempt to breach this division; the planet-moving force had to be shown to be like something that was subject to physics. He expressed this purpose to a certain extent when he came to the explanation of the planets' eccentricities in chapter 38:

> Examples of natural things, and the kinship of celestial things for these terrestrial ones which has hitherto been gathered to exist, cry out that in a simple body the operations which are more general are simpler, while the variables, if any (such as, in the motion of the planets, the varying distance from the sun, or the eccentricity), arise from the concurrence of extrinsic causes.[91]

Ultimately, the only features of the motive force that were important in his analysis of the planets' motions were those he reiterated at the beginning of chapter 39:

> And so, in what has been demonstrated, let us take these axioms, which are of great certainty. First, that the body of a planet is inclined by nature to rest in every place where it is put by itself. Second, that it is transported from one longitudinal position to another by that power which originates in the sun. Third, if the distance of the planet from the sun were not altered, a circular path would result from this motion. Fourth, supposing the same planet to be

in turn at two distances from the sun, remaining there for one whole circuit, the periodic times will be in the duplicate ratio of the distances or magnitudes of the circle.[92] Fifth, the bare and solitary power residing in the body of a planet itself is not sufficient for transporting its body from place to place, since it lacks feet, wings, and feathers by which it might press upon the aethereal air. And nevertheless, sixth, the approach and recession of a planet to and from the sun arises from that power which is proper to the planet. All these axioms are agreeable to nature in themselves, and have been demonstrated previously.[93]

In the final chapters of part 3, Kepler laid the foundations for calculating the motion of a planet based on the planet-moving force. First, he introduced the epicyclic distance model, to discuss how a planet traveling on a circular eccentric changes its distance with respect to the sun, and raised a number of absurdities that arise from trying to interpret epicyclic motion physically. In the final chapter, concluding that though not complete and perfect, the physical theories were complete enough for the case of the sun, he constructed the physical theory of the earth. During the course of this, he introduced the area law as a means of approximating the cumulative effect of the planet-moving force, and discussed the inaccuracy of measuring the sum of the distances from an eccentric point with his analysis of the conchoid.

All of part 3 has a very deliberate and systematic tone. It is thus remarkable when the personal pronoun reappears in its final chapter, and Kepler returns suddenly to the narrative style he had chosen for the book. He had deliberately suspended his narrative, especially for the chapters that laid the foundation of the planet-moving force. As he put it when resuming his account, "Such a long-winded discussion was necessary to prepare a way for a natural form for the equations, on which I am going to be very busy in Part IV."[94] The discussion of the principles, and the analysis with the epicyclic distance model and the conchoid, were necessary to set the stage for the final assault on Mars's orbit. As noted above, when Kepler finally came upon the ellipse, he knew that it was right because it crystallized out of a number of considerations he had been concerned with for some time. For his audience to come to the same epiphany, he needed to have these considerations already raised and unresolved.[95] Furthermore, it would have been valuable to present his philosophical doctrines regarding the planet-moving force dispassionately to avoid the suspicion that the basis of his undertaking was idiosyncratic and unjustified.

When he came to part 4, however, he needed to establish the complicated matter of how and why the ellipse was correct, for which he had to lead his readers along his path. Here it was important to take the blame personally for the failure of his principles, especially his first attempts at constructing an oval orbit, starting in chapter 45. As he put it in the summary for that chapter:

> In what follows, the reader should overlook my credulity, since I am judging everything by my own wits. Indeed, the occasions by which people come to understand celestial things seem to me not much less marvellous than the nature of the celestial things itself. I therefore display these occasions scrupulously, with, no doubt, some attendant difficulty for the reader. Nonetheless, that victory is sweeter that was born in danger, and the sun emerges from the clouds with redoubled splendour. Therefore, O reader, pay heed to the dangers of our army, and contemplate the clouds horrifying in their darkness. Contemplate, I say; for beyond these clouds the sun of truth truly lies hidden, and shortly will emerge.[96]

Here again the use of pathos is obvious, as Kepler seeks forgiveness for his errors and promises the triumph that comes with the unveiling of truth.

Kepler had first to demonstrate that Mars's orbit was an oval. He established this soberly in the first four chapters, 41 through 44, of part 4. First, drawing three sets of heliocentric longitudes and distances for Mars from earlier in the book, he fit them to a circular orbit, but then quickly dismissed the orbit when its aphelion disagreed with the vicarious hypothesis and eccentricity came out significantly larger than he would derive in the following chapter. At the same time, he pointed out that different groups of three consistently yielded disparate results. In the next chapter, he carefully assembled groups of observations when Mars was around aphelion and perihelion and the earth was at different positions around its orbit, from which he triangulated to the two positions of Mars. Knowing its position at both ends of the line of apsides, he could find the true eccentricity, which, not surprisingly, was nearly half the total eccentricity of the vicarious hypothesis (he had referred forward to this procedure when discussing the bisection of Mars's eccentricity earlier). With this half-eccentricity and the vicarious hypothesis's aphelion, he quickly tried employing the area law on a circular orbit for quadrants and octants, checking it against the vicarious hypothesis, as he did throughout part 4 (despite its flaws, he knew the vicarious hypothesis would give at least accurate heliocentric longitudes). The area law on a circular orbit failed. In chapter 44, he reviewed the results:

> With the eccentricity and the ratio of the orbs established with the utmost certainty, it must appear strange to an astronomer that there remains yet another impediment in the way of astronomy's triumph. And me, Christ!—I had triumphed for two full years.[97] Nevertheless, by a comparison of the things which have been established in chapters 41, 42, and 43, preceding, it will easily be apparent what we are still lacking.[98]

His failing was that he was still "imitating the ancients"[99]: the orbit was still circular. That the orbit was oval was important and had to be argued forcefully. First Kepler presented an observational demonstration. Bringing back the distances from chapter 41, he overlaid them on the more accurate circular theory derived from the aphelion and perihelion in chapter 42. The further the distances were from apsides, the more they fell within the circle, which indicated some sort of oval orbit, farthest from the circumscribing circle at quadrants:

> If anyone wishes to attribute this difference to the slippery luck of observing, he must surely not have felt nor paid attention to the force of the demonstrations used hitherto, and will be shamelessly imputing to me the vilest fraud in corrupting the observations of Brahe. I therefore appeal to the observations of subsequent years, at least those made by experienced observers. For if in any respect I have given free rein to my inclinations in one direction, I only go so much the farther into error on the other side. But there is no need of this. I am addressing this to you who are experienced in matters astronomical, who know that in astronomy there is no tolerance for the sophistical loopholes that beset other disciplines.[100]

Having staked his claim on Tycho's observations and dismissed the suspicion that he had "given free rein to his inclination," he now proceeded to a physical analysis of the error around octants, which suggested that the distances needed to be shorter around quadrants, so that the planet would spend less time there.

In broad terms, part 4 was fairly simple. After trying a circular orbit for Mars, Kepler constructed an oval orbit (approximated by an ellipse), which was too narrow. The errors were equal and opposite, therefore the correct orbit was an ellipse that fell between them. And yet, the chapters devoted to Kepler's first oval orbit are the messiest ones in the *Astronomia nova*. Once he devoted himself to calculating the motion based solely on the motion being proportional to the distance to the sun and the distance varying according to an epicyclic distance model, in which the distance varied as the mean anomaly, he ran into a series of intractable problems. Much of this section was devoted to a series of different approximations of the physical equations of motion. What was to be gained by recounting these problems? There are two things that appear to be important. First, it was desirable to show that the means by which he approximated the area law on the narrow oval were sufficiently accurate. Since the final ellipse was shown to be correct by falling between the circular and oval (elliptical) orbits, and thus possessing precisely enough area to put Mars in the right position at octants, it was necessary to establish that the oval orbit, which established the lower boundary, was nearly elliptical and that the equation was correctly calculated.[101] Second, it was necessary for the reader to appreciate *that the problem was intractable*. The ellipse turned out to be simple. Its simplicity was a powerful indication of its truth, but that would not be clear unless the reader were led through the brambles. Kepler thus elicited the emotions of his audience after describing an iterative numerical procedure, "I can't imagine anyone reading this not being overcome by the tedium of it even in the reading. So the reader may well judge how much vexation we (my calculator and I) derived hence."[102]

The fault with the first oval orbit had been Kepler's, as he pointed out repeatedly. When he first realized the orbit was oval, he followed a physical intuition of how this might come about based on the epicyclic distance model. He compared the difficulties he encountered to the proverb, "A hasty dog bears blind pups."[103] He later made fun of himself in a charming passage—another good example of pathos—for seizing upon this physical intuition when, after concluding his various approximations of the motions, he prepared to go back to the observations to find out how far Mars actually fell within the oval:

> While I am thus celebrating a triumph over the motions of Mars, and fetter him in the prison of tables and the leg-irons of eccentric equations, considering him utterly defeated, it is announced in various places that the victory is futile, and war is breaking out again with full force. For while the enemy was in the house as a captive, and hence lightly esteemed, he burst all the chains of the equations and broke out of the prison of the tables. That is, no method administered geometrically under the direction of the opinion of ch[apter] 45 was able to emulate in numerical accuracy the vicarious hypothesis of chapter 16 (which has true equations derived from false causes). Outdoors, meanwhile, spies positioned throughout the whole circuit of the eccentric—I mean the true distances—have overthrown my entire supply of physical causes called forth from ch[apter] 45, and have shaken off their yoke, retaking their liberty. And now there is not much to prevent the fugitive enemy's joining forces with his fellow rebels and reducing me to desperation, unless I send new reinforcements of physical reasoning in a hurry to the scattered troops and old stragglers, and, informed with all diligence, stick to the trail without delay in the direction whither the captive has fled. In the following few chapters, I shall be telling of both these campaigns in the order in which they were waged.[104]

The failure of his physical reasoning forced Kepler to return to the surety of Tycho's observations—his "spies"—and distance himself from his physical intuitions.[105]

Having an accurate solar theory (from part 3) and the vicarious hypothesis giving heliocentric longitudes, Kepler needed only one observation of Mars for any one place on its orbit to find its true distance from the sun. In chapter 51, he carefully calculated a number of such positions and distances at symmetric places with respect to the line of apsides. (That these symmetric distances were the same was the basis for the decisive proof in chapter 52 that Mars's line of apsides passed through the body of the sun.) He offered a different method of calculating the distances in chapter 53, which, in the wake of William Donahue's investigation, has become the most scandalous chapter in the *Astronomia nova*. Donahue discovered that the findings Kepler presented for the procedure he outlined in chapter 53 do not in fact represent the results of that procedure.[106] In one sense, there is nothing surprising in this. Kepler himself admitted in chapter 56 that the distances he presented in chapter 53 were calculated using the distance relation of the ellipse.[107] More disturbing are the values for Mars's heliocentric longitude, which are indistinguishably close to being derived from the ellipse and the area law, and are thus most probably calculated using it, and not Kepler's vicarious hypothesis, as he claims.

Since Kepler never claimed a clinical exactness for his narrative account of his research, assessing the seriousness of his "fraud" in chapter 53 must hinge on what it was meant to accomplish. On the one hand, given that he himself acknowledged that the distances were derived from the ellipse, chapter 53 can be seen as a kind of retroactive assessment of the adequacy of the ellipse.[108] It did after all, contain, real observations that were in close agreement with it. On the other hand, there is also a very bad implication. With the failure of the first oval orbit, Kepler took refuge in the observations, claiming that none of his personal physical theories would enter into the distance derivations. And in fact, he claimed in both chapters 51 and 53 that they were from "infallible observations."[109] However, the findings from chapter 53 were employed only in chapter 55 where, together with those from chapter 51, they were used to confirm that Kepler's first oval orbit was in fact too narrow and fell roughly halfway between the circle and the oval (previously, only his physical interpretation of the failure of the oval, which might have been considered suspect, had indicated this). By chapter 56, Kepler refused to compare the real ellipse's distance relation with the distances from chapter 53, admitting that they were derived from the relation in question. Thus Kepler's irregularity might be dismissed because he used the questionable distances only to disprove his own theory.

Ultimately, Donahue's own understanding of chapter 53 suggests the best account of how these false distances found their way into the *Astronomia nova*. Donahue's analysis shows that the chapter probably underwent multiple revisions, some of which did not fully erase the traces of what had gone before. When Kepler undertook the final revision, it must have been overly burdensome to recalculate the data while pretending that he did not know the true orbit, so he "laundered" the final table, although we must bear in mind that he admitted this openly soon afterward and did not use the distances in an essential demonstration. But if the chapter did exist in previous drafts, why was it there? What did it accomplish that chapter 51 did not? There are many places in the *Astronomia nova* where redundant demonstrations are used to establish the same point. The method in chapter 53 was elegant, if hypersensitive. Kepler thus included it as an independent alternative to chapter 51.[110] Interestingly, though, the only place where the distances from chapter 53 are emphasized over those from chapter 51 is in chapter 55, where Kepler points out that, if incorrect, distances around opposition (such as those from chapter 53) near middle distances can cause an error of 20′ or more:

> Thus, David Fabricius was able to use his observations to charge my hypothesis of chapter 45, which I had communicated to him as true, with this error of having distances that are too short at the middle distances, writing at the very time when I was labouring to seek out the true hypothesis through repeated, careful operations.[111]

Whether this unusual reference to Fabricius's criticism of the first oval influenced the retention of the analysis presented in chapter 53, if not the original values, is unclear. Nor is the value of this reference to Fabricius obvious, unless perhaps Kepler's overgenerous statement that "[h]e was, in fact, quite close to arriving at the truth before me" was meant to imply that Kepler's findings were not idiosyncratic.[112]

Kepler had set his argument up so well that when he finally arrived at the correct distance relation, and thus the ellipse, it came as a flash. Having long before introduced the epicyclic distance model and the analysis of the conchoid, then having revisited both problems with regard to the oval and the lunula between the circular and oval orbits, and having had both the physical and observational indications that the true orbit fell somewhere between them, Kepler had prepared the reader for his nearly mystical revelation:

> While I was anxiously turning this thought over in my mind, reflecting that absolutely nothing was accomplished by chapter 45, and consequently my triumph over Mars was futile, quite by chance I hit upon the secant of the angle 5° 18′, which is the measure of the greatest optical equation [*sic*].[113] And when I saw that this was 100,429, it was as if I were awakened from sleep to see a new light, and I began to reason thus.[114]

The angle 5° 18′ was the angle at Mars between the sun and the center of the orbit at quadrants, where the lunula was widest, 858 parts of 100,000. It should have been half as wide, 429 parts. The number 100,429, the distance from Mars to the sun, was the clue that he should substitute the side, 100,000, for the hypotenuse, thus shortening the distance by the necessary 429 parts. On the epicyclic distance model, it meant that the distance was determined by the projection of the epicycle on the radius. This, in turn, resolved problems he had raised when introducing the epicyclic distance model as to how it regulated its approach to and recession from the sun. Besides providing a table of observed distances that were in good agreement with the reciprocation of the planet along the diameter of the epicyclic distance model, it was almost necessary only to refer the reader back to questions already addressed.

The only physical question that needed to be addressed was how the new distance model could be considered natural. As we saw, the lengthy analysis Kepler offered in chapter 57 was at least partly a response to Fabricius's objection that the variation in distance as the versed sine of the eccentric anomaly was not natural. Kepler's account was heavy going, largely because he had to explain why the planet's distance depended on the eccentric anomaly, which, being the angle at the center of the ellipse, had no real functionality. He began with a general discussion of how a variation as the sine of the eccentric anomaly was natural by introducing the analogy of an oarsman on a circular river. If the oarsman turned his steering oar as the angle about the center of the river, he would in turn approach and recede from an eccentric point in the river (the position of the sun). There was much that was unsatisfying about the analogy, not least of which was that the center of the river had the power of moving the boat, not the point analogous to the sun. Kepler thus turned to the analogy of a magnet, which was much better. Gilbert had shown the body of the earth to be a magnet. If Kepler maintained that the planet naturally maintained its orientation in space and that one pole was attracted to and one repelled from the sun, he could argue that the measure of that attraction and repulsion would go as the sine, but of the true rather than the eccentric anomaly (it would be greatest when the pole was lined up directly with the sun at a true anomaly of 90°, slightly beyond the widest point of lunula). Maintaining that the axis retained its orientation in space was also problematic; Copernicus had given the earth's axis a motion to retain its orientation in space. Furthermore, it was clear that the earth's axis was not perpendicular to the line of apsides, so again the magnetic aspect was merely analogous. Kepler had to concede, "I will be satisfied if this magnetic example demonstrates the general possibility of the proposed mechanism."[115] He then considered how some mental faculty of the planet might be able to bring out the reciprocation along the diameter consistent with the distance model, but this too was fraught with difficulties. Overall, although the chapter is replete with geometrical demonstrations, Kepler often had to resort to handwaving to overcome fundamental obstacles.

With the philosophical problems at least addressed, it remained only to show how the distance law for the ellipse was to be used to calculate the positions (previously, Kepler had shown only that it gave the correct distances for given positions). Here he returned again to his narrative style, and recounted the problem with the *via buccosa*. He began chapter 58 with a quote from Virgil:

> "Galatea seeks me mischievously, the lusty wench:
> "She flees to the willows, but hopes I'll see her first."
>
> It is perfectly fitting that I borrow Vergil's voice to sing this about Nature. For the closer the approach to her, the more petulant her games become, and the more she again and again sneaks out of the seeker's grasp just when he is about to seize her through some circuitous route. Nevertheless, she never ceases to invite me to seize her, as though delighting in my mistakes.[116]

Why did Kepler see fit to include yet another mistake along his journey? Remember that he had not seen fit to inform Fabricius of this particular mistake; he had stricken it through in the draft of the letter that contains this step. Yet he resurrected the mistake for the *Astronomia nova*. It was necessary, he discovered from Fabricius, to include this particular mistake because it was part of the explanation of why it was necessary to go about constructing the position of the planet in the peculiar way he did. Only through the experience with the *via buccosa* had he come to this understanding. When he tried to explain it to Fabricius without this step, Fabricius had become hopelessly mired in a false means of calculating and had raised objections against Kepler's procedure as being so complicated that it would scare people off at first glance. Thus the *via buccosa* had proven itself to be a necessary part of the argument. What ultimately came of Kepler's excursion was the understanding that the means of calculating actually had a deeper meaning—it showed why the area law was actually a precise measure of the planet-moving force.[117] But it was necessary that the reader too learn this from his mistake. It had a twofold purpose: on the one hand, it justified the unusual and strange method of calculating the position, and on the other hand, it tied up a thread in the celestial physics.[118]

Part 4 closes with two chapters in which the properties of an ellipse and the problem of how to calculate the position of a planet on the ellipse are very formally presented in protheorems and demonstrations with proper references to Apollonius. It seems as though Kepler aimed these chapters squarely at mathematical astronomers, to present his unquestionably obscure reformation of astronomical theory clearly in formal mathematical style for their contemplation, and in doing so to make a final plea for a measured judgment of his physical astronomy:

If anyone thinks that the obscurity of this presentation arises from the perplexity of my mind, I shall myself only thus far acknowledge to him my guilt, that I was unwilling to leave anything untested, no matter how utterly obscure, and no matter how irrelevant to the practice of astrology, which many deem the sole end of this celestial philosophy. But as for the subject matter, I urge any such person to read the *Conics* of Apollonius. He will see that there are some matters which no mind, however gifted, can present in such a way as to be understood in a cursory reading. There is need of meditation, and a close thinking through of what is said.[119]

⌘ CONCLUSION ⌘

The title of this book, *The Composition of Kepler's* Astronomia Nova, intentionally plays on two meanings of the word "composition." In that sense, it is a pun and, like all inept wit, needs to be explained. On the one hand, it designates the content of Kepler's *Astronomia nova*, that is, what Kepler's book is made of, what kind of demonstrations Kepler uses, how the argument works, how the book is organized, and so forth. On the other hand, the term also refers to Kepler's process of writing the book, that is, the composing of the book. This second sense of composition would include how he came to write it, the experiences Kepler had while writing it, and how those experiences influenced his writing of the book. It has been my major argument that these two senses of the word "composition" are intimately related in the case of the *Astronomia nova*: what the *Astronomia nova* contains as a narrative and argument is a reflection of his experience while writing the book. Far from being a stream-of-consciousness narrative, it is a subtly composed argument painstakingly tuned to the criticisms and concerns of the contemporary astronomical community. In order to make this argument, it has been necessary to examine the *Astronomia nova* in the development of Kepler's thought, as well as to identify the more specific influences on it.

The *Astronomia nova* cannot be understood without situating it with respect to Kepler's astronomical interests. From the beginning, as a student at Tübingen, Kepler was only marginally interested in the details of Copernican planetary theory, those details that most contemporary astronomers found particularly compelling. It was heliocentric cosmology that captured his attention, and—in particular—the way in which the heliocentric system could be understood as a manifestation of God's providential design. Because of this fundamental religious stimulus, the heliocentric system could be of interest only if it were the physically true representation of how the universe was put together. Kepler went right to the root of this question, and from the beginning argued how the system could be understood and analyzed physically. Thus, in his student disputation he suggested that the planets' differing periods were to be understood as a function of their distances from their source of motive power in the sun.

The assignment to teach mathematics in Graz provided the circumstance for Kepler to devote his full attention to proving the physical truth of heliocentrism. Continuing in the vein of considering the heliocentric system as the product of God's intelligent design, he asked himself some fundamental questions about this design—why were there six and only six planets, and why were they arranged at their particular distances from the sun? The answer that came to him, that God had constructed the universe using the five Platonic solids as an archetype, was for him a compelling physical proof of the truth of heliocentrism, which he presented to the world in his *Mysterium cosmographicum*. But this architectonic scheme was not the only physical argument in the *Mysterium*. Kepler also developed the idea from his student days about the planets' receiving their motive power from the sun. His flawed analysis (which he himself reconsidered within a few years) provided a more rigorous relationship between the planet's distances and periods. But more importantly, he extended the analysis to include the variation in a planet's speed around its own orbit as it changed its distance from the sun. Kepler's eventual success in the *Astronomia nova* can be traced back directly to this insight.

The analysis of motive power for a planet as it moved around the sun was only a single element of Kepler's broad attempt to provide a physical proof of heliocentrism in the *Mysterium*. But complicated circumstances forced him in the short term to pursue that line at the expense of all others. Kepler's interaction with Tycho Brahe was, of course, of monumental importance in the history of astronomy. It was only with the Dane's enormous catalogue of accurate observations that Kepler was able to undertake a real reformulation of astronomical theory. However, when Kepler first came into contact with Tycho Brahe, he was still primarily concerned with the broader argument of the *Mysterium*, the polyhedral hypothesis, which he hoped he might refine and improve with Tycho's observations. Tycho's secretiveness with his data blocked Kepler from that avenue of research. Using only observations of Mars, however, Kepler could still pursue the motive force hypothesis, and he soon experienced remarkable success. Using the Mars observations, he was able to conclude quite quickly that the orbit of the earth had a bisected eccentricity, bringing its theory into line with the theories of the other planets and, at the same time, resolving certain problems with the theory of Mars. The consistency in the theories of all the planets was a gratifying discovery that tended to support the truth of the heliocentric system.

In channeling his efforts into the elaboration of the motive power hypothesis, however, Kepler was pursuing the most controversial of his claims from the *Mysterium*. The notion that a planet's theory might derived from physical principles was a highly suspect proposition at that time. Astronomers generally regarded the details of planetary theory as a province of pure mathematics. Both Tycho and Kepler's own teacher, Maestlin, objected to Kepler's physical mode of reasoning in astronomical theory. It is important to remember, though, that Kepler's primary motivation remained the physical proof of the truth of heliocentrism. A conventional fictionalist stance toward astronomical theory that made no claims about the reality of mechanisms in planetary theory would have held no meaning for him. It was therefore of the utmost importance to pursue and present his vision of a celestial physics.

Tycho's death and Kepler's appointment as imperial mathematician complicated the situation. On the one hand, Kepler could now pursue his physical line of research without fear of Tycho's censure, and his responsibility for the prestigious *Rudolphine Tables* would provide a most suitable venue for presenting his new physical planetary theories. On the other hand, the claims of Tycho's heirs threatened Kepler's new found "philosophical freedom." Anxious to exploit fully the economic potential of Tycho's legacy, the Tychonics regained control over Tycho's posthumous publications. More importantly, since much of the future value of the estate hinged on Tycho's reputation, the heirs took steps to ensure that reputation, and obtained the right to censor Kepler's publications. This wrangling put Kepler in an impossible bind, since his disagreements with Tycho's principles were well known.

The greatest result of losing his responsibility for the *Rudolphine Tables* was that Kepler was obliged to name other works that he would compose to justify his employment. Grasping among his unfinished projects, he named the *Astronomia pars optica* and the *Astronomia nova*. The *Astronomia nova* was thus conceived under pressure, years before the discovery of the true form of Mars's orbit. Still, Kepler had good reasons for thinking that a commentary on the motion of Mars—an unusual genre for an astronomical book—would be a significant step in his physical proof of heliocentrism. With his discoveries concerning the orbit of the earth, he was prepared to put forth an elegant and simplified theory of the planets. All the planets would have the same general form of theory with bisected eccentricities and with their speed a function of their distance from the sun. To Kepler's mind, this was sufficient in itself to rule out Ptolemaic theory, for it would require that the epicycles in the theories of the superior planets, for instance, could no longer be simple circles. Instead, they would have to be eccentrics, each with its own line of apsides, every one of which would maintain the same orientation in space and have the same eccentricity as the earth's orbit. This coincidence would reveal the absurdity of the Ptolemaic system. Because the Tychonic system was an inversion of the Copernican system, it was immune from this criticism; in principle, one could accommodate Kepler's findings simply by adapting the theory of the sun. But if Kepler were successful in showing that the motion of the planets could be derived from consideration of the sun's motive power, that would weigh heavily against the Tychonic system, for how could the sun distribute motion to the planets while it was itself being moved around the earth?

Kepler had several reasons for framing the presentation of his findings in the form of a commentary. First, he had been working on the orbit of Mars almost exclusively for several years. The pressure to present his findings meant that he had to argue from this limited base. The claims of Tycho's heirs might also have complicated extending his analysis to the other planets. But in any case, that would have been unnecessary. The analysis of Mars had been sufficient to make his discoveries; it would also be sufficient to present them. More importantly, the changes in astronomical theory that Kepler was going to present were revolutionary. Since the time of Hipparchus, the earth had always been assigned a simple, nonbisected eccentricity. The unusual procedures Kepler employed to discover that it needed to be bisected would have to be explained. Even more significant and unusual, however, was Kepler's physical analysis of Mars's motion. He knew that this would meet with severe criticism from some members of the astronomical community. But it was central to his proof of heliocentrism. He would therefore require an extended and careful treatment to present his findings, even if they derived only from the analysis of the motion of one planet.

The constraints under which Kepler conceived the *Astronomia nova* and the challenges he faced in presenting his findings were reflected in the form in which he composed the book. Around the time he was facing severe criticism from Longomontanus, the most talented astronomer among the Tychonics, Kepler made his first references to the unusual narrative form of the *Astronomia nova*. The reason he decided to present his results in this form was clearly justification. He felt that it was essential for his readers to understand how he had come to the conclusions he had, and why he had pursued this controversial line of research. At the same time, the scheme was in part a conceit. Facing the accusation from the Tychonics that he was deliberately seeking novelty and attempting to turn Tycho's observations to his own idiosyncratic ends, Kepler presented himself as driven to the conclusions he had reached by the failure of every attempt he had made to represent Mars's motion using the standard mathematical tools of uniform circular motion.

Kepler also discovered that this justificatory narrative could be put to good didactic use. His correspondence with David Fabricius was most significant in this regard. It revealed the great skepticism with which even a sympathetic friend regarded Kepler's innovations, and, more importantly, the manifold ways in which a contemporary astronomer could challenge his findings. Kepler used this information to tailor his narrative. He attempted to forestall the most common objections during the course of his presentation by including alternative or redundant demonstrations. Thus he used Fabricius to map out alternate routes, which Kepler showed were likewise dead ends.

Taking into account this complicated context, we can at last begin to make sense of Kepler's masterful, iconoclastic *Astronomia nova*. Its foundation is a religiously based commitment to the truth of the heliocentric system. Its reality could be established by appeal to physical principles, either in the architectonic design argument from the *Mysterium* or through the success of Kepler's "celestial physics" in describing the motion of the planets. Within the context of his association with Tycho, Kepler was able to develop and refine only the latter. Kepler's research was thus fundamentally redirected and the *Astronomia nova* began to take on a distinct shape, very different from the *Mysterium cosmographicum*. The suspicion with which astronomers viewed the introduction of physics into planetary theory and the machinations of the Tychonics likewise exercised their influence as Kepler became convinced of the need of a careful presentation of his findings. The mode of exposition he chose was avowedly rhetorical. He would take his readers with him during his metaphorical "battle with Mars," leading them along a fictional path in which he first attempted in good faith to construct a theory for Mars using conventional means and resorted to physical principles only when this failed. At the same time, he would repeat his demonstrations in all three major cosmologies to minimize the perceived influence of his commitment to heliocentrism. The final fine structure of the *Astronomia nova* was provided by Kepler's interaction with Fabricius. Here Kepler found precise objections and confusions that could be considered representative of his audience, and he incorporated them into his narrative. The specific correlation of Fabricius's objections and Kepler's demonstrations shows the manifest influence of their discussions.

The complexity of this context and its unseen influence on the *Astronomia nova* have not been previously recognized. The notion that the content of a book might be influenced by the author's expectation of how that book will be read by his contemporaries is surely uncontroversial. Yet why has the *Astronomia nova* evaded this kind of analysis until now? Several factors seem to have obstructed our path to this kind of understanding.

To begin with, the greatness of Kepler's achievement has in some ways obscured a broader contextual analysis. There is no question that the *Astronomia nova* is one of the canonical works of the Scientific Revolution. And works of great originality and brilliance can be approached (to some extent) on their own terms. Nor do I believe that anything I have said diminishes the *Astronomia nova*'s standing as an intellectual tour de force. In fact, I believe that uncovering Kepler's intentionality in his choice of how the book was composed reveals yet another level on which the book can be appreciated. As long as his reasons for doing so remained obscure, Kepler was destined to be viewed as Koestler saw him, a "sleepwalker" who was mentally incapable of presenting his ideas systematically. In the light of our analysis, Kepler is revealed as being—on the contrary—far more clever than that. He presented his research as seemingly aimless intentionally to depict himself as blameless for overthrowing Tycho's legacy and to shield himself against the accusation that his new physical astronomy was dangerously idiosyncratic. If we take into account how long this conceit has remained unchallenged, Kepler gets the last laugh.

A related reason why the *Astronomia nova* has evaded contextualization is that its argument seems to lie right out in the open. Kepler does not hesitate, it seems, to include any failure or blind alley that he encountered. If anything, Kepler's argument in the *Astronomia nova* has bedazzled us by including too much information. With this stultifying surfeit of detail, is there any possibility but that it is a comprehensive account of his research? Here again, our study of the context of the book's composition gives us a better insight as to how the book functions. There is a reason for the great number of seemingly redundant demonstrations that Kepler includes. Far from being a self-indulgent, undisciplined recitation of his experience, the *Astronomia nova* is a carefully constructed argument. Why are so many demonstrations tediously repeated for both the true and the mean sun? Because Fabricius continually objected that Kepler's innovation was itself responsible for the difficulties he encountered. Kepler anticipated the many ways that astronomers might evade the necessity of his conclusions, and he blocked as many of those avenues as possible. I have not provided an explanation for every demonstration that Kepler includes, but I have demonstrated the most important trends and that such explanations are possible. Kepler did not gratuitously include superfluous detail; on the contrary, the book's content represents an active anticipation of its readers' responses.

Finally, it seems that another reason that the *Astronomia nova* has not been read in the context of its composition is that it is, quite simply, a complex technical book. During his many years of work on the theory of Mars, Kepler grew to become a mathematical astronomer of the first caliber. It has taken the efforts of a series of distinguished historians of astronomy simply to describe what Kepler was doing in the book. The nonspecialist faces almost insurmountable difficulties in trying to approach it. Copernicus wrote, "Astronomy is written for astronomers."[1] In a similar vein, Kepler began the *Astronomia nova* by lamenting, "It is extremely hard these days to write mathematical books, especially astronomical ones."[2] As he saw it, the rigorous presentation of mathematics made it tedious to read, and there were too few suitably prepared readers. If in his day Kepler expected few to be able to be equipped follow his argument, what hope would he have held for the general reader of four centuries later? Worse still, in the type of analysis we have attempted, the historian must read deeper than Kepler's presentation to try to understand not only what he was saying but also why he was saying it. And, although there is much of interest in the context in which the book was written and in Kepler's interaction with his peers, this was expressed ultimately in extremely detailed technical issues. Without understanding these and being able to work back from them, the bulk of the meaning in the *Astronomia nova* remains beyond our recognition.

I have not presented a thorough, comprehensive analysis of the *Astronomia nova*, neither as a book of technical astronomy nor as a contextualized product of Kepler's experience. What I have done is to attack the problem already glimpsed by others, of trying to separate Kepler's narrative of discovery from his actual experience. In doing so, I have also attempted to answer the question of why he chose to present his findings in such a narrative. I am certain that there is much that I have overlooked, but I am confident that the general outlines of my answer provide a new a framework in which to view the *Astronomia nova*.

⌘ NOTES ⌘

FREQUENTLY CITED SOURCES

GW Johannes Kepler. *Gesammelte Werke*, ed. Max Caspar et al. 20 vols. to date. Munich: C. H. Beck, 1937– . I will follow the citation standard developed by Owen Gingerich for the Dover edition of Max Caspar, *Kepler*, trans. C. Doris Hellman (New York: Dover, 1993). In the case of Kepler's printed work, I will give the work, chapter, *Gesammelte Werke* volume, page and line numbers. In the case of correspondence, I will give sender, recipient, date, *Gesammelte Werke* volume, letter number and line numbers.

Frisch *Joannis Kepleri astronomi opera omnia*, ed. Christian Frisch. 8 vols. Frankfurt and Erlangen, 1858–1871.

Donahue Johannes Kepler. *New Astronomy*, trans. William H. Donahue. Cambridge: Cambridge University Press, 1992.

Duncan Johannes Kepler. *Mysterium Cosmographicum: The Secret of the Universe*, trans. A. M. Duncan. Introduction and commentary by E. J. Aiton. New York: Abaris Books, 1981.

TBOO *Tychonis Brahe Dani Opera Omnia*, ed. J. L. E. Dreyer. 15 vols. Copenhagen: Gyldendal, 1913–1929.

Unless otherwise indicated, all translations are my own.

PREFACE

1. Owen Gingerich, "The Computer versus Kepler," in *The Eye of Heaven: Ptolemy, Copernicus, Kepler* (New York: The American Institue of Physics, 1993), 357–66. Originally published in *American Scientist* 52 (1964): 218–26.

2. Owen Gingerich, "Kepler's Treatment of Redundant Observations or, the Computer versus Kepler Revisited," in *The Eye of Heaven*, 370. Originally published in *Internationales Kepler-Symposium, Weil der Stadt 1971*, ed. Fritz Krafft, Karl Meyer, and Bernhard Sticker, Arbor scientiarum, ser. A, vol. 1 (Hildesheim: Verlag Dr. H. A. Gerstenberg 1973).

3. Max Caspar, ed., *Nachbericht* to *Astronomia nova*, vol. 3 of Johannes Kepler, *Gesammelte Werke* (Munich: C.H. Beck, 1937), 439.

4. Ibid., 446.

5. Cf. "The history of cosmic theories, in particular, may without exaggeration be called a history of collective obsessions and controlled schizophrenias." Arthur Koestler, *The Sleepwalkers* (New York: Macmillan, 1959), 15.

6. Ibid., 314.

7. Alexandre Koyré, *The Astronomical Revolution: Copernicus—Kepler—Borelli*, trans. R.E.W. Maddison (London: Methuen, 1980), 165–66. It is an indication of Koyré's awareness of the rhetorical nature of the *Astronomia nova* that in a note to this passage he quotes the beginning of the chapter summaries, where

Kepler states this purpose explicitly (see p. 214), but Koyré quotes it in Latin (392, n. 23).

8. Curtis Wilson, "Kepler's Derivation of the Elliptical Path," *Isis* 59 (1968): 5–25; and Eric J. Aiton, "Kepler's Second Law of Planetary Motion," *Isis* 60 (1969): 75–90.

9. D. T. Whiteside, "Keplerian Planetary Eggs, Laid and Unlaid, 1600–1605," *Journal for the History of Astronomy* 5 (1974): 1–21.

10. Bruce Stephenson, *Kepler's Physical Astronomy* (New York: Springer-Verlag, 1987), 2–3.

11. "On the other hand, many will wonder whether Stephenson has found enough evidence to sustain his more general thesis (pp. 3–22) that, contrary to the view of most previous commentators, 'The *Astronomia nova* is arranged not chronologically but rather for didactic reasons.'" Victor E. Thoren, review of *Kepler's Physical Astronomy*, by Bruce Stephenson, *Journal for the History of Astronomy* 19 (1988): 281.

12. William H. Donahue, "Kepler's Fabricated Figures: Covering up the Mess in the *Astronomia nova*," *Journal for the History of Astronomy* 19 (1988): 234.

13. William H. Donahue, translator's introduction to Johannes Kepler, *New Astronomy* (Cambridge: Cambridge University Press, 1992), 3.

INTRODUCTION

1. It should be noted that Copernicus's planetary theory *does* retain uniform motion around the point occupied by Ptolemy's equant. Indeed, it was crafted specifically to reproduce this feature of Ptolemaic theory. But Copernicus recast the motion described by Ptolemy's eccentric and equant into individual motions, each of which was uniform and circular around its own center. See N. M. Swerdlow and O. Neugebauer, *Mathematical Astronomy in Copernicus's* De revolutionibus (New York: Springer-Verlag, 1984), 293–94.

CHAPTER 1

1. Nicolaus Copernicus, *On the Revolutions*, trans. Edward Rosen (1978; reprint, Baltimore: The Johns Hopkins University Press, 1992), 4. I have modified Rosen's translation by substituting the word "commensurability" for "symmetry" (the original reads "*symmetriam*"), and by forcing a new paragraph in the middle for clarity.

2. See James Evans, "On the function and the probable origin of Ptolemy's equant," *American Journal of Physics* 52(1984): 1080–89.

3. In the case of the superior planets, the epicycle represents the motion induced by a moving earth. For the inferior planets, it is the deferent that represents the motion of the earth.

4. See Swerdlow and Neugebauer, *Mathematical Astronomy in Copernicus's* De Revolutionibus, 41–48. For evidence that Marāgha methods might have been known to others, see Jerzy Dobrzycki and Richard L. Kremer, "Peurbach and Marāgha Astronomy? The Ephemerides of Johannes Angelus and Their Implications," *Journal for the History of Astronomy* 27 (1996): 187–237.

5. Swerdlow and Neugebauer, *Mathematical Astronomy in Copernicus's* De Revolutionibus, 85.

6. See Bernard R. Goldstein, *The Arabic Version of Ptolemy's* Planetary Hypotheses, Transactions of the American Philosophical Society, n.s. 57, pt. 4 (1967), and E. J. Aiton, "Peurbach's *Theoricae novae planetarum*: A Translation with Commentary," *Osiris* 3 (1987): 5–44.

7. See Albert Van Helden, *Measuring the Universe: Cosmic Dimensions from Aristarchus to Halley* (Chicago: University of Chicago Press, 1985), ch. 3.

8. Nicolaus Copernicus, *On the Revolutions of the Heavenly Spheres*, trans. Charles Glenn Wallis, Great Books of the Western World, 16 (Chicago: Encyclopædia Britannica, 1952), 528.

9. Robert S. Westman, "The Melanchthon Circle, Rheticus and the Wittenberg Interpretation of the Copernican Theory," *Isis* 66 (1975): 165–93.

10. Owen Gingerich, "Erasmus Reinhold and the Dissemination of Copernican Theory," in *The Eye of Heaven*, 221–51.

11. Quoted from Edward Rosen, *Three Imperial Mathematicians: Kepler Trapped between Tycho Brahe and Ursus* (New York: Abaris Books, 1986), 98–100.

12. See Christine Jones Schofield, *Tychonic and Semi-Tychonic World Systems* (New York: Arno Books, 1981).

13. For a full account of this episode, see Victor E. Thoren, *The Lord of Uraniborg: A Biography of Tycho Brahe* (Cambridge: Cambridge University Press, 1990), 432–39; N. Jardine, *The Birth of History and Philosophy of Science: Kepler's* A Defence of Tycho Against Ursus *with Essays on its Provenance and Significance* (Cambridge: Cambridge University Press, 1984), ch. 1; Rosen, *Three Imperial Mathematicians*; and Owen Gingerich and Robert S. Westman, *The Wittich Connection: Conflict and Priority in Late Sixteenth-Century Cosmology*, Transactions of the American Philosophical Society, 78, pt. 7 (Philadelphia: The American Philosophical Society, 1988), 42–76.

14. Robert S. Westman, "Three Responses to the Copernican Theory: Johannes Praetorius, Tycho Brahe, and Michael Maestlin," in *The Copernican Achievement*, ed. Robert S. Westman (Berkeley: University of California Press, 1975), 289–305.

15. Aristotle, *Physics*, bk.2, ch. 2, trans. R. P. Hardie and R. K. Gaye, quoted from Richard McKeon, ed., *The Basic Works of Aristotle* (New York: Random House, 1941), 238.

16. See Peter Barker and Bernard R. Goldstein, "Realism and Instrumentalism in Sixteenth-Century Astronomy: A Reappraisal," *Perspectives on Science* 6 (1998): 232–58.

17. See Robert S. Westman, "The Astronomer's Role in the Sixteenth Century: A Preliminary Study," *History of Science* 18 (1980): 105–47.

18. Barker and Goldstein, "Realism and Instrumentalism in Sixteenth-Century Astronomy," 253.

19. Ibid., 244.

20. Ibid., 252.

21. For details of Maestlin's life and work, see Richard A. Jarrell, *The Life and Scientific Work of the Tübingen Astronomer Michael Maestlin, 1550–1631* (Ph.D. diss., University of Toronto, 1971).

22. Robert S. Westman, "The Comet and the Cosmos: Kepler, Mästlin, and the Copernican Hypothesis," *Studia Copernicana*, V; *Colloquia Copernicana*, I (1972): 7–30.

23. Richard A. Jarrell, "Astronomy at the University of Tübingen: The Work of Michael Mästlin," in *Wissenschaftsgeschichte um Wilhelm Schickard*, ed. Friedrich Seck (Tübingen: J.C.B. Mohr, 1981), 15.

CHAPTER 2

1. Copernicus's theories were *not* more accurate than Ptolemy's. See Owen Gingerich, "'Crisis' versus Aesthetic in the Copernican Revolution," in *The Eye of Heaven*, 193–204.

2. *Mysterium cosmographicum*, preface, *GW* i, 9:11–21; trans. Duncan, 63. I have altered the translation by rendering "seu mauis" as "or if you prefer" in place of "or rather."

3. *GW* xx.1, 147–49, 491–94. It was originally published in Frisch, vol. 8, 266, where Frisch supplied the title "Fragmentum orationis de motu terrae." In notes to the 1621 edition of the *Mysterium*, Kepler adds, "You will find [the disputation on the first motion] augmented in Book I of the *Epitome of Astronomy*," and "This disputation [ascribing to the earth the motion of the sun] was inserted into my *Commentaries on the motion of Mars*, particularly in the introduction; however it is found in detail in Book IV of the *Epitome*, page 542. Arguments which plainly indicate this were constructed from the total reformation of astronomy" (*GW* viii, 28:23–28, trans. Duncan, 71).

4. *GW* xx.1, 147:25.

5. Ibid., 147:35–39.

6. Ibid., 148:1.

7. Ibid., 148:4–5.

8. Ibid., 148:13–18.

9. Identified by the editors of the *Gesammelte Werke* as John 1:5, *GW* xx.1, 574.

10. *GW* xx.1, 148:19–32.

11. Cf. Copernicus, *De revolutionibus*, I,10; "At rest, however, in the middle of everything is the sun. For in this most beautiful temple, who would place this lamp in another or better position that that from which it can light up the whole thing at the same time? For, the sun is not inappropriately called by some people the lantern of the universe, its mind by others, and its ruler by still others. [Hermes] the Thrice Greatest labels it a visible god, and Sophocles' Electra, the all-seeing. Thus indeed, as though seated on a royal throne, the sun governs the family of planets revolving around it." From Nicolaus Copernicus, *On the Revolutions*, trans. Edward Rosen, 22:3–10.

12. *GW* xx.1, 148:32–47.

13. The terms that Kepler uses here, *actus primus* and [*actus*] *secundus*, are identical to those employed by Hunnius in his commentary on the Epistle to the

Ephesians to describe the omnipresence of the body of Christ. The *actus primus* refers to the passive possession of godly qualities, while the *actus secundus* refers to their only occasional actualization. In the "Notæ ad epistolam D. D. Matthiæ Hafenrefferi" (1625), *GW* xii, 47–62, Kepler writes that he had been much interested in this distinction in Hunnius's commentary while studying theology between 1591 and 1594 (ll. 28–37; see also Johannes Kepler, *Selbstzeugnisse*, ed. Franz Hammer, trans. Esther Hammer (Stuttgart-Bad Cannstatt: Friedrich Frommann Verlag, 1971), 62). These terms are especially suggestive in the light of Kepler's later thoughts identifying the sun, the fixed stars, and the intermediate space with God the Father, the Son, and the Holy Spirit. These same terms are also used in the *Mysterium cosmographicum*, ch. 20, *GW* i, 70:26–27.

14. *GW* xx.1, 148:47–149:18.

15. Cf. Copernicus, *De revolutionibus*, I, 10: "Consequently, with the first principle remaining intact, for nobody will propound a more suitable principle than that the size of the spheres is measured by the length of the time, the order of the spheres is the following, beginning with the highest. The first and highest of all is the sphere of the fixed stars, which contains itself and everything, and is therefore immovable. It is unquestionably the place of the universe, to which the motion and position of all the other heavenly bodies are compared. . . . [The sphere of the fixed stars] is followed by the first of the planets, Saturn, which completes its circuit in 30 years. After Saturn, Jupiter accomplishes its revolution in 12 years. Then Mars revolves in 2 years. The annual revolution takes the series' fourth place, which contains the earth, as I said [earlier in I, 10], together with the lunar sphere as an epicycle. In the fifth place Venus returns in 9 months. Lastly, the sixth place is held by Mercury, which revolves in a period of 80 days." Trans. Rosen, *On the Revolutions*, 21:4–22:2.

16. Copernicus, *De revolutionibus*, I, 4; trans. Rosen, *On the Revolutions*, 11:15–19.

17. *Mysterium*, preface, *GW* i, 9:20–23; trans. Duncan, 63.

18. Georg Joachim Rheticus, *Narratio Prima*, ed. Henri Hugonnard-Roche and Jean-Pierre Verdet, *Studia Copernicana* 20, 60:84–87; trans. Edward Rosen, *Three Copernican Treatises* (New York: Dover, 1959), 146.

19. See above, note 11. Some of these themes extend further back into the medieval period when the sun by virtue of its size and central position among the spheres was characterized as a king in the middle of his kingdom and as the heart in the middle of the body (see Edward Grant, *Planets, Stars, and Orbs: The Medieval Cosmos, 1200–1687* (Cambridge: Cambridge University Press, 1994), 452). With the advent of heliocentrism, the sun's important role in the theories of each of the planets suggested an elaboration of the idea of the sun as the governor of the solar system. Rheticus argued for heliocentrism by pointing out that it was more appropriate for the sun as ruler of the planets to stay put, for "in human affairs the emperor need not himself hurry from city to city in order to perform the duty imposed on him by God" (Rheticus, *Narratio prima*, 56:78–80; trans. Rosen, *Three Copernican Treatises*, 139).

20. *GW* xx.1, 149:43–47.

21. Copernicus, *De revolutionibus*, I, 9; trans. Rosen, *On the Revolutions*, 18:5–8. On the nonviolent character of this motion, see also I, 8, "Yet if anyone

believes that the earth rotates, surely he will hold that its motion is natural, not violent. But what is in accordance with nature produces effects contrary to those resulting from violence, since things to which force or violence is applied must disintegrate and cannot long endure. . . . Ptolemy has no cause, then, to fear that the earth and everything earthly will be disrupted by a rotation created through nature's handiwork" (trans. Rosen, 15:26–32). And on the center being merely a geometrical point, see I, 8, "Surely Aristotle's division of simple motion into three types, away from the middle, toward the middle, and around the middle, will be construed merely as a logical exercise." (trans. Rosen, 17:24–26).

22. In the *Mysterium*, see *GW* i, 56:9–11, and ch. 20, esp. 70:18–32. In the *Astronomia nova*, see esp. *GW* iii, 23:25–25:4.

23. See Robert S. Westman, *Johannes Kepler's adoption of the Copernican hypothesis* (Ph.D. diss., University of Michigan, 1971); and "The Comet and the Cosmos."

24. See Job Kozhamthadam, *The Discovery of Kepler's Laws: The Interaction of Science, Philosophy, and Religion* (Notre Dame: University of Notre Dame Press, 1994); and Peter Barker, "Kepler's Epistemology," in *Method and Order in Renaissance Philosophy*, ed. Daniel A. DiLiscia, Eckhard Kessler, and Charlotte Methuen (Aldershot, Hampshire: Ashgate, 1997), 355–68, and "The Role of Religion in the Lutheran Response to Copernicus," in *Rethinking the Scientific Revolution*, ed. Margaret Osler (Cambridge: Cambridge University Press, 2000), 59–88.

25. For an excellent treatment of the context of Kepler's education, see Charlotte Methuen, *Kepler's Tübingen: Stimulus to a Theological Mathematics* (Brookfield, Vermont: Ashgate, 1998), ch. 2.

26. *Astronomia nova*, ch. 7, *GW* iii, 108:22–24; trans. Donahue, 183–84.

27. Kepler to Maestlin, 3 October 1595, *GW* xiii, nr. 23:253–57.

28. *Mysterium*, preface to the reader, *GW* i, 9:23–34; trans. Duncan, 63.

29. Kepler to Maestlin, 9 April 1597, *GW* xiii, nr. 64:148.

30. Kepler to Maestlin, 2 August 1595, *GW* xiii, nr. 21:7–11.

31. Kepler to Maestlin, 2 August 1595, *GW* xiii, nr. 21:11.

32. 20 July is the date he gives Maestlin for his discovery (Kepler to Maestlin, 2 August 1595, *GW* xiii, nr. 21:39–42), although he gives 19 July in the second edition of the *Mysterium* (*GW* viii, 20:2–5). In the latter, he probably conflates the date of the discovery of the polyhedra with the previous day, when he began working with plane figures.

33. *Mysterium*, preface, *GW* i, 13:16–23; trans. Duncan, 69.

34. Kepler to Maestlin, 14 September 1595, *GW* xiii, nr. 22:24–38.

35. *Mysterium*, preface, *GW* i, 10:38–11:30; trans. Duncan, 65.

36. *Mysterium*, preface, *GW* i, 11:24–26; trans. Duncan, 65.

37. Kepler to Maestlin, 3 October 1605, *GW* xiii, nr. 23:39–52.

38. See chapter 4.

39. See pp. 123–129, and Jardine, *The birth of history and philosophy of science*, esp. 139–44.

40. For the importance of the comet of 1577, see Westman, *Johannes Kepler's adoption of the Copernican hypothesis*, and "The Comet and the Cosmos."

41. See above note 2.

42. Kepler to Maestlin, 3 October 1605, *GW* xiii, nr. 23:82.

43. Kepler to Maestlin, 3 October 1605, *GW* xiii, nr. 23:84–87.

44. Kepler to Maestlin, 3 October 1605, *GW* xiii, nr. 23:107–13. Trans. Anthony Grafton, "Michael Maestlin's Account of Copernican Planetary Theory," *Proceedings of the American Philosophical Society*, 1973, *117*:524.

45. Take, for example, the first case. Kepler had periods for Saturn and Jupiter of 10759;12^d and 4332;37^d respectively. Therefore, $\frac{10759;12 - 4332;87}{2} + 4332;37 =$ 7545;54. This will be the distance of Saturn. The distance of Jupiter is simply 4332.62, so, normalizing to a distance of 1000 for Saturn, $\frac{7545;54}{4332;37} = \frac{1000}{x}$, the distance for Jupiter will be 574.

46. The errors are 2.0, 6.8, 5.5, 5.8, and 2.6 percent.

47. Kepler to Maestlin, 3 October 1605, *GW* xiii, nr. 23:176–205.

48. See p. 180. Although he continued to regard the motive virtue as inversely proportional to the distance from the sun, Kepler did not maintain that its action was simply confined to the plane of the ecliptic in the *Astronomia nova*; see Stephenson, *Kepler's Physical Astronomy*, 68–75. However, his thought was considerably more developed by then. In his first thoughts about the propagation of motor virtue, he appears to have thought only about the plane of the orbit.

49. The cube and the octahedron have smaller errors, the dodecahedron and icosahedron have larger errors, and the unique tetrahedron alone has a negative error (this argument is repeated in the *Mysterium*, ch. 21). Kepler elucidated his concept of similarity of solids in the *Mysterium*, ch. 7. Among other things, the primary solids, the cube and dodecahedron, can generate their secondary counterparts, the octahedron and icosahedron, by making the centers of their faces the vertices of the inscribed solids. Thus, for example, the cube has six faces and the octahedron six vertices. The tetrahedron, on the other hand, having four vertices and four faces, generates itself. On the classification of the solids, see *Mysterium*, chs. 3 and 7; and J. V. Field, *Kepler's Geometrical Cosmology* (Chicago: University of Chicago Press, 1988), 53–60.

50. Kepler to Maestlin, 3 October 1605, *GW* xiii, nr. 23:214–17.

51. Kepler's values from the *Mysterium* (*GW* i, 69) for the periods of Saturn, Mars, the Earth, Venus, and Mercury are in error by less than 0.01 percent. The error in his period for Jupiter is the largest by far at 0.1 percent.

52. Kepler to Maestlin, 3 October 1605, *GW* xiii, nr. 23:227–29.

53. Cf. Copernicus, *De Revolutionibus*, I, 5; "Moreover since the heavens, which enclose and provide the setting for everything, constitute the space common to all things, it is not at first blush clear why motion should not be attributed rather to the enclosed than to the enclosing, to the thing located in space rather than to the framework of space." Trans. Rosen, *On the Revolutions*, 12:11–14. The question of whether the sphere of the fixed stars was a place at all was problematic according to Aristotle's definition of place. On the attempts of medieval commentators to come to grips with the problem, see Grant, *Planets, Stars and Orbs*, 122ff.

54. Kepler to Maestlin, 3 October 1605, *GW* xiii, nr. 23:233–36.

55. Maestlin to Kepler, 27 February 1596 (o.s.), *GW* xiii, nr. 29:1–10.

56. For a translation and an analysis of this document, see Grafton, "Michael Maestlin's Account of Copernican Planetary Theory." The significance of this doc-

ument in the development of Kepler's ideas about the planet-moving force is discussed on pp. 55–58.

57. See Barker, "The Role of Religion in the Lutheran Response to Copernicus," 80–81.

58. Maestlin to Duke Friedrich von Württemberg, 12 March 1596 (o.s.), *GW* xiii, nr. 31.

59. Kepler to Duke Friedrich von Württemberg, 17 February 1596 (o.s.), *GW* xiii, nr. 28; and Kepler to Duke Friedrich von Württemberg, 17 February 1596 (o.s.), *GW* xiii, nr. 30. Kepler also applied for funds so that a replica of his invention could be built in the form of nested copper spheres with different gems denoting the planets, which he even suggested might be constructed with taps so that the spaces between the spheres might be filled with beverages (beer or cheap wine in the large volume between Saturn and Jupiter, good wine between Jupiter and Mars, and so on).

60. Maestlin to Duke Friedrich von Württemberg, 12 March 1596 (o.s.), *GW* xiii, nr. 31:20–37.

61. Maestlin to Duke Friedrich von Württemberg, 12 March 1596 (o.s.), *GW* xiii, nr. 31:58–67 and 73–76.

62. See pp. 67–69.

CHAPTER 3

1. *Prodromus dissertationum cosmographicarum, continens mysterium cosmographicum, de admirabili proportione orbium coelestium, deque causis coelorum numeri, magnitudinis, motuumque periodicorum genuinis & proprijs, demonstratum, per quinque regularia corpora Geometrica.*

2. Kepler to Maestlin [March 1596 (o.s.)], *GW* xiii, nr. 32:32–37.

3. Kepler to Herwart von Hohenburg, 26 March 1598, *GW* xiii, nr. 91:63–87.

4. On the disciplinary boundaries of astronomy during this period, see Robert S. Westman, "The Astronomer's Role."

5. Johannes Praetorius to Herwart von Hohenburg, 23 April 1598 (o.s.), *GW* xiii, nr. 95:7–8. For Kepler's statement in the preface, see *Mysterium cosmographicum*, preface to the reader, *GW* i, 9:19–20; trans. Duncan, 63.

6. For a fuller account, see Field, *Kepler's Geometrical Cosmology*, 30–72.

7. For a discussion of this argument see Nicholas Jardine, "The Forging of Modern Realism: Clavius and Kepler against the Sceptics," *Studies in History and Philosophy of Science* 10 (1979): 141–73, esp. 156–59.

8. *Mysterium*, ch. 1, *GW* i, 15:11–12; trans. Jardine, "The Forging of Modern Realism," 157.

9. *Mysterium*, ch. 1, *GW* i, 15:24–30; trans. Jardine, ibid.

10. Jardine notes that attributing a precise statement of kinematic equivalence to Kepler is problematic. Ibid., 158–59.

11. Strictly speaking, the Ptolemaic and Copernican systems can be observationally distinguished through diurnal parallax, since Mars at opposition, for instance, approaches closer to the earth than the sun in the Copernican system. This effect, however, is undetectable with the naked eye. Nevertheless, its detection was thought to observationally feasible, and Tycho Brahe made various claims to

have measured it. See Owen Gingerich and James R. Voelkel, "Tycho Brahe's Copernican Campaign," *Journal for the History of Astronomy* 29 (1998): 1–34.

12. *Mysterium*, ch. 1, *GW* i, 16:16–21; trans. Duncan, 77.

13. Kepler briefly discussed the argument from the comet of 1577 in chapter 1.

14. He made this remark in the notes to the 1621 edition, *GW* viii, 59:2–3 and 62:25–26; trans. Duncan, 119 and 125.

15. *Mysterium*, ch. 9, *GW* i, 34:28–31 and *Mysterium*, ch. 11, *GW* i, 37:33; trans. Duncan, 115 and 123.

16. See pp. 68–69.

17. *Mysterium*, ch. 13, *GW* i, 43:34–37; trans. Duncan, 149.

18. Note the distinction of the activity of the physicist from that of the astronomer: "For the physicists believe that from the inner surface of the heaven of the Moon right up to the tenth sphere there is no empty space in the celestial spheres, but sphere is always touched by sphere, and the inner surface of the upper is completely united with the higher surface of the lower. Thus if anyone asks, for example, what the position of the heaven of Mars is, physically, they reply: the inner surface of Jupiter's. And from Ptolemy and the customary astronomical description they may be able to draw the following reason: that in them there is no opportunity for investigating the proportions of the spheres, and no assistance. For just as nobody easily refutes those who have written about the new Indies if he has not inspected those regions for himself, so an astronomer cannot reject the physicists' feeble figuring about the contact of the spheres unless the test of observation and the agreement of hypotheses has transported him out to the actual sky and between the spheres" (*Mysterium*, ch. 13, *GW* i, 47:9–20; trans. Duncan, 155).

19. Maestlin to Kepler, 27 February 1596 (o.s.), *GW* xiii, nr. 29:62–428.

20. The moon raised another problem: whether the thickness of the earth's sphere should be calculated including or excluding the moon's sphere. If including it increased the disagreement from observations, Kepler asserted his willingness to measure from the earth's sphere, but he was disturbed by the notion of one sphere carrying "a lump, like a jewel on a ring," and confessed, "Nor, great heavens, do I know in which direction the cosmographical or even the metaphysical arguments tend more" (*Mysterium*, ch. 16, *GW* i, 55:19 and 16–18; trans. Duncan, 165). He finally concluded that including the moon produced better agreement with observation.

21. *Mysterium*, ch. 14, *GW* i, 48:16–18.

22. *Mysterium*, ch. 15, *GW* i, 50:12–21; trans. Duncan, 159.

23. Kepler had been preparing for this contingency since his first letter to Maestlin regarding the polyhedral hypothesis, that of 2 August 1595, when he had asked how accurate the values of the *prosthaphaereses* were.

24. Chapter 16 concerned the case of the moon (see above, note 20). Chapter 17 is a comment on the case of Mercury. Mercury manifestly did not fit the octahedron correctly. Kepler explained this away by allowing Mercury to be inscribed to the sides, rather than the surfaces, of the octahedron.

25. *Mysterium*, ch. 18, *GW* i, 59:18–31; trans. Duncan, 177.

26. *Mysterium*, ch. 18, *GW* i, 60:17–20; trans. Duncan, 179.

27. See pp. 43–45.
28. *Mysterium*, ch. 18, *GW* i, 61:16–21; trans. Duncan, 181.
29. Kepler to Ursus, 15 November 1595, *GW* xiii, nr. 26:29–30.
30. Ibid., nr. 26:30–32.
31. *Mysterium*, ch. 20, *GW* i, 68:27–34; trans. Duncan, 197.
32. *Mysterium*, ch. 20, *GW* i, 70:19–24; trans. Duncan, 199.
33. *Mysterium*, ch. 20, *GW* i, 70:24–32; trans. Duncan, 199–201. Note that the "noble epithets" are from Copernicus; see ch. 2, note 11 above.
34. *Mysterium*, ch. 20, *GW* i, 71:3–4; trans. Duncan, 201.
35. Alternately,

$$\frac{D_S}{D_I} = \frac{\frac{1}{2}(P_S - P_I) + P_I}{P_I} = \frac{\frac{1}{2}(P_S + P_I)}{P_I}$$

where P is the period and D the distance of the superior (S) or inferior (I) planet.
36. The results are comparable, although the value for the Saturn-Jupiter distance is noticeably better and the Venus-Mercury value is worse. From the former to the latter, the errors are 0.3, 5.5, 5.5, 6.0, and 12.6 percent.
37. *Mysterium*, ch. 21, *GW* i, 72:29–30; trans. Duncan, 209.
38. *Mysterium*, ch. 21, *GW* i, 73:3–8; trans. Duncan, 209.
39. Kepler modified the Copernican distance of Mercury by adding half the eccentricity (which is itself half the difference between the maximum and minimum distances) to the mean distance (see Duncan, 250, n. 1). His justification was obscure.
40. *Mysterium*, ch. 21, *GW* i, 74:9–10; trans. Duncan, 211.
41. On Kepler's confusion on this point, see Grafton, "Michael Maestlin's Account of Copernican Planetary Theory," 526–29.
42. On the near equivalence between Ptolemaic and Copernican models, see N. M. Swerdlow and O. Neugebauer, *Mathematical Astronomy in Copernicus's* De Revolutionibus, 295–97.
43. *Mysterium*, ch. 22, *GW* i, 76:17–29; trans. Duncan, 217.
44. *Mysterium*, ch. 22, *GW* i, 77:19–30; trans. Duncan, 219. I have modified Duncan's translation slightly, rendering *particularis* as "individual" rather than "peculiar."
45. *Mysterium* (1621 ed.), dedication, *GW* viii 9:25–27; trans. Duncan, 39.

CHAPTER 4

1. *Mysterium*, ch. 1, *GW* i, 14:21–27; trans. Duncan, 75.
2. Kepler to Maestlin, 3 October 1595, *GW* xiii, nr. 23:43–45.
3. For a full description of this episode, see Edward Rosen, "Kepler and the Lutheran Attitude Towards Copernicanism in the Context of the Struggle Between Science and Religion," in *Kepler: Four Hundred Years* (*Vistas in Astronomy*, 18), ed. Arthur Beer and Peter Beer (Oxford: Pergamon Press, 1975), 317–38.
4. Kepler to the University of Tübingen, 1 May 1596 (o.s.), *GW* xiii, nr. 40.
5. Maestlin to the Prorector of the University of Tübingen [end of May 1596 (o.s.)], *GW* xiii, nr. 43:24–26.
6. Matthias Hafenreffer to Kepler, 6 June 1596 (o.s.), *GW* xiii, nr. 44.

7. Kepler to Maestlin, 9 April 1597, *GW* xiii, nr. 64:2–4.

8. "While I know of no one in the brilliant light of your academy who would intentionally assail my doctrines, one is found in this dark and gloomy wilderness who, since he is not able to understand more, has brought before the estates [of Austria] only the one malicious charge that very much is taken from Copernicus and was already said by him many years ago. What should I do? Neither the hypotheses are mine, nor the solids. I have only joined them together." Kepler to Maestlin, 9 April 1597, *GW* xiii, nr. 64:146–50.

9. Maestlin to Kepler, 30 October 1597 (o.s.), *GW* xiii, nr. 80:11–32; lines 19–32, trans. Rosen, "Kepler and the Lutheran Attitude," 326.

10. Kepler to Maestlin, 6 January 1598, *GW* xiii, nr. 85:121–33; lines 129–33, trans. Rosen, "Kepler and the Lutheran Attitude," 327.

11. *ab initio statim*—the same words as the first line of chapter 1 of the *Mysterium*, cited above.

12. Hafenreffer to Kepler, 12 April 1598 (o.s.), *GW* xiii, nr. 93:22–37; lines 24–35, trans. Rosen, "Kepler and the Lutheran Attitude," 327.

13. See Kepler to Maestlin, 11 June 1596 (o.s.), *GW* xiii, nr. 47:11–13; and Maestlin to Kepler, 15/16 November 1596 (o.s.), *GW* xiii, nr. 52:93–96.

14. Hafenreffer to Kepler, 14 June 1596 (o.s.), *GW* xiii, nr. 48:5–6.

15. Hafenreffer to Kepler, 12 April 1598 (o.s.), *GW* xiii, nr. 93:37–68; trans. Rosen, "Kepler and the Lutheran Attitude," 328–29.

16. Kepler to Maestlin, 1/11 June 1598, *GW* xiii, nr. 99:492–93; trans. Rosen, "Kepler and the Lutheran Attitude," 329.

17. Kepler to Maestlin, 1/11 June 1598, *GW* xiii, nr. 99:502–30; trans. Rosen, "Kepler and the Lutheran Attitude," 329.

18. Kepler to Maestlin, 1/11 June 1598, *GW* xiii, nr. 99:507–12; trans. Rosen, "Kepler and the Lutheran Attitude," 329.

19. Kepler to Maestlin, 1/11 June 1598, *GW* xiii, nr. 99:515–17; trans. Rosen, "Kepler and the Lutheran Attitude," 329.

20. Kepler to Maestlin, 1/11 June 1598, *GW* xiii, nr. 99:505–07; trans. Rosen, "Kepler and the Lutheran Attitude," 329. Cf. Copernicus, *De revolutionibus*, dedication, "Therefore I debated with myself for a long time whether to publish the volume which I wrote to prove the earth's motion or rather to follow the example of the Pythagoreans and certain others, who used to transmit philosophy's secrets only to kinsmen and friends, not in writing but by word of mouth" Trans. Rosen, *On the Revolutions*, 3:15–19.

21. Kepler to Herwart, 12 September 1597, *GW* xiii, nr. 74:19–21.

22. *Mysterium* (1621 ed.), dedication, *GW* viii, 9:19–22; trans. Duncan, 39.

23. *Mysterium* (1621 ed.), notes to the original dedication, *GW* viii, 20:38–40.

24. Kepler wrote in March 1596 that he had neglected to bring the papers with him when he visited Maestlin (Kepler to Maestlin [March 1596 (o.s.)], *GW* xiii, nr. 32:68–70).

25. For details of the correction, see Anthony Grafton, "Michael Maestlin's Account of Copernican Planetary Theory," 526–29. Kepler's error shows that although he apparently understood the equivalence of Copernicus's and Ptolemy's theories with respect to uniform motion around the equant (Kepler to Maestlin

[March 1596 (o.s.)], *GW* xiii, nr. 32:70–73), he actually did not. Kepler's faulty argument justified uniform motion around the center of Copernicus's deferent, reasoning that it was located halfway between the center of Ptolemy's deferent and the equant because one-half of the difference in motion (that caused by the increase of the sizes of the orbits) was not present. Actually, the position of the planet on the epicycle in Copernicus's theory does move uniformly around the equant, as we saw above. Maestlin's masterful revision of Kepler's argument was true to Kepler's physical ideas while explaining the equivalence of uniform motion around the equant in Ptolemy and Copernicus.

26. Maestlin to Kepler, 9 March 1597 (o.s.), *GW* xiii, nr. 63:127–38.

27. Kepler was suitably chastened. He responded to the criticism, writing, "Nor less seriously do I also [take] that advice about moderating those speculations." Kepler to Maestlin, 9 April 1597, *GW* xiii, nr. 64:209–10.

28. Michael Maestlin, *Epitome astronomiae: qva brevi explicatione omnia, tam ad Sphæricam quàm Theoricam eius partem pertinentia, ex ipsius scientiæ fontibus deducta, perspicuè per quæstiones traduntur* (Tübingen, 1597).

29. Maestlin's textbook is geocentric, although he was a Copernican. He explained in a letter to the duke of Württemberg that even though all specialists adhered to Copernicus, the older doctrines were taught because they were easier for the students to understand (Maestlin to Duke Friedrich von Württemberg, 12 March 1596 (o.s.), *GW* xiii, nr. 31:45–48).

30. Maestlin, *Epitome astronomiae*, 42.

31. Ibid., 30. Compare the last sentence "Hoc modo. Κόσμ[ος] ἐστίν ἡ τῶν ὅλων τάξις τε κ[αὶ] διακόσμησις ὑπὸ θε[οῦ] τε κ[αὶ] διὰ θεὸν πεποιημένη, κ[αὶ] διαφυλαττομένη. Mundus est vniversorum ordinatio, & distributio, quæ à Deo & propter Deum, creata est & conseruatur" to Pseudo-Aristotle, *De mundo*, trans. D. J. Furley, *The Loeb Classical Library*, no. 400 (Cambridge, Massachusetts: Harvard University Press, 1978), 391b11–12; "In another sense, *cosmos* is used to signify the orderly arrangement of the universe, which is preserved by God and through God [λέγεται δὲ καὶ ἑτέρως κόσμος ἡ τῶν ὅλων τάξις τε καὶ διακόσμησις, ὑπὸ θεοῦ τε καὶ διὰ θεὸν φυλαττομένη]." This statement is attributed to Posidonius; see, *Poseidonios: Die Fragmente*, ed. Willy Theiler, *Texte und Kommentare*, 10 (Berlin: Walter de Gruyter, 1982), vol. 2, 156.

32. Jarrell, *The Life and Scientific Work of . . . Michael Maestlin*,, 140–41.

33. See chapter 8, note 1, below.

34. Maestlin to Kepler, 1 October 1616, *GW* xvii, nr. 744:24–29; trans. from Owen Gingerich, "Mästlin's, Kepler's, and Schickard's Copies of *De revolutionibus*," in *Zum 400. Geburtstag von Wilhelm Schickard*, ed. Friedrich Seck [*Contubernium*, 41] (Sigmaringen: Jan Thorbecke Verlag, 1994), 171.

35. Kepler to Maestlin [beginning of October 1597], *GW* xiii, nr. 75:121.

36. See Stillman Drake, *Galileo Studies: Personality, Tradition, and Revolution* (Ann Arbor: The University of Michigan Press, 1970), 123–25.

37. Galileo to Kepler, 4 August 1597, *GW* xiii, nr. 73:16–25.

38. Kepler to Galileo, 13 October 1597, *GW* xiii, nr. 76:25–46.

39. There is evidence that Kepler did make such use of letters. In the 1621 edition of the *Mysterium*, he mentions the delayed receipt of a letter from Tycho Brahe: "In the end, I regaled myself on it abundantly, and added it to the public

joy which then prevailed in Styria." *Mysterium* (1621 ed.), notes on the original dedication, *GW* viii, 20:42–43; trans. Duncan, 59.

40. Kepler to Galileo, 13 October 1597, *GW* xiii, nr. 76:46–50.

41. Ibid., nr. 76:50–53.

42. Ibid., nr. 76:64–69. On Tycho's attempt to measure stellar parallax, see Thoren, *Lord of Uraniborg*, 279.

43. Kepler to Galileo, 13 October 1597, *GW* xiii, nr. 76:55–56.

44. Kepler to Herwart, 26 March 1598, *GW* xiii, nr. 91:162–68.

45. Drake, *Galileo Studies*, ch. 10; and *Galileo at Work: His Scientific Biography* (Chicago: University of Chicago Press, 1978), 36–38.

46. *Astronomia nova*, introduction, *GW* iii, 26; Donahue, 56–57.

47. Galileo Galilei, *Dialogue Concerning the Two Chief World Systems—Ptolemaic and Copernican*, trans. Stillman Drake (2d ed.; Berkeley: University of California Press, 1967), 462.

48. Drake, *Galileo at Work*, 34, 56.

49. Alexandre Koyré, *Galileo Studies*, trans. John Mepham (Atlantic Highlands, New Jersey: Humanities Press, 1978), 28–38.

50. Drake, *Galileo at Work*, 12, 51.

51. Quoted from Drake, *Galileo at Work*, 52.

52. Koyré, *Galileo Studies*, 201–9.

53. Drake, 62–66; and "Galileo's 'Platonic' Cosmology and Kepler's *Prodromus*," *Journal for the History of Astronomy* 4 (1973): 174–91.

54. Drake, "Galileo's 'Platonic' Cosmology," 184. For his earlier, less plausible explanation, see *Galileo Studies*, 127–30.

55. Drake has pointed out that the closest Galileo ever came to acknowledging elliptical orbits was a comment in the *Sidereus nuncius* that even a "nearly straight" (presumably highly eccentric) "oval" motion would be incapable of explaining the apparent change of size in Jupiter's satellites, though even this reference is questionable. Stillman Drake, ed. and trans., *Discoveries and Opinions of Galileo* (New York: Doubleday, 1957), 58.

56. Cesi to Galileo, 21 July 1612, *Le opere di Galileo Galilei*, Edizione nazionale (Florence, 1901), vol. 11, 365–67; trans. Rosen (as corrected by Drake), *Isis* 48 (1957): 66.

57. See Erwin Panofsky, "Galileo as a Critic of the Arts: Aesthetic Attitude and Scientific Thought," *Isis* 47 (1956):3–15. This article is an abridged version of Erwin Panofsky, *Galileo as a Critic of the Arts* (The Hague: Martinus Nijhoff, 1954); revised in accordance with suggestions in Alexandre Koyré, "Attitude esthetique et pensée scientifique," *Critique*, IX, vol. 12, no. 100–01 (1955):835–47. See also Edward Rosen's review of the 1954 version, *Isis* 47 (1956):78–80; and Panofsky's reply, ibid., 182–85.

58. Koyré, *Galileo Studies*, 222, n. 115.

59. See Galileo to Castelli, 21 December 1613, and "Letter to the Grand Duchess Christina" in *The Galileo Affair: A Documentary History*, ed. and trans. Maurice A. Finocchiaro (Berkeley: University of California Press, 1989), 49–54, 87–118.

60. Galileo to Castelli, 21 December 1613; Finocchiaro, *The Galileo Affair*, 54.

61. Finocchiaro, *The Galileo Affair*, 71, 97.

62. ibid., 116–17.

63. Owen Gingerich, foreword to Donahue, xii.

64. Drake, *Galileo Studies*, 125.

65. Ibid., 138.

66. For biographical details, see Westman, "Three Responses," 290–91; and "The Melanchthon Circle," 171.

67. Quoted from Westman, "Three Responses," 292–93.

68. Praetorius, through his personal encounter with Rheticus, was well aware of Copernicus's realist intention, despite Osiander's anonymous instrumentalist preface. Notes in his hand in a copy of *De revolutionibus* indicate, "Rheticus affirmed that this preface was added by Andreas Osiander. But it was rejected by Copernicus." Ibid., 304.

69. Ibid., 293–301.

70. "A first glance at Kepler pleases me very much, the attentiveness employed was the greatest, in the context itself and the style," Praetorius to Herwart, 14/24 February 1598, *GW* xiii, nr. 86:177.

71. Praetorius to Herwart, 23 April 1598 (o.s.), *GW* xiii, nr. 95:5–21.

72. See the *Mysterium*, ch. 8, "On the disagreement between the equations derived from the solids and those of Copernicus in general, and on the precision of astronomy"; trans. Duncan, 177–89.

73. Praetorius to Herwart, 23 April 1598 (o.s.), *GW* xiii, nr. 95:21–39.

74. Quoted from Westman, "Three Responses," 303.

75. On Roeslin, see C. Doris Hellman, *The Comet of 1577: Its Place in the History of Astronomy* (New York: Columbia University Press, 1944), 159–73.

76. Tycho addressed Roeslin's work in his book on the comet of 1577 (Hellman, *The Comet of 1577*, 163), and Kepler mentioned him in the same breath as Lansberg, Ursus, Magini, and Tycho in a letter to Herwart (Kepler to Herwart, 26 March 1598, *GW* xiii, nr. 91:174–80).

77. Roeslin to Herwart, 4/14 May 1597, *GW* xiii, nr. 68:1–21.

78. *De opere Dei creationis sev de mundo Hypotheses . . .* (1597). See Schofield, *Tychonic and Semi-Tychonic World Systems*, 136–44.

79. Schofield, *Tychonic and Semi-Tychonic World Systems*, 140–41.

80. These early letters between Roeslin and Kepler have not survived.

81. Roeslin to Herwart, 20/30 July 1597, *GW* xiii, nr. 72:4–16.

82. Text reads "*Corpus*."

83. Roeslin to Herwart, 20/30 July 1597, *GW* xiii, nr. 72:17–45.

84. Herwart to Kepler, 12 March 1598, *GW* xiii, nr. 88:11–13. It is not clear to whose letter Herwart is referring. This letter is dated 12 March 1598. It follows Roeslin's letters by over seven months, though such a delay was not unheard of. It could not have referred to Praetorius's letters, for those follow this letter.

85. Kepler to Maestlin [beginning of October 1597], nr. 75:127–29; trans. Rosen, *Three Imperial Mathematicians*, 90–91.

86. For a full account of this dispute see Rosen, *Three Imperial Mathematicians*. On Tycho Brahe, see Thoren, *The Lord of Uraniborg*.

87. Bürgi (1552–1632) was a watchmaker and instrument maker to Wilhelm IV of Hesse and later to Emperor Rudolf II. Ursus credited him with being his teacher in astronomy (Ursus to Kepler, 29 May 1597, *GW* xiii, nr. 69:16–17).

88. Ursus to Kepler, 29 May 1597, *GW* xiii, nr. 69:22–24; trans. Rosen, *Three Imperial Mathematicians*, 90.

89. Ursus to Kepler, 29 May 1597, *GW* xiii, nr. 69:29–32; trans. Rosen, *Three Imperial Mathematicians*, 90.

90. Kepler to Ursus, 15 November 1595, *GW* xiii, nr. 26:12.

91. Kepler to Tycho, 13 December 1597, *GW* xiii, nr. 82:7–16; trans. Rosen, *Three Imperial Mathematicians*, 106.

92. Tycho Brahe, *Astronomiæ instauratæ progymnasmatum secunda pars*, in *TBOO*, vol. 2, 421. According to Plato's description of the composition of the polyhedra in the *Timeaus*, each face would be made up of a number of constituent triangles, see Field, *Kepler's Geometrical Cosmology*, 9–13.

93. Tycho to Kepler, 1 April 1598 (o.s.), *GW* xiii, nr. 92:13–18.

94. Ibid., nr. 92:171–72.

95. Ibid., nr. 92:18–24.

96. Copernicus was able to eliminate the Ptolemaic equant by replacing it with an eccentric epicyclet. The center of Copernicus's deferent was halfway between the equant and the center of the Ptolemaic deferent, thus the distance from the sun to the center of the Copernican deferent was 3/2 e, or the distance from the equant to the new center of the deferent is 1/3 of the whole Copernican eccentricity. The epicycle was of size 1/2 e—1/3 of the Copernican eccentricity—and rotated in the same direction and with the same angular velocity as the deferent in such a way that the planet moved uniformly with respect to the equant and did not stray sensibly from Ptolemy's eccentric. See Swerdlow and Neugebauer, *Mathematical Astronomy in Copernicus's De Revolutionibus*, 295–97.

97. Copernicus, in fact, relied on the property of his model that the planets moved uniformly around the equant in his derivation. Without this assumption it could not have worked. See Swerdlow and Neugebauer, *Mathematical Astronomy in Copernicus's De Revolutionibus*, 307–8.

98. He repeats this argument about Copernicus's theories being too flawed to support Kepler in a letter to Maestlin (Tycho to Maestlin, 21 April 1598 (o.s.), *GW* xiii, nr. 94:27–36).

99. This idea is expressed in Maestlin's preface to Rheticus's *Narratio prima*, which was appended to the *Mysterium*, *GW* i, 82. See also Maestlin to Duke Friedrich von Württemberg, 12 March 1596 (o.s.), *GW* xiii, nr. 31:58–67; and Maestlin to the Prorector of the University of Tübingen, [end of May 1596], *GW* xiii, nr. 43:24–33.

100. Tycho to Maestlin, 21 April 1598 (o.s.), *GW* xiii, nr. 94:16–26; trans. Rosen, *Three Imperial Mathematicians*, 118–19.

101. In a Tychonic system the distance to the fixed stars is determined by the outermost distance of Saturn. In his *De hypothesibus astronomicis tractatus* (1597), Ursus mercilessly ridiculed Roeslin for the effrontery of so determining "the altitude of the heavens" (Jardine, *The Birth of History and Philosophy of Science*, 50–51).

102. Tycho to Kepler, 1 April 1598 (o.s.), *GW* xiii, nr. 92:90–96.

103. Tycho to Kepler, 9 December 1599, *GW* xiv, nr. 145:197–99.

104. Ibid., nr. 145:202–09.

105. Cf. Copernicus, *De revolutionibus*, I, 4, "We must acknowledge, nevertheless, that their motions are circular or compounded of several circles, because these nonuniformities recur regularly according to a constant law. This could not happen unless the motions were circular, since only the circle can bring back the past. Thus, for example, by a composite motion of circles the sun restores to us the inequality of days and nights as well as the four seasons of the year. Several motions are discerned herein, because a simple heavenly body cannot be moved by a single sphere nonuniformly. For this nonuniformity would have to be caused either by an inconstancy, whether imposed from without or generated from within, in the moving force or by an alteration in the revolving body. From either alternative, however, the intellect shrinks." Trans. Rosen, *On the Revolutions*, 11:10–19.

106. Tycho to Kepler, 9 December 1599, *GW* xiv, nr. 145:192–93.

107. Tycho to Kepler, 1 April 1598 (o.s.), *GW* xiii, nr. 92:67–71.

108. Ibid., nr. 92:180–81.

109. Kepler to Galileo, 13 October 1597, *GW* xiii, nr. 76:57–70; Kepler to Herwart, 26 March 1598, *GW* xiii, nr. 91:90–161; and Kepler to Herwart, 16 December 1598, *GW* xiii, nr. 107:90–174.

110. Tycho to Kepler, 9 December 1599, *GW* xiv, nr. 145:228–42.

111. Thoren, *Lord of Uraniborg*, 432–42.

112. One might conclude that Tycho's kindness was not motivated by the Ursus affair based on the fact that all mention of Ursus in this first letter is restricted to a postscript. This fact, and Tycho's mention that Kepler's letter and Ursus's book came via the same courier, might conjure up the image of Tycho writing to Kepler, then examining Ursus's book, and then writing the angry postscript. One must note, however, that the postscript is intentionally written on a separate sheet, which Tycho asks be "consigned to the flames" (Tycho to Kepler, 1 April 1598 (o.s.), *GW* xiii, nr. 92:163).

113. Little is known about Limnaeus [Wirn]. He entered the University of Jena in 1571, became a *magister* in 1681, and was named professor of mathematics in 1588. He was also prorector at Jena in 1593 and 1609. See Johann Christoph Adelung, *Fortsetzung und Ergänzungen zu Christian Gottlieb Jöchers allgemeinem Gelehrten-Lexikon* (7 vols., Leipzig: 1784–1897), vol. 3, 1836; and Rosen, *Three Imperial Mathematicians*, 104 and 345, n. 1.

114. This letter has not survived.

115. A Babylonian astronomer of the 3d century B.C.

116. Limnaeus to Kepler, 24 April 1598 (o.s.), *GW* xiii, nr. 96:3–28.

117. Herwart to Kepler, 12 March 1598, *GW* xiii, nr. 88:13–24.

118. See pp. 46–47.

119. See pp. 71–72.

120. Kepler to Herwart, 26 March 1598, *GW* xiii, nr. 91:180–92.

121. Ibid., nr. 91:194–97; trans. Rosen, *Three Imperial Mathematicians*, 101.

CHAPTER 5

1. *Astronomia nova*, ch. 7, *GW* iii, 108–10.
2. See, for instance, Arthur Koestler, *The Sleepwalkers*, 303.
3. Max Caspar, *Kepler*, trans. C. Doris Hellman (Dover edition; New York: Dover, 1993), 77–85.
4. Kepler to Maestlin, 16/26 February 1599, *GW* xiv, nr. 113:118–23; trans. Rosen, *Three Imperial Mathematicians*, 149.
5. Kepler to Maestlin, 16/26 February 1599, *GW* xiii, nr. 113:45–93 (emphasis in original).
6. Kepler to Herwart, 12 July 1600, *GW* xiv, nr. 168:105–07; Kepler to Magini, 1 June 1601, *GW* xiv, nr. 190:18–21; and *Astronomia nova*, ch. 7, *GW* iii, 109:8–10.
7. On Longomontanus, see pp. 153–54.
8. Thoren, *Lord of Uraniborg*, 431–32. For biographical details on all of these figures, see John Robert Christianson, *On Tycho's Island: Tycho Brahe and His Assistants, 1570–1601* (Cambridge: Cambridge University Press, 2000). See also J.L.E. Dreyer, *Tycho Brahe: A Picture of Scientific Life and Work in the Sixteenth Century* (Edinburgh: Adam & Charles Black, 1890), 287–88.
9. Thoren, *Lord of Uraniborg*, 447.
10. Ibid., 440.
11. Kepler to Herwart, 12 July 1600, *GW* xiv, nr. 168:21–23; Kepler to Maestlin, 9 September 1600, *GW* xiv, nr. 175:7–10.
12. Christianson, *On Tycho's Island*, 327.
13. Ibid.
14. On the dispute between Tycho and Ursus, see Thoren, *Lord of Uraniborg*, ch. 13; trans. Rosen, *Three Imperial Mathematicians*; and Jardine, *The Birth of History and Philosophy of Science*.
15. Tycho to Kepler, 1 April 1598 (o.s.), *GW* xiii, nr. 92:156–59; and Thoren, *Lord of Uraniborg*, 437.
16. Thoren, *Lord of Uraniborg*, 459. The document, titled "Quarrel between Tycho and Ursus over Hypotheses," is translated in Rosen, *Three Imperial Mathematicians*, 290–96.
17. Thoren, *Lord of Uraniborg*, 437.
18. On Tycho's doubts about Kepler, see Thoren, *Lord of Uraniborg*, 441. On his reaction to his dealings with Wittich, see Gingerich and Westman, *The Wittich Connection*.
19. Gingerich and Westman, *The Wittich Connection*, 20.
20. Kepler to Herwart, 12 July 1600, *GW* xiv, nr. 168:105–16.
21. Kepler writes in the *Astronomia nova* that had Longomontanus been working on another planet he would have been assigned to that planet too, *GW* iii, 109:16–17.
22. *GW* xix, doc. 2.5, 48.
23. Kepler mss., Pulkova XIV (Russian Academy of Sciences, St. Petersburg), fol. 68. Fol. 68r is reproduced in Owen Gingerich, *The Eye of Heaven*, 335. My

thanks to Prof. Gingerich for giving me access to a microfilm of Kepler's Mars notebook.

24. *Astronomia nova*, ch. 7, *GW* iii, 109:22–29; trans. Donahue, 185.

25. Tycho Brahe, *Triangulorum planorum et sphaericorum praxis arithmetica* (Prague, 1886), fol. 6.

26. *Astronomia nova*, ch. 7, *GW* iii, 109:39–41; trans. Donahue, 185.

27. *Mysterium cosmographicum*, chs. 15, 16, and 18.

28. The reason for this dependence is complex. Remember that in Ptolemaic theory, for the superior planets, the deferent represented the planet's motion and the epicycle the earth's motion around the sun, and for the inferior planets, the deferent represented the earth's motion around the sun and the epicycle represented the planet's motion. Copernicus's conversion of Ptolemaic theory into heliocentric form did not resolve the situation, but further complicated it. See Swerdlow and Neugebauer, *Mathematical Astronomy in Copernicus's* De revolutionibus, 369–71.

29. *Mysterium cosmographicum*, *GW* i, 77:31–37; trans. Duncan, 219.

30. *Astronomia nova*, ch. 7, *GW* iii, 109:10–13; trans. Donahue, 184.

31. *Astronomia nova*, ch. 7, *GW* iii, 109:41–110:2; trans. Donahue, 185.

32. Pulkova XIV, fol. 68. For a description of these manuscripts, see Owen Gingerich, "Kepler's Treatment of Redundant Observations; or, the Computer Versus Kepler Revisited," 307–14.

33. See Stephenson, *Kepler's Physical Astronomy*, 31–38.

34. *Astronomia nova*, part 2, title page, *GW* iii, 107.

35. *Astronomia nova*, summaries of the individual chapters, *GW* iii, 36: 11–15.

36. Gingerich, "Kepler's Treatment of Redundant Observations," 310.

37. Thoren, *Lord of Uraniborg*, ch. 7, 235.

38. For a fuller account of this procedure, see Koyré, *The Astronomical Revolution*, 173–75, or Stephenson, *Kepler's Physical Astronomy*, 42–44.

39. See pp. 130–134. The analysis of this document will be based on the selection excerpted by Frisch under this title; Frisch, viii:229–334.

40. Frisch, viii:229.

41. Ibid.

42. For Kepler's first thoughts about the harmonic spacing of the planets, see Kepler to Edmund Bruce, 18 July 1599, *GW* xiv, nr. 128; and Kepler to Maestlin, 19/29 August 1599, *GW* xiv, nr. 132.

43. This statement apparently indicates his realization that the demonstration in chapter 22 of the *Mysterium* was only accurate around apsides.

44. The periods for the planets are in sexigesimal notation. Kepler multiplies by sixty in order to take the root of a whole number. Similarly, he multiplies this result by ten before taking the root below.

45. Frisch, viii:229.

46. Ibid.

47. *Astronomia nova*, ch. 16, *GW* iii, 152:39–153:8; trans. Donahue, 251–52.

48. *Astronomia nova*, ch. 16, *GW* iii, 152:21–34; trans. Donahue, 251.

49. For an analysis of Copernicus's inexact bisection, see Swerdlow, *Mathematical Astronomy in Copernicus's* De revolutionibus, 356–61. Brahe probably

followed Copernicus's precedence in the Mars theory. His ratio of the eccentricity of the eccentric to the eccentricity of the equant converted into Ptolemaic form is given by Kepler as 9,900:7,560 (*Astronomia nova*, ch. 16, *GW* iii, 152:34–38; trans. Donahue, 251), although this appears to be a mistake. When Kepler arrived at Benatky, he found Longomontanus working with a Mars theory in the Copernican form with a total eccentricity of 20,160, whose two epicycles were in the ratio 16,380:3,780 (*Astronomia nova*, ch. 7, *GW* iii, 109:22–29). Converted into Ptolemaic form, this theory would have a ratio of the eccentricity of the eccentric to the eccentricity of the equant of 12,300:7,560. The ratio of 9,900:7,560 given in the *Astronomia nova* derives from a value for the size of Tycho's larger epicycle of 13,680 and for the smaller of 3,780. It seems that Kepler was guilty of a numerical spoonerism for misremembering 16,380 as 13,680.

50. Frisch, viii:229.

51. See pp. 155–57.

52. Gingerich, "Kepler's Treatment of Redundant Observations," 311.

53. Kepler to Herwart, 12 July 1600, *GW* xiv, nr. 168:119–26; trans. Koyré, *The Astronomical Revolution*, 396.

54. Kepler to Herwart, 12 July 1600, *GW* xiv, nr. 168:126–32.

55. Not knowing that this was caused by the earth's eccentricity not being bisected, Kepler's initial reaction had been simply to transfer the annual change in the size of Mars's orbit to the earth's orbit, reasoning that this was preferable physically (or rather metaphysically). In a letter to Herwart of 30 May 1599, he wrote, "But I can say in a similar way that the earth's annual orbit expands, and the same effects will follow. . . . In physical or cosmological discussions, however, the approach of the single earth to all five planets as its orb expands will be far more probable than the assertion that the spheres of the five planets, with the sun at their center, have risen away from the earth in an expanded circular framework" (*GW* xiii, nr. 123:306–14; trans. Rosen, *Three Imperial Mathematicians*, 211–12).

56. Kepler to Herwart, 12 July 1600, *GW* xiv, nr. 168:132–34.

57. Thoren, *Lord of Uraniborg*, 313.

58. *Astronomia nova*, chapter summaries, *GW* iii, 43:11–12; trans. Donahue, 89.

59. Kepler to Archduke Ferdinand, beginning July 1600, *GW* xiv, nr. 166:51–67 (emphasis mine). Kepler refers here to "another element of motion." This is the librating force that causes the planet alternately to approach and recede from the sun.

60. Kepler to Archduke Ferdinand, beginning July 1600, *GW* xiv, nr. 166: 129–31.

61. Ibid., nr. 166:230–31.

62. Tycho to Kepler, 9 December 1599, *GW* xiv, nr. 145:206–9.

63. Kepler to Herwart, 12 July 1600, *GW* xiv, nr. 168:184–87.

64. Gingerich, "Kepler's Treatment of Redundant Observations," 311.

65. Kepler to Maestlin, 9 September 1600, *GW* xiv, nr. 175:11.

66. Ibid., nr. 175.

67. Caspar, *Kepler*, 111–15. Maestlin did respond to Kepler's pleas, but not until Kepler had left Graz. See Maestlin to Kepler, 9 October 1600, *GW* xiv, nr. 178.

68. Kepler to Longomontanus, beginning 1605, *GW* xiv, nr. 323:187–93.

69. On the significance of Venus and Mercury in this context, see below.

70. Kepler to Longomontanus, beginning 1605, *GW* xiv, nr. 323:193–98.

71. Gingerich, "Kepler's Treatment of Redundant Observations," 311.

72. Pulkova XIV, fol. 71v. This folio corresponds to page 11 of Kepler's Mars notebook. By July 1600, he was past page 200.

73. Gingerich, "Kepler's Treatment of Redundant Observations," 312.

74. Ibid.

75. Ibid., 313.

76. Kepler to Magini, 1 June 1601, *GW* xiv, nr. 190:142–50.

77. Ibid., nr. 190:36–53.

78. Kepler to Herwart, 26 March 1598, *GW* xiii, nr. 91:177–78.

79. Kepler to Magini, 1 June 1601, *GW* xiv, nr. 190:32–34.

80. Kepler to Magini, 1 June 1601, *GW* xiv, nr. 190:9–21 and 53–62. On the confusion over whether Kepler actually sent a copy explicitly for Magini, see Drake, *Galileo Studies*, 127–31.

81. Kepler to Magini, 1 June 1601, *GW* xiv, nr. 190:59–66.

82. Ibid., nr. 190:66–68.

83. Ibid., nr. 190:77–78.

84. Swerdlow, *Mathematical Astronomy in Copernicus's* De Revolutionibus, 369–71.

85. Kepler to Magini, 1 June 1601, *GW* xiv, nr. 190:88–90.

86. Ibid., nr. 190:75.

87. Ibid., nr. 190:127–41.

88. Ibid., nr. 190:256–57.

89. Ibid., nr. 190:251–54.

90. Ibid., nr. 190:420–24.

91. See Kepler to Magini, 22 March 1610, *GW* xvi, nr. 560; and Magini to Kepler, 20 June 1610, *GW* xvi, nr. 569.

92. He told Magini that he intended to stay three more weeks. Kepler to Magini, 1 June 1601, *GW* xiv, nr. 190:455–56.

93. Kepler to Longomontanus, beginning 1605, *GW* xv, nr. 323:201–3; Gingerich, "Kepler's Treatment of Redundant Observations," 313.

94. Kepler to Longomontanus, beginning 1605, *GW* xv, nr. 323:218–20; and Kepler to Maestlin, 10/20 December 1601, *GW* xiv, nr. 203:7–9.

95. See Ch. 6.

96. Kepler to Maestlin, 10/20 December 1601, *GW* xiv, nr. 203:16–40.

97. Kepler to Maestlin, 8 February 1601, *GW* xiv, nr. 183:161–62; trans. Rosen, *Three Imperial Mathematicians*, 283.

98. See Jardine, *The Birth of History and Philosophy of Science*, 25–27.

99. For a complete account, see Jardine, *The Birth of History and Philosophy of Science*; for Tycho's plan, 26.

100. Cf. Kepler to Maestlin [16/26 February 1599], written as a note on a copy of Kepler to Tycho [19 February 1599], *GW* xiii, nr. 112:140–43: "Finally, I added that it did not seem worthy of Tycho's stature to be so violently upset by this

disparagement. He should rather allow their works to speak for both of them, confident of what the learned would think" (trans. Jardine, *The Birth of History and Philosophy of Science*, 19).

101. "De lite causa hypothesium D. Tychonem inter et Ursum," *GW* xx.1, 66:8–9; trans. Jardine, 67.

102. Jardine, 75.

103. Ibid., 74.

104. See Jardine, "The Forging of Modern Realism," 141–73.

105. Jardine, *The Birth of History and Philosophy of Science*, 211–13.

106. Ibid., 211–12, 214–15.

107. Ibid., 212.

108. *Mysterium*, ch. 1, *GW* i, 15:5–16:38. He had planned to confront this argument from the planning stages of the *Mysterium*; see Kepler to Maestlin, 3 October 1595, *GW* xiii, nr. 23:46–49.

109. Jardine, *The Birth of History and Philosophy of Science*, 147, n. 64.

110. *Apologia*, ch. 1, *GW* xx.1, 20:12–16; trans. Jardine, *The Birth of History and Philosophy of Science*, 139.

111. *Apologia*, ch. 1, *GW* xx.1, 20:23–25; trans. Jardine, 139.

112. *Apologia*, ch. 1, *GW* xx.1, 28:28–34; trans. Jardine, 153.

113. *Mysterium*, ch. 22, *GW* i, 77:24; trans. Duncan, 219.

114. Kepler to Maestlin [beginning of October 1597], *GW* xiii, nr. 75:5–25. Kepler repeated this claim publicly on the verso of the title page of the *Astronomia nova*, *GW* iii, 6.

115. *Apologia*, ch. 1, *GW* xx.1, 28:40–29:2; trans. Jardine, 154.

116. *Apologia*, ch. 1, *GW* xx.1, 21:5–11; trans. Jardine, 140.

117. *Apologia*, ch. 1, *GW* xx.1, 21:16–18; trans. Jardine, 140.

118. *Apologia*, ch. 1, *GW* xx.1, 21:37–40; trans. Jardine, 141.

119. In his commentary on Aristotle's *Physics*, Simplicius quotes Geminus's comment "inspired by the views of Aristotle": "[T]he physicist will in many cases reach the cause by looking to creative force; but the astronomer, when he proves facts from external conditions, is not qualified to judge of the cause, as when, for instance, he declares the earth or the stars to be spherical; sometimes he does not even desire to ascertain the cause, as when he discourses about an eclipse; at other times he invents by way of hypothesis, and states certain expedients by the assumption of which the phenomena will be saved." Quoted from Heath, *Aristarchus of Samos* (Oxford, 1913; Dover ed., New York: Dover, 1981), 275–76.

120. *Apologia*, ch. 1, *GW* xx.1, 23:44–24:3; trans. Jardine, 144–45.

121. *Astronomia nova*, introduction, *GW* iii, 20:2–3; trans. Donahue, 48.

122. *Astronomia nova*, introduction, *GW* iii, 20:21–22; trans. Donahue, 48.

CHAPTER 6

1. See, for instance, Kepler to Herwart, 12 July 1600, on pp. 111–12.

2. William H. Donahue, "Kepler's First Thoughts on Oval Orbits: Text, Translation, and Commentary," *Journal for the History of Astronomy* 24 (1993): 73.

3. Again, the analysis of this document will be based on the selection excerpted by Frisch under this title, Frisch, viii: 229–334. The "Consideratio" begins on folio K435=P307 of the Mars notebook. A related section on the area law runs

through K451=P315. K381=P276 is dated 23 November 1601. K474=P326[v] contains the date 29 March 1602, but this was probably added with a later annotation. In addition, Kepler to Maestlin, 10/20 December 1601, *GW* xiv, nr. 203, discusses very similar material. See Donahue, "Kepler's First Thoughts on Oval Orbits," 73.

4. Frisch, viii, 229.

5. It is not clear to me what solar theory Kepler used in his research. In chapter 20 of the *Astronomia nova*, he used Tycho's solar theory, with the qualification that it would suffice for this purpose. This may have been a necessary conceit, however, for he had not yet begun the revision of the theory of the earth in part 3. Kepler knew when undertaking the research that the theory of the earth needed an equant, so he may have had a solar theory with a bisected eccentricity. Future research on the Mars manuscripts may resolve this question.

6. Frisch, viii, 229–30.

7. Named by E. S. Kennedy after the medieval Arab astronomer Nasir ad-Din at-Tusi (1201–1274), the Tusi couple consists of two equal circles, one of which acts as an epicycle on the other. If the epicycle rotates in a contrary sense to the deferent with twice the speed, a point on the epicycle will periodically traverse a straight line whose length is twice the diameter of the circles. Copernicus made use of this principle in his elimination of the equant.

8. Frisch, viii, 230.

9. Ibid.

10. Ibid.

11. Kepler is speaking of minutes only in principle; in practice, he divided the circle only into individual degrees.

12. Frisch, viii, 237–40.

13. *Astronomia nova*, ch. 30, *GW* iii, 229:15–16; trans. Donahue, 364.

14. *Astronomia nova*, ch. 30, *GW* iii, 228:24–35; trans. Donahue, 363.

15. Kepler to Maestlin, 10/20 December 1601, *GW* xiv, nr. 203.

16. See chapter 8, note 1, below.

17. Kepler to Maestlin, 10/20 December 1601, *GW* xiv, nr. 203:42–43.

18. By this time, Kepler had determined that bisecting the earth's eccentricity introduced no more than a 1′ change in Tycho's solar theory. This pleased him, since he recognized the solar theory as the summit of Tycho's work, which, along with the catalogue of stars and the lunar theory, "exude pure ambrosia" (Kepler to Maestlin, 10/20 December 1601, *GW* xiv, nr. 203:29).

19. See Swerdlow, *Mathematical Astronomy in Copernicus's* De revolutionibus, 483–91, specifically 489.

20. Kepler to Maestlin, 10/20 December 1601, *GW* xiv, nr. 203:52–54.

21. Ibid., nr. 203:54–55.

22. Ibid. nr. 203:55–56; trans. Koyré, *The Astronomical Revolution*, 397.

23. Kepler to Maestlin, 10/20 December 1601, *GW* xiv, nr. 203:61.

24. Ibid., nr. 203:64–68.

25. Ibid., nr. 203:71–72.

26. Ibid., nr. 203:83–86.

27. Ibid., nr. 203:94–102ff.

28. Donahue, "Kepler's First Thoughts on Oval Orbits."

29. Ibid., 73.
30. Ibid., 71, 75.
31. Ibid., 84.
32. Pulkova, XIV, fol. [K531=P357]; trans. Donahue, "Kepler's First Thoughts on Oval Orbits," 84. Donahue points out the delightful analogy Kepler uses to explain this situation in the *Astronomia nova*, ch. 44: "Therefore, the times, when they are abstracted from the plane and adjusted upward and downward, will be accumulated at aphelion and perihelion in much the same manner as, if one were to squeeze a fat-bellied sausage at its middle, he would squeeze and squash the ground meat, with which it is stuffed, outwards from the belly towards the two ends, emerging above and below from beneath his hand."
33. *Mysterium*, ch. 22, *GW* i, 77:22–24.
34. Donahue, "Kepler's First Thoughts on Oval Orbits," 88. These faculties are more deeply considered in Donahue, ibid., esp. 85–88.
35. Pulkowa, XIV, fol. [K532=P358]; trans. Donahue, "Kepler's First Thoughts on Oval Orbits," 92.
36. Pulkova, XIV, fol. [K532=P358]-[K533(?)=P358v]; trans. ibid., 95, interpolations are mine.
37. Pulkova, XIV, [K533(?)=P358v] ; trans. ibid., 96–97.
38. *Astronomia nova*, ch. 45, *GW* iii, 288:4–31; trans. Donahue, 455–56.

CHAPTER 7

1. Kepler to Longomontanus [beginning of 1605], *GW* xv, nr. 323:216–20; and *GW* xix, doc. 2.8.
2. The most detailed account of Kepler's conflict with the Tychonics is in Frisch, vol. 3, 11–13, 35 and 444. See also Caspar, *Kepler*, 139–41; *GW* iii, 436–37; and Dreyer, *Tycho Brahe*, appendix, passim.
3. Kepler to Maestlin, 10/20 December 1601, *GW* xiv, nr. 203:7–15. For Kepler's responsibility for Tycho's instruments, see also *GW* xix, doc. 2.9.
4. Uraniborg was the name of Tycho's castle observatory on the Danish island of Hven.
5. Christianson, *On Tycho's Island*, 316.
6. Ibid., 367.
7. Thoren, *Lord of Uraniborg*, 460.
8. *GW* xx.1, 95: 11–14.
9. Pierre Gassendi, *Tychonis Brahei equitis dani, astronomorum coryphi, vita* (Paris, 1654), 205.
10. *TBOO*, vol. 14, 259.
11. Thoren, *Lord of Uraniborg*, 462–63.
12. *TBOO*, vol. 14, 265.
13. Kepler to Heydon [October 1605], *GW* xv, nr. 357:22–27.
14. *TBOO*, vol. 14, 249; Thoren, *Lord of Uraniborg*, 472.
15. On Kepler's refusal to turn over the observations, see Kepler to Fabricius, 1 October 1602, *GW* xiv, nr. 226:31–34. On the lawsuit, see Kepler to Longomontanus [beginning of 1605], *GW* xv, nr. 323:223–28.
16. Kepler to Fabricius, 1 October 1602, *GW* xiv, nr. 226:29–31 and 34–38.

17. Dreyer, *Tycho Brahe*, 368–69; Kepler to Longomontanus [beginning of 1605], *GW* xv, nr. 323: 204f.

18. Kepler to Herwart [12 November 1602], *GW* xiv, nr. 232:337–39.

19. Kepler to Herwart, 7 October 1602, *GW* xiv, nr. 228:16–31.

20. Fabricius to Kepler, 13 March 1602 (o.s.), *GW* xiv, nr. 211.

21. On Fabricius's views, see ch. 8.

22. *GW* x, 10*. (Vol. x of the *GW* has two paginations, one distinguished with an asterisk.)

23. Kepler had this responsibility initially; see note 3, above. In the 7 October 1602 letter to Herwart, Kepler says Tengnagel "is obtruding himself in anticipation of the *Rudolphine Tables*." (note 19, above). Tengnagel's responsibility for the tables is mentioned in a document of 8 July 1604 (*GW* xix, 189, doc. 5.1). The *Rudolphine Tables* had undoubtedly been the most lucrative of Kepler's responsibilities. It appears that he was left with the less attractive task of looking after the instruments; see *GW* xix, doc. 2.9.

24. Kepler to Longomontanus [beginning of 1605], *GW* xv, nr. 323:227–32.

25. Kepler to Herwart, 7 October 1602, *GW* xiv, nr. 228:79–83.

26. Kepler to Herwart [12 November 1602], *GW* xiv, nr. 232:108–11.

27. *Astronomia nova*, part 3 title page, *GW* iii, 189; trans. Donahue, 303.

28. *TBOO*, vol. 14, 252, 265.

29. Kepler to Fabricius [February 1604], *GW* xv, nr. 281:225–42. Kepler was apparently being ironic about Tycho's wife—the phrase *genere nobilem* also means "noble by birth," although it was a well-known fact that she was not.

30. Kepler to Maestlin, 10/20 December 1601, *GW* xiv, nr. 203:7–15.

31. See *GW* xix, docs. 2.7, 2.10, 2.12, 2.13, etc.

32. Max Caspar, *Bibliographia Kepleriana* (2nd ed.; Munich: C.H. Beck, 1968), 89.

33. *GW*, xix, doc. 5.4, 194.

34. *GW*, xix, docs. 5.1 and 5.2, 189–90.

35. Kepler to Pistorius [15 June 1607], *GW*, xv, nr. 431:124–28.

36. *GW*, xix, doc. 5.1, 189.

37. *GW*, xix, doc. 5.2, 190.

38. As he wrote to his former teacher Michael Maestlin, "High honors and offices I do not have. I live here on the stage of the world as a private man. If I can squeeze out a portion of my salary at the court I am happy not to have to live entirely on my own means. As for the rest, I take the attitude that I serve not the emperor but rather the whole human race and posterity. In this confident hope, I scorn with secret pride all honors and offices and, in addition, if it is necessary, also those things which they bestow. I count as the only honor the fact that by divine decree I have been put near the Tychonic observations" (Kepler to Maestlin, 5 March 1605, *GW*, xv, nr. 355:5–12; trans. from Caspar, *Kepler*, 159).

39. Kepler to Heydon [October 1605], *GW* xv, nr. 357:28–44.

40. Kepler to the University of Tübingen, 12 December 1604, *GW* xv, nr. 304:28–29.

41. Ibid., nr. 304: 32–44.

42. University of Tübingen to Kepler, 26 January 1605 (o.s.), *GW* xv, nr. 324:20–21. For the record of the senate's decision, see *GW* xix, doc. 7.46; note the statement of this qualification: "(do es nitt wider die Kay. Maitt. sei)."

43. Kepler to Herwart von Hohenburg, 13 February 1605, *GW* xv, nr. 325: 69–72.

44. Herwart von Hohenburg to Kepler, 8 March 1605, *GW* xv, nr. 336:11–13.

45. On Longomontanus, see John Christianson, *On Tycho's Island*, 313–19; Kristian-Peder Moesgaard, "Longomontanus," *Dansk biografisk leksikon* (Copenhagen: Gyldendal, 1979–1984), vol. 9, 109–10; Victor E. Thoren, "Severin, Christian," *Dictionary of Scientific Biography*, ed. Charles Coulston Gillispie (18 vols.; New York: Scribner, 1970–1990), vol. 12, 332; Thoren, *Lord of Uraniborg*, passim.; and J.-B.-J. Delambre, *Histoire de l'astronomie moderne* (Paris, 1821), 262–87.

46. Thoren, *Lord of Uraniborg*, 397–98, 407–8, 419, 431, 449.

47. Ibid., 449–50.

48. Longomontanus to Kepler, 3 August 1600, *GW* xiv, nr. 170.

49. Ibid., nr. 170:60–70.

50. Longomontanus apparently heard about the reformation of lunar tables through Tengnagel shortly after it was composed, but then did not write until a year later (Longomontanus to Kepler, 6 May 1604 (o.s.), *GW* xv, nr. 287:3–7). Kepler's transformation of Tycho's (actually, it was primarily Longomontanus's) lunar theory was composed in March or April 1603. Although its purpose was primarily to simplify the use of Tycho's lunar theory, the way that Kepler set about this was clearly based on the same kind of physical considerations as his early Mars work. Kepler reintroduced equants into the theory in place of Tycho's carefully crafted system of uniform circular motions, in the same way that he initially reintroduced equants into planetary theory because they were the simplest mathematical analogues of natural motions he was trying to model. More significantly, he explained the annual variation in the moon's motion, one of Tycho's greatest discoveries, in terms of the force for moving the moon that the earth receives from the sun. This force was more active when the moon is near the earth-sun axis, and thus the moon moved more quickly there. This explanation of the variation was recounted in the *Astronomia nova*, chapter 37. The development of Kepler's lunar theory was complex and passed through a number of phases. Only around 1616 did he begin to consider how the effect of the solar force on the earth's rotation could physically explain the variation. On the development of Kepler's lunar theory generally, see *GW* xx.1, 516–38; on the "Transformatio," 516–23.

51. Longomontanus to Kepler, 6 May 1604 (o.s.), *GW* xv, nr. 287:19–20.

52. Ibid., nr. 287:31–35.

53. There was apparently a story about Rheticus's experiences with the theory of Mars circulating in the astronomical community. Kepler recounts it in the dedication to the *Astronomia nova*: "It is said that Georg Joachim Rheticus . . . when he was brought up short in amazement by the motion of Mars, and did not disentangle himself, fled to the oracle of his familiar Genius, either intending (the gods willing) to explore that being's erudition, or driven by a headstrong desire for the truth, whereupon that stern patron, exasperated, alternately caught the importunate inquirer by the hair and stretched his head towards the low-hanging panelled

ceiling, and then threw him down, flattening him on the paved floor, adding the reply: 'This is the motion of Mars.' The story is a bad thing: there is nothing else more injurious to good reputation, for it is as tenacious of deception and distortion as it is informative of the truth. It is nevertheless not unbelievable that Rheticus himself, when his speculations were not succeeding and his spirit was in turmoil, leapt up in fury and pounded his head against the wall." (*Astronomia nova*, dedication, *GW* iii, 8:30–9:3; trans. Donahue, 32).

54. Longomontanus to Kepler, 6 May 1604 (o.s.), *GW* xv, nr. 287:49–86.

55. Longomontanus's reference to "Platonic music" may be more than a stab at the *Mysterium cosmographicum*. Recall that when Kepler initially reassessed his work in the *Mysterium*, he lost faith in his model for the relation between the periods and the distances and pursued the idea that they were governed by "harmonic music" (see above, pp. 109–10). This was about the time Kepler and Longomontanus were both in Prague, so it is possible that they discussed the idea.

56. Longomontanus to Kepler, 6 May 1604 (o.s.), *GW* xv, nr. 287:125–43.

57. Ibid., nr. 287:168–69.

58. Ibid., nr. 287:169.

59. Kepler to Longomontanus [early 1605], *GW* xv, nr. 323. Kepler had not received Longomontanus's letter until 2 January 1605, see ll. 2–4.

60. Ibid., nr. 323:6–11.

61. Ibid., nr. 323:15–18.

62. Ibid., nr. 323:23–26.

63. Ibid., nr. 323:26–38.

64. Kepler to Herwart, 22 April 1603, *GW* xiv, nr. 255. It is so identified by the editors in *GW* xx.1, 516, 556. There are two existing copies. *GW* xiv, nr. 255 is based on a draft in Kepler's hand, rather than the copy in Matthias Seiffart's hand to which Kepler referred in his letter to Longomontanus, but a copy of the letter also exists in the Pulkowa manuscripts (*GW* xiv, 502, note to line 103). I have not determined whether the copy is in Seiffart's hand. Even if it is, there is no evidence that it was the copy in Tengnagel's possession, so my analysis depends on assuming that the letters are the same.

65. Kepler to Herwart, 22 April 1603, *GW* xiv, nr. 255:27–48.

66. Kepler to Longomontanus [early 1605], *GW* xv, nr. 323:44–46 and 55–56.

67. Ibid., nr. 323:93–109.

68. Ibid., nr. 323:183–87.

69. Ibid., nr. 323:251–72 (emphasis mine).

70. By spirals, Kepler means the complex path of a planet through space caused by its retrograde motion in a geocentric (or geo-heliocentric) system. For an illustration, see *Astronomia nova*, ch. 1; Donahue, 119.

71. Kepler to Longomontanus [early 1605], *GW* xv, nr. 323:277–95.

72. Ibid., nr. 323:362–68.

73. *Astronomia nova*, ch. 6, *GW* iii, 89:7–11; trans. Donahue, 157.

74. For example, Caspar, *Kepler*, 121–22; and Donahue, "Translator's introduction," 3–4.

75. Kepler to Fabricius [February 1604], *GW* xv, nr. 281:107–8.

76. Kepler to Herwart, 10 February 1605, *GW* xv, nr. 325:55–69 and 73–76.
77. Ibid., nr. 325:82–83.
78. Ibid., nr. 325:84–85.
79. Friedrich Seck, "Johannes Kepler und der Buchdruck," *Archiv für Geschichte des Buchwesens* 11 (1970), 643.
80. *GW* iii, 436.
81. Kepler to Odontius, 5 August 1605, *GW* xv, nr. 351:47–49.
82. *GW* iii, 438.
83. Kepler to Rudolf II, 4 December 1606, *GW* xv, nr. 402.
84. *GW* xix, 52–54, docs. 2.16 and 2.18. For Kepler's expectation of 800 florins, see Seck, "Johannes Kepler und der Buchdruck," 643.
85. Seck, "Johannes Kepler und der Buchdruck," 644–45.
86. Kepler to Pistorius [15 June 1607], *GW* xv, nr. 431:120–30.
87. Seck, "Johannes Kepler und der Buchdruck," 644.
88. Kepler to Fabricius, 1 August 1607, *GW* xvi, nr. 438:292–94.
89. Frisch, vol. 3, 454.
90. *GW* xix, doc. 2.24.
91. Seck, "Johannes Kepler und der Buchdruck," 645.
92. Christianson, *On Tycho's Island*, 369.
93. *Astronomia nova*, to the reader, *GW* iii, 17:3–21; trans. Donahue, 43–44, with my modifications based on Max Caspar and Walther von Dyck, *Johannes Kepler in seinen Briefen* (Munich: R. Oldenbourg, 1930), vol. 1, 325.
94. Donahue, 43, n. 16.

CHAPTER 8

1. After Maestlin's letter to Kepler of 9 October 1600 (o.s.) (*GW* xiv, nr. 178), despite Kepler's pleas for a reply, he did not write again until 28 January 1605 (o.s.) (*GW* xv, nr. 322), and even then, Maestlin's next letter was not until 7 September 1610 (o.s.) (*GW* xvi, nr. 592). Maestlin's reasons for allowing their correspondence to lapse are obscure. He lamely offered that he had nothing to contribute (nr. 322:8–9), and certainly his alienation from his former student was, to some extent, due to Kepler's increasingly unconventional research. However, it was also clear that there was some misunderstanding about Kepler's intention to publish some letter of Maestlin's (Rosen, *Three Imperial Mathematicians*, 284–85).
2. On Fabricius, see Diedrich Wattenberg, "David Fabricius: Der Astronom Ostfrieslands (1564–1617)," *Vorträge und Schriften der Archenhold-Sternwarte*, 19 (Berlin-Treptow, 1964); A. W. Lang, "Die 'Nie vnd warhafftige Beschrivinge des Ostfreslandes' des David Fabricius von 1589," *Nordseeküste* 8 (Juist: Die Bake, 1963); Georg Sello, *Des David Fabricius Karte von Ostfriesland und andere Fabriciana des Oldenburger Archivs* (Norden und Norderney: Herm, Braams, 1896); Gerhard Berthold, *Der Magister Johann Fabricius und die Sonnenflecken, nebst einem Excurse über David Fabricius* (Leipzig: Veit & Comp., 1894); W. Olbers, "Materialien zu einer Lebensbeschreibung der beiden Astronomen, David und Johannes Fabricius," *Astronomische Nachrichten*, 31 (1851), no.

729:129–42; and E. F. Apelt, *Die Reformation der Sternekunde* (Jena: Friedrich Mauke, 1852), 267–434.

3. E. F. Apelt, for instance, does not hesitate to call Fabricius a member of Tycho's school; *Die Reformation der Sternekunde*, 270.

4. Kepler to Fabricius, 11 October 1605, *GW* xv, nr. 358:613.

5. Cf. *Astronomia nova*, introduction, "I myself, who am known as a mathematician, find my mental forces wearying when, upon rereading my own work, I recall from the diagrams the sense of the proofs, which I myself had originally introduced from my own mind into the diagrams and the text." *GW* iii, 18: 12–14; trans. Donahue, 45–46.

6. Kepler to Fabricius, 11 October 1605, *GW* xv, nr. 358:660–66.

7. *De stella nova in pede serpentarii* (1606), ch. 12, *GW* i, 210:33–37.

8. Fabricius to Kepler, 13 March 1602 (o.s.), *GW* xiv, nr. 211:2–21.

9. Ibid., nr. 211:23–27.

10. Ibid., nr. 211:41–42.

11. Kepler mentions that he had written such a letter in Kepler to Fabricius, 1 October 1602, *GW* xiv, nr. 226:687.

12. Fabricius to Kepler, 1/5 August 1602 (o.s.), *GW* xiv, nr. 221:33–35.

13. Kepler to Fabricius, 1 October 1602, *GW* xiv, nr. 226:685–86.

14. Fabricius to Kepler, 8 November 1602 (o.s.), *GW* xiv, nr. 234:9–11. For a more explicit statement of their understanding, see also Fabricius to Kepler, 27 February 1608 (o.s.), *GW* xvi, nr. 481:204–7.

15. Fabricius to Kepler, 28 April 1602 (o.s.), *GW* xiv, nr. 214:34–38.

16. Fabricius to Kepler, 1/5 August 1602 (o.s.), *GW* xiv, nr. 221:56–72.

17. Kepler to Fabricius, 1 October 1602, *GW* xiv, nr. 226:266–67.

18. Cf. Kepler to Fabricius, 1 October 1602, *GW* xiv, nr. 226:294–95: "This, therefore, is the reason why we can be fully adequate with a false hypothesis in the place of the true one."

19. Fabricius's confusion apparently arose after his own investigations, using different triplets of observations, produced different values for the eccentricity. I infer, since Fabricius did not understand Kepler's iterative procedure using all four observations, that he believed that Kepler's method of four acronychals should similarly have produced four different values from the four permutations.

20. Fabricius to Kepler, 1/5 August 1602, *GW* xiv, nr. 221:63–72 (emphasis mine).

21. Kepler to Fabricius, 2 December 1602, *GW* xiv, nr. 239:654–719.

22. Fabricius to Kepler, 1/5 August 1602, *GW* xiv, nr. 221:138–45.

23. Kepler to Fabricius, 2 December 1602, *GW* xiv, nr. 239:720–51.

24. For a further example, see Fabricius to Kepler, 7 February 1603 (o.s.), *GW* xiv, nr. 248:436–37; "Give me some probable cause, and you will persuade me."

25. Koyré, *The Astronomical Revolution*, 182.

26. Wilson, "Kepler's Derivation of the Elliptical Path," 5–10.

27. Fabricius to Kepler, 13 March 1602 (o.s.), *GW* xiv, nr. 211:35–53; and Fabricius to Kepler, 1/5 August 1602 (o.s.), *GW* xiv, nr. 221:73–145.

28. Fabricius to Kepler, 1/5 August 1602 (o.s.), *GW* xiv, nr. 221:101–2.

29. Ibid., nr. 221:103–4.

30. Ibid., nr. 221:108. After Copernicus's invention, Kepler cogently responded, "Why should one [planet] prescribe laws to another?" (Kepler to Fabricius, 1 October 1602, *GW* xiv, nr. 226:406).

31. *Astronomia nova*, ch. 29, *GW* iii, 225:4–7.

32. See, for example, Fabricius to Kepler, 1/5 August 1602 (o.s.), *GW* xiv, nr. 221:99–145.

33. In his second letter, Kepler writes, "In this way that oval figure is produced, which I mentioned in my previous letter" (Kepler to Fabricius, 1 October 1602, *GW* xiv, nr. 226:594–95).

34. Fabricius to Kepler, 4 November 1602 (o.s.), *GW* xiv, nr. 233:50–60.

35. Fabricius to Kepler, 1/5 August 1602 (o.s.), *GW* xiv, nr. 221:186–93.

36. "In an astronomy constructed without hypotheses God would be with you with His divine will. There will be the greatest possible honors and glories, if you bring it about" (Fabricius to Kepler, 1/5 August 1602 (o.s.), *GW* xiv, nr. 221:205–7).

37. Kepler to Archduke Ferdinand [beginning of July 1600], *GW* xiv, nr. 166. See above, pp. 113–14.

38. Kepler to Fabricius, 1 October 1602, *GW* xiv, nr. 226:228–29.

39. Fabricius to Kepler, 13 March 1602 (o.s.), *GW* xiv, nr. 211:38–39; Fabricius to Kepler, 1/5 August 1602 (o.s.), *GW* xiv, nr. 221:104–6; and Fabricius to Kepler, 4 November (o.s.), *GW* xiv, nr. 233: 56–57.

40. Kepler to Fabricius, 1 October 1602, *GW* xiv, nr. 226:370–71. The passage is garbled by the copyist, although it is clear that this has something to do with the sun's eccentricity. One month later, Kepler wrote to Herwart that he had found in Mars the "key to a universal astronomy" (see above, p. 147).

41. Kepler to Fabricius, 1 October 1602, *GW* xiv, nr. 226:557–59.

42. Ibid., nr. 226:590–92.

43. Cf. Kepler to Maestlin, 3 October 1595, *GW* xiii, nr. 23:184–91.

44. "For it comes to H and reaches the distance AH more rapidly, than it reaches the quadrate point from B around the sun. And thus to the sides or in the middle distances shorter distances are produced than the proportion of a perfect circle requires, and the difference around perigee is larger than around apogee, which is a proper definition of an oval figure" (Kepler to Fabricius, 1 October 1602, *GW* xiv, nr. 226:595–99).

45. Ibid., nr. 226:607–8.

46. Ibid., nr. 226:572–80. Also partially translated in Koyré, *The Astronomical Revolution*, 227–28.

47. Kepler to Fabricius, 1 October 1602, *GW* xiv, nr. 226: 616–24.

48. Ibid., nr. 226:638–40.

49. For a discussion of the difficulty of translating this term, see Donahue, 23–24. On Kepler's concept of the solar *species*, see also Stephenson, *Kepler's Physical Astronomy*, 68–77.

50. Stephenson, *Kepler's Physical Astronomy*, 68, n. 57.

51. See above, pp. 108–11.

52. Kepler to Fabricius, 1 October 1602, *GW* xiv, nr. 226:649–66.

53. Ibid., nr. 226:666–75.

54. Ibid., nr. 226:675–80.

55. Fabricius to Kepler, 8 November 1602 (o.s.), *GW* xiv, nr. 234:17–19.

56. Ibid., nr. 234:22–26.

57. Ibid., nr. 234:49–50.

58. Fabricius to Kepler, 8 December 1602 (o.s.), *GW* xiv, nr. 240:12–34.

59. Ibid., nr. 240:35–36.

60. Ibid., nr. 240:38–40.

61. Ibid., nr. 240:41–42; Kepler's marginal note.

62. In *De revolutionibus*, the general model Copernicus offers for the planets is an eccentric with a small epicycle, which nearly reproduces the inequality of motion Ptolemy produced with the equant. He also mentions the alternative of a double epicycle, where motion on a large epicycle replaces the eccentric, while the small epicycle retains its original purpose (*De revolutionibus*, V, 4; trans. Rosen, *On the Revolutions*, 242:7–12). Tycho preferred the latter form, as was illustrated above in Longomontanus's theory of Mars, p. 101. In either case, the inequality being modeled is that of the planet alone. See Swerdlow and Neugebauer, *Mathematical Astronomy in Copernicus's* De revolutionibus, 291–301, specifically 299.

63. Fabricius to Kepler, 8 December 1602 (o.s.), *GW* xiv, nr. 240:54–55; Kepler's marginal note.

64. Kepler to Fabricius, 2 December 1602, *GW* xiv, nr. 239:128–30.

65. Ibid., nr. 239:158–62.

66. For a diagram of this theory, see Fabricius to Kepler, 7 February 1603 (o.s.), *GW* xiv, nr. 248, p. 374. Fabricius's theory is clarified and corrected to some extent in Fabricius to Kepler, 24 June 1603 (o.s.), *GW* xiv, nr. 261.

67. Fabricius to Kepler, 7 February 1603 (o.s.), *GW* xiv, nr. 248:499–500 and 509–10.

68. Ibid., nr. 248:448–49; Kepler's marginal note.

69. Ibid., nr. 248:479–86.

70. Kepler to Fabricius, 4 July 1603, *GW* xiv, nr. 262:866–71.

71. See Fabricius to Kepler, 8 December 1602 (o.s.), *GW* xiv, nr. 240:12–31; 30 January 1603 (o.s.), *GW* xiv, nr. 246: 6–9; 7 February 1603 (o.s.), *GW* xiv, nr. 248: 430–34 and 503–5; 7 May 1603 (o.s.), *GW* xiv, nr. 257:27–31; 24 June 1603 (o.s.), *GW* xiv, nr. 261:93–95; and 11 August 1603 (o.s.), *GW* xiv, nr. 266:17–20.

72. Fabricius to Kepler, 7 February 1603 (o.s.), *GW* xiv, nr. 248:434; Kepler's marginal note.

73. After trying to clear up the confusion, Kepler continued to suggest that Fabricius use the vicarious hypothesis as an interim solution; see, for instance, Kepler to Fabricius [February 1604], *GW* xv, nr. 281:376.

74. Kepler to Fabricius, 4 July 1603, *GW* xiv, nr. 262:29–50.

75. *Astronomia nova*, ch. 55, *GW* iii, 345: 8–13; Donahue, 542.

76. For an account of Fabricius's finding, see Apelt, *Die Reformation der Sternkunde*, 301–13.

77. Fabricius to Kepler, 27 October 1604 (o.s.), *GW* xv, nr. 297:53–79.

78. Kepler to Fabricius [February 1604], GW xv, nr. 281:135–38. Also translated in Koyré, *The Astronomical Revolution*, 490, n. 12.

79. See, most importantly, Curtis Wilson, "Kepler's Derivation of the Elliptical Path."

80. Kepler to Fabricius [February 1604], *GW* xv, nr. 281:141–42.

81. Fabricius acknowledged the receipt of this installment in Fabricius to Kepler, 23 September 1605, *GW* xv, nr. 355: 1, but he was clearly awaiting further replies to a backlog of queries (nr. 355: 10). It is possible that there was also an earlier installment of this letter. Fabricius later mentioned Kepler's "last letter from the spring [1605]," which had the somewhat comical fate of being dropped in a puddle by the courier, rendering it partially illegible (Fabricius to Kepler, 20 January 1607 (o.s.), *GW* xv, nr. 408:8–13). No letter from spring 1605 is known. Fabricius lamented that figures for determining Saturn's eccentricity were obliterated. Kepler had provided acronychal observations of Saturn in nr. 358:1231–39. But if the letter in the spring had been an excerpt of nr. 358, which included these figures, it would have represented very much of nr. 358, leaving little for Fabricius to acknowledge in nr. 355.

82. *GW* xv, 527.

83. Like all of Kepler's letters to Fabricius, the surviving manuscript is a draft. One of Kepler's celebrated announcements that the orbit is an ellipse was, in fact, crossed through with a notation for the copyist on the following paragraph, "These things remain." (*GW* xv, 528, note to line 377. The paragraph in question is Kepler to Fabricius, 11 October 1605, *GW* xv, nr. 358:377–400. It is quoted in Koyré, *The Astronomical Revolution*, 261.) Also, the previous paragraph, in which Kepler tries to explain the reasoning that led to the ellipse, bears the marginal annotation, "This was totally useless. See below." (*GW* xv, 528, note to line 377).

84. D. T. Whiteside has shown that this asymmetry is in fact very small, and actually opposite to what Kepler intuited; see D. T. Whiteside, "Keplerian Planetary Eggs, Laid and Unlaid," 10.

85. Kepler to Fabricius, 11 October 1605, *GW* xv, nr. 358:304–27. Also translated in Koyré, *The Astronomical Revolution*, 259–60.

86. *Astronomia nova*, ch. 56, *GW* iii, 346:2–3; trans. Donahue, 543.

87. The optical equation is the angle viewed from the planet between the center of the orbit and the sun. Stephenson correctly points out that Kepler misstates this point in the *Astronomia nova*, where he refers to the maximum optical equation. The point he is actually considering is where the lunula between the circumscribing circle and the ellipsoid is its broadest, at 90° of eccentric anomaly. Stephenson, *Kepler's Physical Astronomy*, 108.

88. After Kepler to Fabricius [February 1604], GW xv, nr. 281, p. 30; and similarly in *Astronomia nova*, chs. 40, 43, and 47.

89. On the conchoid analysis, see Stephenson, *Kepler's Physical Astronomy*, 82–87.

90. Kepler to Fabricius [February 1604], *GW* xv, nr. 281:529–31.

91. Stephenson (*Kepler's Physical Astronomy*, 123) rightly takes issue with Koyré's and Aiton's criticism of this assumption.

92. After Stephenson, *Kepler's Physical Astronomy*, 122.

93. See note 83, above. These annotations are noted in *GW* xv, 528. Their exact placement was determined by examination of the manuscript, designated Pulkovo X, in the Archive of the Russian Academy of Sciences in St. Petersburg.

94. *Astronomia nova*, ch. 58, *GW* iii, 364: 18–21; trans. Donahue, 573.

95. Kepler to Fabricius, 11 October 1605, *GW* xv, nr. 358: 453–59.

96. Ibid., nr. 358:462–71.

97. Ibid., nr. 358:851–58.

98. Fabricius to Kepler, 11 January 1606 (o.s.), *GW* xv, nr. 371.

99. Fabricius to Kepler, 20 January 1607 (o.s.), *GW* xv, nr. 408:15–30.

100. Ibid., nr. 408:82–88, 95–102, and 107–11.

101. Fabricius to Kepler, 1 June 1607 (o.s.), *GW* xv, nr. 430:49–56.

102. *Astronomia nova*, ch. 57, *GW* iii, 348–64; trans. Donahue, 547–72.

103. It should be noted, however, that as early as the *Mysterium*, when Kepler is describing the action that causes the variation in a planet's distance, he describes it in terms of a *remigium*, a set of oars. *Mysterium cosmographicum*, ch. 22, *GW* i, 77:23.

104. Fabricius to Kepler, 20 January 1607 (o.s.), *GW* xv, nr. 408:45–57.

105. Fabricius to Kepler, 11 January 1606 (o.s.), *GW* xv, nr. 371:59–61.

106. Fabricius to Kepler, 5 April 1607 (o.s.), *GW* xv, nr. 419:20–310; and 1 June 1607 (o.s.), *GW* xv, nr. 430: 66–281.

107. Kepler to Fabricius, 1 August 1607, *GW* xvi, nr. 438:8–27.

108. Ibid., nr. 438:61.

109. Ibid., nr. 438:72–79.

110. Ibid., nr. 438:82–83.

111. Ibid., nr. 438:572–75.

112. Thomas S. Kuhn, *The Structure of Scientific Revolutions* (2d ed.; Chicago: The University of Chicago Press, 1970), 149: "Since new paradigms are born from old ones, they ordinarily incorporate much of the vocabulary and apparatus, both conceptual and manipulative, that the traditional paradigm had previously employed. But they seldom employ these in quite the traditional way. Within the new paradigm, old terms, concepts, and experiments fall into new relationships one with the other. The inevitable result is what we must call, though the term is not quite right, a misunderstanding between two competing schools."

113. See Curtis Wilson, "Predictive Astronomy in the Century after Kepler," in *Planetary Astronomy from the Renaissance to the Rise of Astrophysics*, ed. René Taton and Curtis Wilson (Cambridge: Cambridge University Press, 1989), 161–206.

114. Kepler to Fabricius, 1 August 1607, *GW* xvi, nr. 438:163–64.

115. Kepler to Herwart, 12 July 1600, *GW* xiv, nr. 168:171–315.

116. Kepler to Fabricius, 4 July 1603, *GW* xiv, nr. 262:831–35. Also partially translated in Koyré, *The Astronomical Revolution*, 230. Fabricius did indeed try to enlist one of his countrymen to help Kepler, but had to report that he had, unfortunately, died (Fabricius to Kepler, 27 October 1604 (o.s.), *GW* xv, nr. 297:104–5).

117. Kepler to Fabricius, 1 August 1607, *GW* xvi, nr. 438:113–19 and 342–44.

118. Ibid., nr. 438:450–52.

119. This characterization of Fabricius's theory as "thought up" is the same as that of his earlier theory, which Kepler had refused to assess for precisely that reason. See above, p. 185.

120. Kepler to Fabricius, 1 August 1607, *GW* xvi, nr. 438:215–31.

121. Ibid., nr. 438:435–45.

122. Ibid., nr. 438:48–58.

123. For the most comprehensive treatment of Fabricius's theory, see Apelt, *Die Reformation der Sternekunde*, 313–24.

124. For a more detailed explanation see Apelt, ibid., especially figure 17.

125. Ibid., 315.

126. Fabricius to Kepler, 27 February 1608 (o.s.), GW xvi, nr. 481:177–204.

127. Ibid., nr. 481:204–11.

128. Ibid., nr. 481:187–91.

129. Apelt, *Die Reformation der Sternekunde*, 324.

130. Kepler to Fabricius, 10 November 1608, *GW* xvi, nr. 508:392–11.

131. Ibid., nr. 508:419–25.

132. Fabricius to Kepler, 12 March 1609 (o.s.), *GW* xvi, nr. 524:330–429.

133. Kepler did finally write to Fabricius again on 1 October 1616 (*GW* xvii, nr. 746), after Fabricius had taken to addressing Kepler publically in his published calendars. This letter marked the end of their correspondence; Fabricius was murdered on 7 May 1617.

134. Although Fabricius's libration was motion along a straight line, such motion was easily reproduced using uniform circular motion by means of a Tusi couple. Even if not modeled in this way, Fabricius's model was repugnant by virtue of being purely mathematical.

135. Kepler to Fabricius, 10 November 1608, *GW* xvi, nr. 508: 114–16.

CHAPTER 9

1. See above, pp. 32–33.

2. *Harmonice mundi*, bk. 5, ch. 9, *GW* vi, 362:40–363:5; trans. from Caspar, *Kepler*, 375.

3. See Kozhamthadam, *The Discovery of Kepler's Laws*, esp. ch. 1; and Barker, "The Role of Religion in the Lutheran Response to Copernicus."

4. In 1619, Johannes Remus Quietanus wrote to Kepler that Galileo wanted his "Copernican book" (Quietanus to Kepler, 23 July 1619, *GW* xvii, nr. 845:9–12). Kepler answered, "All of my books are Copernican," but surmised that the book in question was the *Epitome*. Cited from Robert S. Westman, "Two Cultures or One? A Second Look at Kuhn's *The Copernican Revolution*," *Isis* 85 (1994): 107, n. 91.

5. *Astronomia nova*, chapter summaries, *GW* iii, 36: 4–15; trans. Donahue, 78.

6. The Latin words *oratoria* and *rhetorice* are more or less synonyms, with the former deriving from Latin and the latter from Greek. Cicero's major work on rhetoric was entitled *De oratore* [*On the orator*].

7. Aristotle, *On Rhetoric*, bk. 3, ch. 16 is devoted to the use of narrative.

8. *Astronomia nova*, ch. 7, *GW* iii, 111:5–7; trans. Donahue, 187.

9. Even something as abstract as a mathematical proof can be considered rhetorical in some sense. See Philip Kitcher, "Persuasion," in *Persuading Science: The Art of Scientific Rhetoric*, ed. Marcello Pera and William R. Shea (Canton, Mass.: Science History Publications, 1991), 3–27.

10. Caspar, *Kepler*, 56.

11. Jardine, *The Birth of History and Philosophy of Science*, 74–79.

12. For a good summary of the relation between dialectic and rhetoric in seventeenth-century science, see Jean Dietz Moss, *Novelties in the Heavens: Rhetoric and Science in the Copernican Controversy* (Chicago: Chicago University Press, 1993), ch. 1. She writes, "'Scientific rhetoric' would be for [Renaissance scholars] a contradiction in terms" (22).

13. Aristotle, *On Rhetoric*, bk. 1, ch. 1; trans. George A. Kennedy (Oxford: Oxford University Press, 1991), 34.

14. Aristotle, *On Rhetoric*, bk. 1, ch. 2; trans. Kennedy, 37.

15. *Astronomia nova*, ch. 16, *GW* iii, 156:4–8; trans. Donahue, 256. Delambre, the astronomer and historian of astronomy, rejected Kepler's plea for compassion, noting, "Kepler was sustained by his desire to have a case against Tycho, Copernicus, Ptolemy, and all the astronomers in the world; he has tasted this satisfaction, and I don't believe he deserves our pity for making all these calculations" (quoted from Gingerich, "The Computer versus Kepler," 357–58).

16. *Astronomia nova*, chapter summaries, *GW* iii, 36:15–26; trans. Donahue, 78–79.

17. For an excellent treatment of the astronomical argument, see Stephenson, *Kepler's Physical Astronomy.*

18. *Astronomia nova ΑΙΤΙΟΛΟΓΗΤΟΣ, seu physica coelestis, tradita commentariis de motibus stellæ Martis, ex observationibus G.V. Tychonis Brahe: jussu & sumptibus Rudolphi II Romanorum imperatoris &c: Plurimum annorum pertinaci studio elaborata Pragæ a Sæ. Cæ. M.tis Sæ. Mathematico Joanne Keplero*; trans. Donahue, 27.

19. The *Rudolphine Tables* also provides a particularly good example of the seriousness with which the title page was regarded; see *GW* x, 28*–31*.

20. Jardine, *The Birth of History and Philosophy of Science*, 229.

21. See Swerdlow and Neugebauer, *Mathematical Astronomy in Copernicus's* De revolutionibus, 41–48.

22. See Jardine, *The Birth of History and Philosophy of Science*, ch. 7.

23. Fabricius to Kepler, 1/5 August 1602 (o.s.), *GW* xiv, nr. 221:194–202; Detlef Forst to Kepler, 9 September 1608 (o.s.), *GW* xvi, nr. 500:28–30.

24. *Astronomia nova*, title page verso, *GW* iii, 6:22–23; trans. Donahue, 28.

25. *Astronomia nova*, title page verso, *GW* iii, 6:24–31; trans. Donahue, 28.

26. Kepler was able to name Osiander as the author of this preface because he had been identified by Hieronymous Schreiber in an annotation in Kepler's copy of *De revolutionibus*, now preserved at the Universitätsbibliothek, Leipzig. See Owen Gingerich, "Mästlin's, Kepler's, and Schickard's Copies of *De revolutionibus*," 173–74.

27. *Astronomia nova*, title page verso, *GW* iii, 6:37–38; trans. Donahue, 28.

28. See above, pp. 126 and 181.

29. We have not had time to examine Kepler's critique of traditional astrological theory, *De fundamentis astrologiæ certioribus* (1601); however, its emphasis on the importance of physical causes and the efficacy of faculties of light must at least be mentioned. See Gerard Simon, *Kepler: astronome, astrologue* (Paris: Gallimard, 1979).

30. *Astronomia nova*, dedication, *GW* iii, 8:13–15; trans. Donahue, 31.

31. *Astronomia nova*, dedication, *GW* iii, 8:16–22; trans. Donahue, 31–32.

32. Ibid., 8:23–27; trans. Donahue, 32.

33. Ibid., 9:7–14; trans. Donahue, 32–33.

34. Ibid., 9:15–20; trans. Donahue, 33.

35. Ibid., 9:22–26; trans. Donahue, 33. The "Brahean [siege] machines" must refer to Tycho's observations rather than his instruments; Kepler self-deprecatingly referred to his own observing efforts in the *Astronomia nova* as "a clown show" (*Astronomia nova*, ch. 11, *GW* iii, 124:19; trans. Donahue, 207).

36. *Astronomia nova*, dedication, *GW* iii, 9:27–10:7; trans. Donahue, 33–34.

37. Ibid., 10:8–12; trans. Donahue, 34.

38. Ibid., 10:13–18; trans. Donahue, 34.

39. On Kepler as a humanist, see Anthony Grafton, "Humanism and Science in Rudolphine Prague: Kepler in Context," in *Literary Culture in the Holy Roman Empire, 1555–1720*, ed. James A. Parente, Jr., et al. (Chapel Hill: University of North Carolina Press, 1991), 19–45.

40. *Astronomia nova*, epigrams, *GW* iii, 14:32–33 and 36–37; trans. Donahue, 40–41.

41. *Astronomia nova*, epigrams, *GW* iii, 15:1–11; trans. Donahue, 41. The quotation in the last line is from Tycho's "Hortatory Ode" (*TBOO*, vol. 2, 302).

42. Donahue identifies this as Phoebe, the moon, but from the context and meter it is clearly Phoebus Apollo, the sun. Both are *Phoebe* in the vocative, but the vocative form of *Phoebus* has a long final e, which is required in the dactyl. I would like to express my appreciation to Roger Ceragioli for helping me with the analysis of the meter.

43. These four lines refer to the triangulation to the earth's orbit and the bisection of its eccentricity. I have translated the lines from "Without you . . ." through "And how many parts . . ." based on the elegant translation in Johannes Kepler, *Neue astronomie*, trans. Max Caspar (Munich-Berlin: R. Oldenbourg, 1929), 17. Donahue's translation reads "Only by 'looking up' could you rightly adjust your dioptra, / I would 'look down' upon earth's course from there. / I would measure swift paces, Libra confronting the Goat, and, / Phoebus, the ratio giving to you the path's centre." It seems that Donahue has fallen prey to one of Kepler's puns. *Jugum* means "a yoke" and also the constellation Libra. Since it is meant to contrast with the sun's "swift paces," I have translated it loosely as "a reining in." The "yoke confronting the goat" is perhaps also a play on a goat's obstinacy.

44. *Astronomia nova*, epigrams, *GW* iii, 15:37–16: 20; trans. Donahue, 42–43.

45. *Astronomia nova*, introduction, *GW* iii, 18:2–3; trans. Donahue, 45.

46. Ibid., 35:26–28; trans. Donahue, 68–69.

47. Ibid., *GW* iii, 18:12–14; trans. Donahue, 45–46. Cf. Kepler to Fabricius, "even I now scarcely grasp the sense of what I have written, unless I reread it carefully" (p. 171 above).

48. *Astronomia nova*, chapter summaries, *GW* iii, 29–33; Donahue, 79.

49. *Astronomia nova*, introduction, *GW* iii, 20:4–22; trans. Donahue, 48.

50. Whether Kepler ever reassessed Fabricius's theory is unclear. It did not adequately reproduce the area law, but it did produce an ellipse. The production of an ellipse by natural causes was an important element of Kepler's argument, so he was suppressing this aspect, at least.

51. *Astronomia nova*, introduction, *GW* iii, 20:32–33; trans. Donahue, 49.

52. Ibid., 20:39–41; trans. Donahue, 49.

53. Ibid., 21:8–13; trans. Donahue, 49.

54. See pp. 111–14.

55. *Astronomia nova*, introduction, *GW* iii, 22:23–38; trans. Donahue, 51.

56. "The astronomical things are set forth in the triple form of the hypotheses [i.e., Ptolemy, Copernicus, Brahe], but the physical things to some extent conflict directly with the Tychonic form. And I would not want to bear Tengnagel himself as a censor, and thus to say something publicly that I believe differently in private." See above, pp. 165–66.

57. The note reads: "Of whom, in all fairness, most honest and grateful mention is made, and recognition given, since I build this entire structure from the bottom up upon his work, all the materials being borrowed from him" (*Astronomia nova*, introduction, *GW* iii, 22:38–23:5; trans. Donahue, 51).

58. *Astronomia nova*, introduction, *GW* iii, 23:16–17; trans. Donahue, 52.

59. This argument was present already in Kepler's student disputation; see above, pp. 28–29.

60. *Astronomia nova*, introduction, *GW* iii, 24:1–11; trans. Donahue, 53.

61. Ibid., 24:14–18; trans. Donahue, 53.

62. Ibid., 24:18–20; trans. Donahue, 53.

63. Aristotle, *On Rhetoric*, bk. 3, ch. 14: "The most necessary and specific function of the prooemion is this: to make clear what is the 'end' [telos] for which the speech [is being given]. . . . The other kinds that are used are *remedies* and [are] common [to all species of rhetoric]. These are derived from the speaker and the hearer and the subject and the opponent: from the speaker and the opponent [are derived] whatever refutes or creates a prejudicious attack [*diabole*]. . . . [T]he defendant, when he is going to introduce himself, has to remove whatever hinders his case, and thus must first counteract the attack" (trans. Kennedy, 262–63).

64. *Astronomia nova*, introduction, *GW* iii, 34:19–28; trans. Donahue, 67.

65. Ibid., 34:34–35:2; trans. Donahue, 67–68.

66. Ibid., 35:2–4; trans. Donahue, 68.

67. Ibid., 35:5–9; trans. Donahue, 68.

68. Ibid., 35:10–12; trans. Donahue, 68.

69. Ibid., 35:12–14; trans. Donahue, 68.

70. Ibid., 35:15–17; trans. Donahue, 68.

71. Ibid., 35:22–23; trans. Donahue, 68.

72. For example: "I, to be sure, have laid all my prejudices and preferences out in the open here, so that I am more afraid of appearing to the reader to be importunate than I am of seeming untrustworthy" (*Astronomia nova*, ch. 26, *GW* iii, 211:6–7; trans. Donahue, 334).

73. *Astronomia nova*, ch. 6, *GW* iii, 103:37–38; Donahue, trans. 177.

74. *Astronomia nova*, ch. 16, *GW* iii, 152:39–153: 4; trans. Donahue, 251–52.

75. Ibid., 152:21–26; trans. Donahue, 251.

76. *Astronomia nova*, ch. 26, *GW* iii, 215:5–10; trans. Donahue, 339–40.

77. *Astronomia nova*, ch. 20, *GW* iii, 181:6–10; trans. Donahue, 291. For another example, see ch. 22: "Let us suppose that the second inequality is measured starting from the line of the sun's mean motion, as the theorists have hitherto been pleased to hold (lest anyone hold suspect here my innovation of using the sun's apparent motion)" (*Astronomia nova*, ch. 22, *GW* iii, 192:19–21; trans. Donahue, 306–7).

78. *Astronomia nova*, introduction, *GW* iii, 19:2–3; trans. Donahue, 46.

79. *Astronomia nova*, ch. 21, *GW* iii, 187:9–22; trans. Donahue, 300–1.

80. *Astronomia nova*, part 3, *GW* iii, 189; trans. Donahue, 303.

81. *Astronomia nova*, ch. 22, *GW* iii, 191:14–18; trans. Donahue, 305.

82. Ibid., 191:24–26; trans. Donahue, 305–6.

83. See above, p. 85.

84. *Astronomia nova*, ch. 22, *GW* iii, 192:6–18; trans. Donahue, 306.

85. Ch. 32 is similarly presented in the chapter summaries as being the decisive point of departure for physical astronomy: "Physicists, prick up your ears! For here is raised a deliberation involving an inroad to be made into your province" (*Astronomia nova*, chapter summaries, *GW* iii, 43:23–24; trans. Donahue, 89).

86. Kepler included the note, "Now more than that." This indicates that the chapter was composed in 1605 (or 1604, if one takes the nominal date 1596 from the title page rather than the actual completion of printing in 1597). Kepler disclosed his plan to recount all of his efforts in early 1605.

87. *Astronomia nova*, ch. 32, *GW* iii, 233:22–31; trans. Donahue, 372.

88. Ibid., 238:15–20; trans. Donahue, 379.

89. Ibid., 238:9–10; trans. Donahue, 378–79.

90. See Koyré, *The Astronomical Revolution*, 197–214.

91. *Astronomia nova*, ch. 38, *GW* iii, 255:6–10; trans. Donahue, 404 (emphasis mine).

92. That is, because the length of the orbit and the weakness of the force are both proportional to distance, the total effect goes with the inverse square of the distance.

93. *Astronomia nova*, ch. 39, *GW* iii, 256:10–23; trans. Donahue, 407.

94. *Astronomia nova*, ch. 40, *GW* iii, 263:5–6; trans. Donahue, 417.

95. Cf. end of ch. 40 (after the conchoid analysis): "Granted, these general considerations of the computation of the physical equation are not yet well enough supported by the geometrical apparatus. Nevertheless, I wished to give them a preliminary treatment here, so that when all the planetary inequalities are determined . . . , this operation will not be so sharply divorced from its basis in physical theory. . . . But the very things which have been rejected in this chapter

as paralogisms will be taken up again below, when we shall have come to a perfectly correct way of expressing the equations, when the thing that gave rise to the paralogism will have been eliminated from that hypothesis" (*Astronomia nova*, ch. 40, *GW* iii, 269:27–39; trans. Donahue, 426).

96. *Astronomia nova*, chapter summaries, *GW* iii, 47:18–25; trans. Donahue, 95.

97. I.e., early 1602.

98. *Astronomia nova*, ch. 44, *GW* iii, 285:20–26; trans. Donahue, 451.

99. In the synoptic table of contents, these chapters were described with the oxymoronic phrase, "The first [inequality in longitude], proper to the planet and specifically to its eccentric, which is established by physical causes / with a preconceived opinion of the orbit (which the planet describes, whether / *circular, in imitation of the ancients, where I presuppose the manner in which the physical causes act*, as in ch. 39" (emphasis mine; *Astronomia nova*, synoptic table, *GW* iii, between 34 and 35; trans. Donahue, 70 and 74).

100. *Astronomia nova*, ch. 44, *GW* iii, 286:13–21; trans. Donahue, 452.

101. Kepler had already approximated the oval with an ellipse and found the equal and opposite errors already by the end of ch. 47, but continued with a variety of different approximations in chs. 48 through 50. These efforts were justified at the end of ch. 47: "But because the plane of the circle is not equivalent to the aggregate of all the distances, nor is the oval figure that, according to the opinion of ch. 45, Mars describes, a perfect ellipse, as we have assumed, the causes for the departure from the truth are still hidden. For in addition to these two concurring errors in the calculation, there can be a third, relating to its foundation, the opinion expressed in ch. 45. So we have not yet set up equations according to the law of the opinion of ch. 45, and have not yet done justice to the hypothesis we took up there, because we have been abandoned by geometry. Therefore, we still cannot charge it with being erroneous. For the computation which is going to do this will declare for itself the rule by which its innocence will be judged" (*Astronomia nova*, ch. 47, *GW* iii, 303:12–20; trans. Donahue, 478).

102. *Astronomia nova*, ch. 48, *GW* iii, 306:1–3; trans. Donahue, 482.

103. *Astronomia nova*, ch. 45, *GW* iii, 288:13; trans. Donahue, 455.

104. *Astronomia nova*, ch. 51, GW iii, 322:30–323: 10; trans. Donahue, 508.

105. "Now that it is clear that the oval of chapter 45 produces false equations, the question is raised whether it is also in error concerning the distances. Consequently, in this chapter, first the observations and second the distances of the sun from the earth are adopted, such as were established with great certainty in part III. Aside from these, nothing is assumed or admitted as a principle of demonstration" (*Astronomia nova*, chapter summaries, *GW* iii, 49:38–42; trans. Donahue, 99).

106. Donahue, "Kepler's Fabricated Figures: Covering up the Mess in the *Astronomia nova*."

107. It has also been pointed out that Kepler did not necessarily need the ellipse for an original derivation of the procedure. Splitting the difference between the circle and the narrow ellipse would have sufficed. See Volker Bialas, "Keplers komplizierter Weg zur Wahrheit: Von neuen Schwierigkeiten, die '*Astronomia nova*' zu lesen," *Berichte zur Wissenschaftsgeschichte* 13 (1990):167–76.

108. As Kepler wrote when unveiling his trick, "In the observations of chapter 53, there is no need to do the [comparison with distances derived from the ellipse]. For I previously used this same method of reciprocation to find out the distances of Mars from the sun which I presented in order to compute Mars's apparent positions. And since they are in agreement with the observations, they are therefore correct" (*Astronomia nova*, ch. 56, *GW* iii, 347:26–34; trans. Donahue, 546).

109. *Astronomia nova*, ch. 55, *GW* iii, 344: 27; trans. Donahue, 541.

110. "Since we are establishing new hypotheses here, in that we are enquiring into the natural cause of the eccentric equations, it is appropriate that we should explore everything as carefully as possible, lest in neglecting the foundations we build upon them a building doomed to ruin. And so it will further us to explore this same thing (perfectly accurate distances of Mars from the sun) by many methods" (*Astronomia nova*, ch. 53, *GW* iii, 337:24–26; trans. Donahue, 529).

111. *Astronomia nova*, ch. 55, *GW* iii, 345:8–13; trans. Donahue, 542. I have modified Donahue's translation by substituting "middle distances" for "mean longitudes." For a discussion of the preferred translation of "mediae longitudines," see Bruce Stephenson, review of Johannes Kepler, *New Astronomy*, trans. William H. Donahue, *Isis* 85 (1994), 326.

112. *Astronomia nova*, ch. 55, *GW* iii, 345:13–14; trans. Donahue, 542.

113. See chapter 8, note 87, above.

114. *Astronomia nova*, ch. 56, *GW* iii, 345:35–346: 4; trans. Donahue, 543.

115. *Astronomia nova*, ch. 57, *GW* iii, 355:31–32; trans. Donahue, 559.

116. *Astronomia nova*, ch. 58, *GW* iii, 364:22–28; trans. Donahue, 573.

117. "The very truth, and the nature of things, though repudiated and ordered into exile, sneaked in again through the back door, to be received by me under an unwonted guise" (*Astronomia nova*, ch. 58, *GW* iii, 365: 35–37; trans. Donahue, 575).

118. In ch. 59, after explaining very formally the error of the *via buccosa*, Kepler concluded, "And when the fact was established, I was afterwards driven, once I had settled on the principles, to seek the cause of the matter which I have revealed to the reader in this chapter as skillfully and lucidly as possible. And unless the physical causes that I had taken in the place of principles had been good ones, they would never have been able to withstand an investigation of such exactitude" (*Astronomia nova*, ch. 59, *GW* iii, 375:32–37; trans. Donahue, 591).

119. *Astronomia nova*, ch. 59, *GW* iii, 375:38–376:6; trans. Donahue, 591.

NOTES TO CONCLUSION

1. Nicolaus Copernicus, *On the Revolutions*, trans. Edward Rosen, p. 5, ll. 44–45.

2. *Astronomia nova*, introduction, *GW* iii, 18:2–3; trans. Donahue, 45.

⌘ BIBLIOGRAPHY ⌘

Aiton, Eric J. "Johannes Kepler and the Astronomy without Hypotheses." *Japanese Studies in the History of Science* 14 (1975): 49–71.

———. "Kepler's Second Law of Planetary Motion." *Isis* 60 (1969): 75–90.

———. "Peurbach's *Theoricae novae planetarum*: A Translation with Commentary," *Osiris* 3 (1987): 5–44.

Apelt, E. F. *Die Reformation der Sternekunde*. Jena: Friedrick Mauke, 1852.

Applebaum, Wilbur. "Keplerian Astronomy after Kepler: Researches and Problems." *History of Science* 34 (1996): 451–504.

Apt, A. J. *The Reception of Kepler's Astronomy in England: 1596–1650*. Ph.D. diss., Oxford University, 1982.

Aristotle. *The Basic Works of Aristotle*, ed. Richard McKeon. New York: Random House, 1941.

———. *On Rhetoric*, trans. George A. Kennedy. Oxford: Oxford University Press, 1991.

Barker, Peter. "Kepler's Epistemology." In *Method and Order in Renaissance Philosophy*, ed. Daniel A. DiLiscia, Eckhard Kessler, and Charlotte Methuen, 355–68. Aldershot, Hampshire: Ashgate, 1997.

———. "The Role of Religion in the Lutheran Response to Copernicus." In *Rethinking the Scientific Revolution*, ed. by Margaret Osler, 59–88. Cambridge: Cambridge University Press, 2000.

Barker, Peter, and Bernard R. Goldstein. "Realism and Instrumentalism in Sixteenth Century Astronomy: A Reappraisal." *Perspectives on Science* 6 (1998): 232–58.

Baumgardt, Carola. *Johannes Kepler: Life and Letters*. New York: Philosophical Library, 1951.

Berthold, Gerhard. *Der Magister Johann Fabricius und die Sonnenflecken, nebst einem Excurse über David Fabricius*. Leipzig: Veit & Comp., 1894.

Bialas, Volker. "Keplers komplizierter Weg zur Wahrheit: Von neuen Schwierigkeiten, die '*Astronomia nova*' zu lesen." *Berichte zur Wissenschaftsgeschichte* 13 (1990): 167–76.

Brahe, Tycho. *Tychonis Brahe Dani Opera Omnia*, ed. J.L.E. Dreyer, 15 vols. Copenhagen: Libraria Gyldendaliana, 1913–1929.

Caspar, Max. *Bibliographia Kepleriana: Ein Führer durch das Gedruckte Schrifttum von Johannes Kepler.* 2d ed. Munich: C. H. Beck, 1968.

———. *Kepler*, trans. C. Doris Hellman. Dover edition with a New Introduction and References by Owen Gingerich, and Bibliographical Citations by Owen Gingerich and Alain Segonds. New York: Dover, 1993.

Caspar, Max, and Walther von Dyck. *Johannes Kepler in seinen Briefen*. 2 vols. Munich: R. Oldenbourg, 1930.

Christianson, John Robert. *On Tycho's Island: Tycho Brahe and His Assistants, 1570–1601*. Cambridge: Cambridge University Press, 2000.

Copernicus, Nicolaus. *On the Revolutions*, trans. Edward Rosen. Reprint, Baltimore: The Johns Hopkins University Press, 1992. Originally published as vol.

2 of *Nicholas Copernicus: Complete Works*, ed. Jerzy Dobrzycki (Warsaw and Cracow: Polish Scientific Publishers, 1978).

———. *On the Revolutions of the Heavenly Spheres*, trans. Charles Glenn Wallis. Great Books of the Western World, 16. Chicago: Encyclopædia Britannica, 1952.

Davis, A.E.L. "Kepler's Resolution of Individual Planetary Motion," "Kepler's 'Distance Law'—Myth not Reality," "Grading the Eggs (Kepler's Sizing Procedure for the Planetary Orbit)," "Kepler's Road to Damascus," and "Kepler's Physical Framework for Planetary Motion." *Centaurus* 35 (1992): 97–191.

Dobrzycki, Jerzy, and Richard L. Kremer. "Peurbach and Marāgha Astronomy? The Ephemerides of Johannes Angelus and Their Implications." *Journal for the History of Astronomy* 27 (1996): 187–237.

Donahue, William H. "Kepler's Fabricated Figures: Covering up the Mess in the *Astronomia nova*." *Journal for the History of Astronomy* 19 (1988): 217–37.

———. "Kepler's First Thoughts on Oval Orbits: Text, Translation, and Commentary." *Journal for the History of Astronomy* 24 (1993): 71–100.

Drake, Stillman. *Galileo at Work: His Scientific Biography*. Chicago: University of Chicago Press, 1978.

———. *Galileo Studies: Personality, Tradition, and Revolution*. Ann Arbor: The University of Michigan Press, 1970.

———. "Galileo's 'Platonic' Cosmology and Kepler's *Prodromus*." *Journal for the History of Astronomy* 4 (1973): 174–91.

Dreyer, J.L.E. *Tycho Brahe: A Picture of Scientific Life and Work in the Sixteenth Century*. Edinburgh: Adam & Charles Black, 1890.

Evans, James. "On the function and the probable origin of Ptolemy's equant." *American Journal of Physics* 52(1984): 1080–89.

Field, J. V. *Kepler's Geometrical Cosmology*. Chicago: University of Chicago Press, 1988.

Finocchiaro, Maurice A., ed. and trans. *The Galileo Affair: A Documentary History*. Berkeley: University of California Press, 1989.

Galilei, Galileo. *Dialogue Concerning the Two Chief World Systems—Ptolemaic and Copernican*, trans. Stillman Drake. 2d ed. Berkeley: University of California Press, 1967.

———. *Discoveries and Opinions of Galileo*, ed. and trans. Stillman Drake. New York: Doubleday, 1957.

Gassendi, Pierre. *Tychonis Brahei equitis dani, astronomorum coryphi, vita*. Paris, 1654.

Gingerich, Owen. "The Computer versus Kepler." In *The Eye of Heaven: Ptolemy, Copernicus, Kepler*, 357–66. New York: The American Institue of Physics, 1993. Originally published in *American Scientist* 52 (1964):218–26.

———. "'Crisis' versus Aesthetic in the Copernican Revolution." In *The Eye of Heaven: Ptolemy, Copernicus, Kepler*, 193–204. Originally published in *Vistas in Astronomy* 17 (1975): 85–95.

———. "Erasmus Reinhold and the Dissemination of Copernican Theory." In *The Eye of Heaven: Ptolemy, Copernicus, Kepler*, 221–51. Originally published as "The Role of Erasmus Reinhold and the Prutenic Tables in the Dissemination of Copernican Theory," *Studia Copernicana, VI; Colloquia Copernicana, II* (1973): 43–62.

———. "Heliocentrism as Model and Reality." In *The Eye of Heaven: Ptolemy, Copernicus, Kepler*, 286–302. Originally published in *Proceedings of the American Philosophical Society* 117 (1973): 513–22.

———. "Kepler's Treatment of Redundant Observations or, the Computer versus Kepler Revisited." In *The Eye of Heaven: Ptolemy, Copernicus, Kepler*, 367–78. Originally published in *Internationales Kepler-Symposium, Weil der Stadt 1971*, ed. Fritz Krafft, Karl Meyer, and Bernhard Sticker, Arbor scientiarum, ser. A, vol. 1 (Hildesheim: Verlag Dr. H. A. Gerstenberg, 1973).

———. "Mästlin's, Kepler's, and Schickard's Copies of *De revolutionibus*." In *Zum 400. Geburtstag von Wilhelm Schickard*, ed. Friedrich Seck, 167–83. Contubernium, 41. Sigmaringen, Germany: Jan Thorbecke Verlag, 1994.

Gingerich, Owen, and James R. Voelkel. "Tycho Brahe's Copernican Campaign." *Journal for the History of Astronomy* 29 (1998): 1–34.

Gingerich, Owen, and Robert S. Westman. *The Wittich Connection: Conflict and Priority in Late Sixteenth-Century Cosmology*. Transactions of the American Philosophical Society, 78, pt. 7. Philadelphia: The American Philosophical Society, 1988.

Goldstein, Bernard R. *The Arabic Version of Ptolemy's* Planetary Hypotheses. Transactions of the American Philosophical Society, n.s. 57, pt. 4. Philadelphia: The American Philosophical Society, 1967.

Grafton, Anthony. "Humanism and Science in Rudolphine Prague: Kepler in Context." In *Literary Culture in the Holy Roman Empire, 1555–1720*, ed. James A. Parente, Jr., et al., 19–45. Chapel Hill: University of North Carolina Press, 1991.

———. "Michael Maestlin's Account of Copernican Planetary Theory." *Proceedings of the American Philosophical Society* 117 (1973): 523–50.

Grant, Edward. *Planets, Stars, and Orbs: The Medieval Cosmos, 1200–1687*. Cambridge: Cambridge University Press, 1994.

Hellman, C. Doris. *The Comet of 1577: Its Place in the History of Astronomy*. Studies in History, Economics, and Public Law, 510. New York: Columbia University Press, 1944.

Holton, Gerald. "Johannes Kepler's Universe: Its Physics and Metaphysics." *American Journal of Physics* 24 (1956): 340–51.

Jardine, N. *The Birth of History and Philosophy of Science: Kepler's* A Defence of Tycho Against Ursus *with Essays on its Provenance and Significance*. Cambridge: Cambridge University Press, 1984.

———. "The Forging of Modern Realism: Clavius and Kepler against the Sceptics." *Studies in History and Philosophy of Science* 10 (1979): 141–73.

Jarrell, Richard A. "Astronomy at the University of Tübingen: The Work of Michael Mästlin." In *Wissenschaftsgeschichte um Wilhelm Schickard*, ed Friedrich Seck, 9–19. Contubernium: Beiträge zur Geschichte der Eberhard-Karls-Universität Tübingen, 26. Tübingen: J.C.B. Mohr, 1981.

———. *The Life and Scientific Work of the Tübingen Astronomer Michael Maestlin, 1550–1631*. Ph.D. diss., University of Toronto, 1971.

Kepler, Johannes. *Gesammelte Werke*, ed. Max Caspar et al. 20 vols. to date. Munich: C. H. Beck, 1937–.

———. *Joannis Kepleri astronomi opera omnia*, ed. Christian Frisch. 8 vols. Frankfurt and Erlangen: Heyder & Zimmer, 1858–1871.

———. *Johannes Kepler in seinen Briefen*, trans. and ed. Max Caspar and Walther von Dyck. 2 vols. Munich and Berlin: R. Oldenbourg, 1930.

———. *Mysterium Cosmographicum: The Secret of the Universe*, trans. A. M. Duncan. Introduction and commentary by E. J. Aiton. New York: Abaris Books, 1981.

———. *Neue astronomie*, trans. Max Caspar. Munich-Berlin: R. Oldenbourg, 1929.

———. *New Astronomy*, trans. William H. Donahue. Cambridge: Cambridge University Press, 1992.

———. *Selbstzeugnisse*. Edited by Franz Hammer, and translated by Esther Hammer. Stuttgart-Bad Cannstatt, Germany: Friedrich Frommann Verlag, 1971.

Koestler, Arthur. *The Sleepwalkers: A History of Man's Changing Vision of the Universe*. New York: Macmillan, 1959.

Koyré, Alexandre. *The Astronomical Revolution: Copernicus—Kepler—Borelli*, trans. R.E.W. Maddison. London: Methuen, 1980.

———. *Galileo Studies*, trans. John Mepham. Atlantic Highlands, N.J.: Humanities Press, 1978.

Kozhamthadam, Job. *The Discovery of Kepler's Laws: The Interaction of Science, Philosophy, and Religion*. Notre Dame: University of Notre Dame Press, 1994.

Kuhn, Thomas S. *The Structure of Scientific Revolutions*. 2d ed. Chicago: The University of Chicago Press, 1970.

Lang, A. W. "Die 'Nie vnd warhafftige Beschrivinge des Ostfreslandes' des David Fabricius von 1589: Erläuterungen zur Lichtdruckausgabe." *Nordseeküste: Volkstümliche Vorträge und Abhandlungen des Küstenmuseums Juist*, Heft 8. Juist: Die Bake, 1963. Originally published in *Emder Jahrbuch* 43 (1963).

Methuen, Charlotte. *Kepler's Tübingen: Stimulus to a Theological Mathematics*. Brookfield, Vt.: Ashgate, 1998.

Moss, Jean Dietz. *Novelties in the Heavens: Rhetoric and Science in the Copernican Controversy*. Chicago: Chicago University Press, 1993.

Panofsky, Erwin. "Galileo as a Critic of the Arts: Aesthetic Attitude and Scientific Thought." *Isis* 47 (1956): 3–15.

Rheticus, Georg Joachim. *Narratio Prima* (1540), ed. Henri Hugonnard-Roche and Jean-Pierre Verdet. Studia Copernicana, 20. Wroclaw: Polish Academy of Sciences, 1982.

Rosen, Edward. "Kepler and the Lutheran Attitude Towards Copernicanism in the Context of the Struggle Between Science and Religion." In *Kepler: Four Hundred Years*, ed. Arthur Beer and Peter Beer, 317–38. Vistas in Astronomy, 18. Oxford: Pergamon Press, 1975.

———. *Three Copernican Treatises: The* Commentariolus *of Copernicus, The* Letter against Werner, *The* Narratio prima *of Rheticus*. 2d ed. New York: Dover, 1959.

———. *Three Imperial Mathematicians: Kepler Trapped between Tycho Brahe and Ursus*. New York: Abaris Books, 1986.

Russell, J. L. "Kepler's Laws of Planetary Motion: 1609–1666." *The British Journal for the History of Science* 2 (1964): 1–24.

Schofield, Christine Jones. *Tychonic and Semi-Tychonic World Systems.* New York: Arno Books, 1981.

Seck, Friedrich. "Johannes Kepler und der Buchdruck: Zur äußeren Entstehungsgeschichte seiner Werke." *Archiv für Geschichte des Buchwesens* 11 (1970), 609–726.

Sello, Georg. *Des David Fabricius Karte von Ostfriesland und andere Fabriciana des Oldenburger Archivs.* Norden und Norderney: Herm. Braams; 1896.

Simon, Gerard. *Kepler: astronome, astrologue.* Paris: Gallimard, 1979.

Stephenson, Bruce. *Kepler's Physical Astronomy.* Studies in the History of Mathematics and Physical Sciences, 13. New York: Springer-Verlag, 1987.

Swerdlow, N. M. and O. Neugebauer. *Mathematical Astronomy in Copernicus's* De revolutionibus. 2 vols. Studies in the History of Mathematics and Physical Sciences, 10. New York: Springer-Verlag, 1984.

Thoren, Victor E. *The Lord of Uraniborg: A Biography of Tycho Brahe.* Cambridge: Cambridge University Press, 1990.

Van Helden, Albert. *Measuring the Universe: Cosmic Dimensions from Aristarchus to Halley.* Chicago: University of Chicago Press, 1985.

Wattenberg, Diedrich. "David Fabricius: Der Astronom Ostfrieslands (1564–1617)." *Vorträge und Schriften der Archenhold-Sternwarte,* 19. Berlin-Treptow: Archenhold-Sternwarte, 1964.

Westman, Robert S. "The Astronomer's Role in the Sixteenth Century: A Preliminary Study." *History of Science* 18 (1980): 105–47.

———. "The Comet and the Cosmos: Kepler, Mästlin, and the Copernican Hypothesis." *Studia Copernicana,* V; *Colloquia Copernicana,* I (1972): 7–30.

———. *Johannes Kepler's Adoption of the Copernican Hypothesis.* Ph.D. diss., University of Michigan, 1971.

———. "The Melanchthon Circle, Rheticus and the Wittenberg Interpretation of the Copernican Theory." *Isis* 66 (1975): 165–93.

———. "Three Responses to the Copernican Theory: Johannes Praetorius, Tycho Brahe, and Michael Maestlin." In *The Copernican Achievement,* ed. Robert S. Westman, 289–305. UCLA Center for Medieval and Renaissance Studies Contributions, 7. Berkeley: University of California Press, 1975.

———. "Two Cultures or One? A Second Look at Kuhn's *The Copernican Revolution.*" *Isis* 85 (1994): 79–115.

Whiteside, D. T. "Keplerian Planetary Eggs, Laid and Unlaid, 1600–1605." *Journal for the History of Astronomy* 5 (1974): 1–21.

Wilson, Curtis. "From Kepler's Laws, So-called, to Universal Gravitation: Empirical Factors." *Archive for the History of Exact Sciences* 6 (1970): 89–170.

———. "Kepler's Derivation of the Elliptical Path." *Isis* 59 (1968): 5–25.

———. "Predictive Astronomy in the Century after Kepler." In *Planetary Astronomy from the Renaissance to the Rise of Astrophysics,* edited by René Taton and Curtis Wilson, 161–206. The General History of Astronomy, 2A. Cambridge: Cambridge University Press, 1989.

⌘ INDEX ⌘

⌘ INDEX OF CORRESPONDENCE ⌘

Numbers refer to Kepler *Gesammelte Werke*